POLLUTION OF THE ARCTIC ATMOSPHERE

Environmental Management Series

Edited by

J. CAIRNS JNR
University Center for Environmental and Hazardous Material Studies, Virginia Polytechnic Institute and State University, Blacksburg, Virginia 24061-0415, USA

R. M. HARRISON
Institute of Public and Environmental Health School of Biological Sciences, University of Birmingham Edgbaston, Birmingham, B15 2TT, UK

Titles published in the *Environmental Management Series*

Instrumental Analysis of Pollutants
C. N. Hewitt

Pollution of the Arctic Atmosphere
W. J. Sturges

Organic Contaminants in the Environment
K. C. Jones

Titles published in the *Pollution Monitoring Series*

Quantitative Aquatic Biological Indicators
D. J. H. Phillips

Biomonitoring Air Pollutants with Plants
W. J. Manning and W. A. Feder

Effect of Heavy Metal Pollution on Plants 1 & 2
N. W. Lepp

Biological Monitoring of Heavy Metal Pollution
M. H. Martin and P. J. Coughtrey

Biological Indicators of Freshwater Pollution and Environmental Management
J. M. Hellawell

Pollution Threat of Heavy Metals in Aquatic Environments
G. Mance

Ecophysiology of Metals in Terrestrial Invertebrates
S. P. Hopkin

Ecology Effects of Thermal Discharges
T. E. L. Langford

Cover Illustration:

The *Glomar Beaufort Sea I* drilling platform in position off Prudhoe Bay. Possible influences on arctic air chemistry from power generation by the oil industry and Prudhoe are discussed in Chapter 9. Plans are under way to use this platform as a floating research station, including atmospheric measurement facilities. After locking it into multi-annual ice it would be allowed to drift slowly around the Arctic Ocean gyre. In the background the sun is turned red as it sets into a bank of 'arctic haze': air pollution transported over thousands of kilometres to the Arctic from mid-latitude sources; the topic of other chapters in this volume. (Photograph courtesy of S. B. Wetmore and reproduced with permission of the Global Marine Drilling Company.)

POLLUTION OF THE ARCTIC ATMOSPHERE

Edited by

W. T. STURGES

Cooperative Institute for Research in Environmental Sciences
University of Colorado, Boulder,
Colorado 80309-0446, USA

ELSEVIER SCIENCE PUBLISHERS
LONDON and NEW YORK

ELSEVIER SCIENCE PUBLISHERS LTD
Crown House, Linton Road, Barking, Essex IG11 8JU, England

Sole Distributor in the USA and Canada
ELSEVIER SCIENCE PUBLISHING CO., INC.
655 Avenue of the Americas, New York, NY 10010, USA

WITH 20 TABLES AND 103 ILLUSTRATIONS

British Library Cataloguing in Publication Data

Pollution of the Arctic atmosphere—(Environmental management series).
1. Arctic. Air, pollution
I. Sturges, W. T. (William T.) II. Series
363.73920998

ISBN 1-85166-619-2

Library of Congress Cataloging-in-Publication Data

Pollution of the Arctic atmosphere/edited by W. T. Sturges.
p. cm.—(Environmental management series)
Includes bibliographical references and index.
ISBN 1-85166-619-2
1. Air—Pollution—Arctic regions. 2. Air—Pollution—Arctic regions—Meteorological aspects. 3. Climatic changes. I. Sturges, W. T. II. Series.
TD883.7.A68P65, 1991
363.73′92′0998—dc20 91-4769
CIP

Typeset by Interprint Ltd, Malta.
Printed in Great Britain at the University Press, Cambridge.

Foreword

Environmental Management Series

The current expansion of both public and scientific interest in environmental issues has not been accompanied by a commensurate production of adequate books, and those which are available are widely variable in approach and depth.

The Environmental Management Series has been established with a view to co-ordinating a series of volumes dealing with each topic within the field in some depth. It is hoped that this Series will provide a uniform and quality coverage and that, over a period of years, it will build up to form a library of reference books covering most of the major topics within this diverse field. It is envisaged that the books will be of single, or dual authorship, or edited volumes as appropriate for respective topics.

The level of presentation will be advanced, the books being aimed primarily at a research/consultancy readership. The coverage will include all aspects of environmental science and engineering pertinent to management and monitoring of the natural and man-modified environment, as well as topics dealing with the political, economic, legal and social considerations pertaining to environmental management.

J. Cairns and R. M. Harrison

Preface

> 'The sun sometimes sweeps two or three times round the Pole, without being for a moment obscured by cloud. Objects the most remote, may be seen perfectly distinct and clear. A ship's top-gallant-mast, at the distance of five or six leagues, may be discerned when just appearing above the horizon...and the summits of some mountains are visible at the distance of sixty to a hundred miles.'

William Scoresby's '*An Account of the Arctic Regions*' published in 1820 (reprinted by Augustus M. Kelly, New York, 1969) reflects the cherished romantic view of the Arctic: remote, pristine, knowing nothing of Man but the fearless adventurer and hardy native.

That picture is beginning to change. Writing in 1956 J. Murray Mitchell noted:

> 'Perhaps the most surprising outcome of the 'Ptarmigan' observations [meteorological reconnaissance flights] has been the high incidence of haze at flight altitudes which restricts horizontal and slant visibility to an estimated range of 2–5 miles. The haze...sometimes stretches over the entire 2000 miles of the reconnaissance track. Arctic haze possesses a grey-blue hue in antisolar directions and a reddish-brown hue in the direction of the sun.' (*Journal of Atmospheric and Terrestrial Physics*, Special Supplement.)

We now know that Mitchell was observing air pollution swept into the Arctic from the industrialised regions of the northern mid-latitudes. The phrase he coined, 'arctic haze', has now passed into the terminology of arctic atmospheric science to describe the layers of polluted air, containing visibility-reducing particles, heavy metals, acidic compounds, toxic organics, and gaseous effluents from industry and fossil-fuel combustion, that are found throughout the entire arctic region every year during winter and spring.

In the past two decades research on the arctic atmosphere has blossomed from humble beginnings and is now intensifying at an unprecedented rate. The governments and universities of Canada, the USA, Norway, Sweden, Germany and Denmark are among those engaged in multi-million dollar national and international research programmes. Some of these programmes have been shining examples of international co-operation.

Arctic air pollution is no longer just a question of simple aesthetics; the marring of pristine environments, although that should be reason enough for concern. The potential for real environmental peril is now being realised. The Arctic is indeed a harsh and hostile world, but it is also one of great fragility and delicate balance. Toxic substances are beginning to appear in the food chains there (Chapter 8), whilst acids and heavy metals are being deposited to the snow and ice (Chapters 3, 4 and 6) with as yet unknown consequences to the arctic ecology following snowmelt. Much attention has recently been focused on the antarctic ozone 'hole': it appears that the arctic stratosphere is itself poised for similar ozone destruction processes to take place (Chapter 7). Perhaps of most concern is the potential for climate change in the Arctic (Chapters 5 and 10). The sensitivity of the arctic climate (Chapter 2) is such that global change may be detected there first, and change there could in turn have wider repercussions on northern hemisphere climate and global sea levels. Finally, many regions of the Arctic contain enormous resources of minerals and fossil fuels. Exploitation of these resources is already underway (Chapter 9): it is important for us to understand more of the complex processes of the arctic environment to be able to predict and ameliorate future impacts from the growing industrialisation of the North.

In addition to those persons named by individual authors who gave their time to review the articles herein, I would also like to thank Russ Schnell, Jack, Winchester, Barry Bodhaine, Pat Sheridan and Jon Kahl for their helpful comments and suggestions, and Karen and Thomas Sturges for their patience and support.

Mitchell was not the first to observe arctic haze. The native people had already observed this harbinger of change and given it a name: 'poo-jok', meaning dark haze (D. B. MacMillan, Crocker Land Expedition of 1914, Geographical Report, American Museum of Natural History, New York). This book is about poo-jok and is dedicated to those people of the Arctic.

William T. Sturges
Cooperative Institute for Research in Environmental Sciences
University of Colorado, USA

Contents

List of Contributors

L. A. BARRIE

Atmospheric Environment Service, 4905 Dufferin Street, Downsview, Ontario, Canada, M3H 5T4

J.-P. BLANCHET

Atmospheric Environment Service, 4905 Dufferin Street, Downsview, Ontario, Canada, M3H 5T4 (Present address: Département de physique, Université du Québec à Montréal, Case Postale 8888, succursale "A", Montréal, Qué, H3C 3P8, Canada)

J. W. BOTTENHEIM

Atmospheric Environment Service, 4905 Dufferin Street, Downsview, Ontario, Canada, M3H 5T4

C. I. DAVIDSON

Department of Civil Engineering, Carnegie Mellon University, Pittsburgh, PA 15213, USA

D. J. GREGOR

Lakes Research Branch, National Water Research Institute, Environment Canada, P. O. Box 5050, Burlington, Ontario, Canada, L7R 4A6

D. A. JAFFE

Geophysical Institute and Department of Chemistry, University of Alaska, Fairbanks, AK 99707, USA

J.-L. JAFFREZO

Department of Civil Engineering, Carnegie Mellon University, Pittsburgh, PA 15213, USA

P. A. MAYEWSKI

Glacier Research Group, Institute for Study of Earth, Oceans and Space, University of New Hampshire, Durham, NH 038424, USA

S. J. OLTMANS

National Oceanic and Atmospheric Administration, Climate Monitoring and Diagnostics Laboratory, 325 Broadway, Boulder, Colorado 80303, USA

J. M. PACYNA

Norwegian Institute for Air Research, P.O. Box 64, N-2001 Lillestrøm, Norway

W. E. RAATZ

Deutscher Wetterdienst, Wallneyer Strasse 10, D-4300 Essen, Federal Republic of Germany

J. G. ROEDERER

Geophysical Institute and Department of Physics, University of Alaska, Fairbanks, AK 99775-0800, USA

G. E. SHAW

Geophysical Institute, University of Alaska, Fairbanks, AK 99707, USA

Chapter 1

Understanding the Arctic: Research Policies and Responsibilities

JUAN G. ROEDERER

Geophysical Institute and Department of Physics
University of Alaska Fairbanks
Fairbanks, Alaska 99775-0800, USA

1.1 THE ENVIRONMENTAL CHALLENGE

Revolutionary transformations are sweeping the world, shaking up relationships between the superpowers, political blocks and military alliances, touching every single nation, large or small. The global change that is happening in the international political scene brings great hope for an enduring peace.

At the same time as the nuclear risks are decreasing, the global environmental challenge is increasing. Human society finds itself in a position of inducing environmental change similar to that of natural forces, but on a much shorter time scale. Six gigatonnes of carbon dioxide are injected per year into the atmosphere as a result of human activity—about one tonne per person per year—and the turnover of minerals in mining and construction projects is becoming comparable to that caused by tectonic processes. Industrial and agricultural activities and large-scale deforestation are perturbing the natural nitrogen and sulfur cycles. The ever-increasing demands for energy and natural resources and their transformation into goods and services are leading to global environmental insult—the degradation of the environment of our planet is no longer a threat but a reality.

Society must learn to cope with new responsibilities of living on Planet Earth: no single nation is anymore in control of the future of its own environment and energy supplies. We must reduce the related uncertainties by urgently addressing the need for global security through sustainable development.

Since the early 1800s society has undergone the transition from an agrarian society to an industrial society and, more recently, to an information society. These transitions were largely spontaneous, unplanned, driven mainly by science when the unpredictable results of scientific research were applied to technological development. Now we are witnessing another transition, driven by the fact that military and national security concerns are being overshadowed by global environmental security concerns. During the industrial revolution, and even during the computer revolution, there was the belief that all problems had a technological solution: there was a characteristic neglect of the human dimension in the long-term. This human dimension now emerges as a most crucial element in the transition to global environmental security. This next transition can be, and must be, planned.[1]

All transitions bring turbulence, inefficiency and instability. The threats of human-caused environmental disasters and the increased vulnerability of today's complex society to natural disasters have created an 'era of anxiety'. This is compounded by the fact that only one-quarter of the world population—the industrialized nations—control 75% of the world's wealth and energy consumption and thus cause three-quarters of all environmental problems (as measured, for instance, by the amounts of anthropogenic carbon dioxide production). In economic terms we have islands of affluence in a sea of despair, a fact that leads to worldwide social unrest and political crisis.

At the root of the demand on natural resources and the concomitant environmental degradation is the exponential growth of world population. Where there is life, there is waste—one cannot wish it away: the Second Law of Thermodynamics demands it!

Global environmental security deals with the twin imperatives of environmental protection and economic development. In this task, the scientific community has a unique responsibility to identify and illuminate the driving forces of global change. This is an international responsibility which demands that scientists develop a common, shared, knowledge base and reach a consensus on pertinent recommendations, on the basis of which the public and its representatives can make decisions.

Nowhere is this requirement more evident and pressing than in the Arctic. As a consequence of the ever-increasing demands for energy, minerals and renewable resources, industrial activity and urbanization are inexorably expanding toward the North, into a pristine and fragile environment. This entails the obligation to reconcile the use of arctic resources and the development of urban centers with the preservation of environmental values and native cultures. The challenge is to devise a course showing that arctic resources *can* be extracted with minimum adverse impact on the environment; profit-motivated industrial operations *are* possible within environmentally conscious communities; indigenous lifestyles *can* be preserved without relinquishing the benefits of modern society; improved transportation systems *can* be built with minimum disturbance of animal migration routes; and a basic understanding of how the arctic system works as a component of our planet *can* be gained while at the same time solving practical problems posed by arctic environmental hazards.

To accomplish this formidable task, it is necessary to have detailed and accurate scientific knowledge of how the arctic system works and how it will evolve in the future under the influence of human activity. Such knowledge does not yet exist—to acquire it will demand a major international effort of research cooperation in the North. Understanding the Arctic will not only benefit the arctic countries: it will benefit human society as a whole.

1.2 ARCTIC RESEARCH: INTERNATIONAL COOPERATION AND NATIONAL POLICIES

The Arctic has a most vulnerable environment, extremely sensitive to perturbations from outside. The delicate balance between its physical, chemical and ecological components, governed by the very low rate of biogenesis and chemical turnover, makes the Arctic an 'early warning system' for global change, where the signatures of climate change are expected to occur first.[2]

The Arctic is a natural storage reservoir for atmospheric and water pollution. Industrial effluents, aerosols and gases from lower latitudes in Europe, the USSR and North America appear in the form of 'arctic haze' over large regions of the Arctic.[3] The absolute concentrations of pollutants are comparatively small and may not pose immediate health hazards to arctic residents. However, their long-term accumulation in the food chain and subsequent release to wildlife and subsistence livestock feeding on them, as well as the magnitude of their effects on the radiation balance and the chemistry of the atmosphere, while still difficult to appraise quantitatively, are of clear potential concern. Similarly, the behavior and trends of stratospheric ozone in the Arctic are only now beginning to be studied systematically.

Concerning water pollution, the Arctic Ocean receives as much as 10% of all the world's riverine input, in spite of representing only 1·2% of the total ocean water mass. Since it is surrounded by populated land and has very limited outflows into the world oceans, it is much more vulnerable to industrial, urban and agricultural pollutants discharged into rivers than any other ocean. Local air and water pollution also exhibit distinct characteristics in the Arctic, due to the climatic conditions, low-altitude inversion layers, inefficient operation of combustion engines, impermeability of frozen soils, etc.

Far from being a passive element, the Arctic is an active component of the global geosphere–biosphere system. Atmosphere–ocean coupling in the Arctic is an important feedback mechanism in the 'thermodynamic machine' that controls the climate of our planet. The world oceans' dominant water mass, i.e. the bottom water in the Atlantic, is formed mainly in the Arctic Ocean; processes involving sea ice determine the temperature and the concentrations of oxygen, carbon compounds and nutrients in this deep sea water reservoir. Highly localized physical, chemical and biological processes in the Arctic Ocean's upper layers play a crucial role in the removal of carbon dioxide and other biogenic and man-made materials from the atmosphere.[4]

The arctic upper atmosphere is a 'window to outer space'. A vast region of the Earth's outer environment, the magnetosphere, is connected to the polar regions by the Earth's magnetic field lines; because of this connection the arctic upper atmosphere acts as a giant TV screen on which the signatures of many deep-space phenomena are projected. The magnificent display of northern lights, or aurora, is a conspicuous example. Space processes can thus be 'remotely sensed' with instruments deployed on the ground in the Arctic.[5] These processes are controlled by solar activity and can cause important disturbances at high latitudes such as blackout of radiowave communications, removal of stratospheric ozone by NO_x from energetic electron reactions, upper atmosphere heating and subsequent satellite orbit decay, electric power line failures, and interference with defense surveillance systems.

Finally, the arctic environment *per se* is a gigantic scientific laboratory, where unique physical, chemical, and biological processes and mechanisms can be studied, giving us better knowledge of how Planet Earth works, how it worked in the past, and where it will be going in the future.[6]

The importance of international scientific cooperation in the polar regions has long been recognized; indeed, the First Polar Year of 1882/83 was historically the first major, multidisciplinary international scientific enterprise ever to be carried out. It was followed by the Second Polar Year fifty years later. Their successor was the International Geophysical Year (IGY) of 1957/58, with a strong arctic research component; the IGY launched both the space age and the international antarctic research program.

Antarctic research cooperation is sometimes pointed out as an ideal to be emulated in the Arctic. However, the differences are profound. First, while there exists an international treaty on Antarctica, the Arctic is one notable area as yet largely untouched by formal international agreements (the Spitsbergen Treaty of 1920 is geographically constrained). Secondly, in contrast to the Arctic, there are no indigenous people in Antarctica, there is as yet no non-renewable resource development there, and military activities are banned and territorial claims suspended as long as the Antarctic Treaty remains in force. Thirdly while antarctic research is mainly driven by scientific curiosity, much of arctic research is directed toward solving practical problems related to the inevitable northward expansion of human activity and to national defense. Even in the geophysical–biospheric aspects there are profound differences, given by the fact that Antarctica is an uninhabited land and ice mass surrounded by oceans, whereas in the Arctic we have an ocean surrounded by densely populated continents.

Many research needs in the Arctic are directly linked to the economic, military and political interests of the 'Arctic Countries'.[7] This makes international cooperation in research difficult;[8] in certain arctic regions and in certain fields of research, international cooperation becomes practically impossible when expected research findings are classified as sensitive or proprietary by governments or industrial firms, respectively, or when research activities have the potential of yielding significant intelligence data. Even within a single country certain arctic research issues are socially or politically sensitive. Because of strategic, political or economic considerations, there will always be a limit to how far governments will go in accepting pertinent arctic research subjects to be included in international cooperative projects.[9]

The question of establishing an international body to coordinate research activities in the Arctic is an old one. Proposals have been formulated on several occasions during the last three decades since the IGY, but they have never been implemented. Most cooperative ventures in arctic research have been conducted on an *ad hoc* basis. In terms of international non-governmental organizations, there is no equivalent to the Scientific Committee on Antarctic Research (SCAR), established thirty years ago by the International Council of Scientific Unions (ICSU). The idea of a 'SCAR of the North' has been advanced several times in recent years, but pertinent discussions never came to fruition. In particular, attempts to draw the USSR into a multinational body of arctic research have systematically failed in the past.

A breakthrough occurred in October 1987, when the then General Secretary Gorbachev, in a public speech[10] in Murmansk, USSR, stated that 'the scientific study of the Arctic is of immense importance for the whole of mankind. We have a wealth of experience here and are prepared to share it. In turn, we are interested in the studies conducted in other arctic and northern countries.'[11]

Because of the changing international climate, an initiative of the United States Arctic Research Commission formulated in 1986 to create an International Arctic Science Committee (IASC) as a non-governmental body is now being implemented. The Committee was established in August 1990 by the eight Arctic Countries; six non-arctic countries which have genuine research interests in the region have now joined as members. This organization promises to become a much needed forum for the planning and coordination of joint scientific projects in the Arctic. The study of global change, while not the only area of research that needs to be conducted in a cooperative fashion in the Arctic, will become a most important activity of the IASC during its first years.

Even at the national level, arctic research coordination and the development of related policies have encountered difficulties in some Arctic Countries. The USA is a case in point. Historically, the responsibility for planning, funding and implementing arctic research has been divided between several federal agencies, the State of Alaska, private industry, universities, and national laboratories. As a result of this pluralistic approach, individual studies are proposed and supported separately, and their costly logistic requirements must be funded in competition with research carried out in the rest of the country under less-demanding environmental conditions.[12] Environmental monitoring and fundamental data gathering are the responsibility of public agencies whose missions are separate and whose budgets may not reflect arctic priorities. Applied research has often been proprietary; as industrial activities increased in the Arctic, much scientific information on arctic processes moved out of the public domain.

There have been several high-level attempts to coordinate federal arctic research programs in the USA during the past two decades. Unfortunately, mechanisms for the adequate implementation of policy decisions were not developed at the same time. A study by the Departments of Interior, Defense, and Energy[13] concluded in 1982 that the USA lacked an explicit arctic research policy.

It was not until 1984 that the US Congress passed the Arctic Research and Policy Act to set policy guidelines and create the presidential Arctic Research Commission and the Interagency Arctic Research Policy Committee. The main tasks of these two bodies are, *inter alia*, to develop arctic research policy, to establish a biennially updated five-year Arctic Research Plan, and to coordinate the implementation of arctic research programs at the federal level.[14]

Turning to the other side of the Bering Strait, the USSR has had a long-standing economic development policy and clearly defined goals for its arctic regions, with 20–25 000 scientists and engineers at work there, no less than 170 scientific institutes linked to offshore oil and gas exploration alone, 37 polar research vessels, and 19 ice breakers, several of them nuclear-powered. But it was only recently that environmental concerns have moved the government and the Soviet Academy of Sciences to establish an executive-level Commission on Arctic Research to develop a comprehensive research policy for the region.

1.3 SOCIAL RESPONSIBILITIES OF INVOLVED SCIENTISTS

Achieving a scientific understanding of the inner workings of the terrestrial environment is one of the most difficult and ambitious endeavors of humankind, rivaling in complexity the harnessing of nuclear energy, the conquest of space, and the understanding of the human brain. Unfortunately, the more we learn about the environment, anthropogenic effects and climate change, the more political problems emerge. In litigation, what used to be 'acts of God' are replaced by 'acts of man'. Lawyers, judges and politicians have trouble coping with the complexities involved because of the technical problems for which they have little training. They often make decisions and render verdicts on the basis of environmental concepts they grasp poorly and without a clear understanding of the scientific method, the inherent uncertainties of scientific measurements, and the natural limitations of scientific predictions.

The task is full of dangers and traps for the participating scientists. We are dealing with a horrendously complex, non-linear, interactive system, which is the scientific way of saying that nothing is proportional to anything, and everything is coupled to everything else. Lessons learned from disciplines such as plasma physics and solid state physics which have been dealing with complex non-linear systems for decades, are not very encouraging. The first thing one learns is that there are intrinsic, physical limits to the 'knowability' and the predictability of a complex system that have nothing to do with the particular techniques employed to describe and model it. There are strong indications that the atmosphere is such a system.

For instance, it is found that the predictability of a complex system can be inherently impossible within certain, limited ranges of time and space. The atmosphere seems fairly well predictable on a time-scale of a few days. Conceivably, it could be predictable on a millennial time-scale, too. But nothing yet can be said about its inherent predictability on the most important decadal or centennial time-scales.

What one learns from one time-scale cannot be simply scaled up or down to other time-scales. The effects of a given increase in atmospheric carbon dioxide concentration over only 200 years may be totally different from the effects caused by the same total increase taking thousands of years to build up in the course of an Ice Age cycle (leaving completely aside the question of whether CO_2 increases in the inter-glacial periods were the cause or the effect of global warming!). Thus there may be severe, inherent restrictions to how we can use the information on past, comparatively slow, natural changes of global climate (such as obtained from sediments and ice cores) to make predictions about anthropogenic effects in a relatively short-term future.

A complex system may evolve in discrete steps with sudden, unpredictable transitions from one state of quasi-equilibrium to another. As an eminently non-linear system, the coupled atmosphere–ocean system has this potential, but we do not know enough about the totality of physical and biogeochemical processes that could trigger sudden change in order to incorporate them into the global climate models.

The only physically and mathematically legitimate characterization of a complex system is by the quantitative, multidimensional description of its total state; examples for the lower atmosphere are global circulation, global distribution of minor constituents, or the global vector wind field. The compression of all the relevant information into just one single scalar variable, such as the mean surface temperature, however intuitive, can lead to serious misrepresentations or misinterpretations of the full internal dynamic state of the complex system (leaving completely aside the difficulties with the unbiased experimental determination of such planetary-average quantities and their long-term variability!).

Another lesson learned from non-linear system studies in other fields of physics, is that temporal and spatial scales of disturbance are linked, even in the most highly non-linear systems. In response to some external forcing, small-scale turbulence is bound to appear and develop earlier than large-scale global change. Translated to the atmosphere, increases in frequency and intensity of violent storms and regional variability may appear well before true global warming, but have the potential of being as devastating for society as the latter. Unfortunately, none of the current supercomputer climate models can handle such smaller-scale features.

Finally, the study of complex systems subjected to the rules of chaos shows that it is not legitimate to neglect inputs or coupling mechanisms simply because they are 'small', 'weak' or 'slow'. A most frustrating consequence of this is that approximations using simple-minded or oversimplified models may give wrong answers, even if such models are self-consistent and make good physical sense, such as the 'greenhouse model' for anthropogenic CO_2-induced global warming.[15] In this connection we should note that it is not enough to confirm the predictions of one model with those of a different one developed by another group, if both models are based on the same simplifying assumptions (e.g. about atmosphere–ocean coupling) and carry the same computational limitations (such as grid size): equally deficient models are bound to give mutually consistent results!

How can we scientists sustain a public sense of the common danger of global environmental change while remaining honest, in view of the realities of scientific uncertainty? How can we nurture this sense of common danger without making statements based on half-baked ideas, statistically unreliable results, or oversimplified models? How can we strike a balance between the need to overstate a case to call the attention of the media, and the obligation to adhere strictly to the ethos of science?

Our problem is compounded by the fact that scientific results are gobbled up, often out of context, by policy makers and politicians as they come out in piecemeal fashion. In this process, scientific results are translated into 'understandable' terms by intermediaries who often bias the translation to further their own political or ideological agendas. This can lead to billion-dollar solutions for million-dollar problems, and possible economic disaster; or to million-dollar solutions for billion-dollar problems, and possible environmental disaster.

Do we need proof beyond reasonable doubt before important environmental policy decisions are made? By the time we have the proof it may be too late. Should costly policy decisions be made on the basis of insufficient scientific information just as a preventive measure? Decisions on how much scientific information is 'enough' should not be left to politicians alone, nor to scientists alone. Both groups must work together: not only must science issues be brought in clearly understandable terms to the policy makers, policy issues must be brought in clearly understandable terms to the science makers!

This situation is not particular to any single nation. It presents both a great dilemma and a great challenge to the scientific community worldwide. It gives an added dimension to the need for international scientific cooperation: a dimension of awesome social responsibility.

Atmospheric chemists share a significant part of this responsibility. After all, it is they who study the most conspicuous form of anthropogenic disturbance to our planet: atmospheric pollution. It is they who deal with the principal biogeochemical cycles that control the functions of the 'terrestrial machine' and its deceptively non-linear response to change. It is they who are called to find out about the Earth's climate in the past and to monitor our planet's health in the future. And in all this, the Arctic is a good place to start!

ACKNOWLEDGEMENT

The author gratefully acknowledges a grant from the John D. and Catherine T. MacArthur Foundation for the support of his studies on arctic research policy issues.

REFERENCES AND NOTES

1. Roederer, J. G., Confidence building. *EOS*, Transactions of the American Geophysical Union, **70**, October 17 (1989) 42.
2. *Arctic Interactions.* Published by the Office for Interdisciplinary Earth Studies, University Corporation of Atmospheric Research, Boulder, Colorado, 1988.

3. Stonehouse, B. (ed.), *Arctic Air Pollution*. Cambridge University Press, Cambridge, UK, 1986. See also Chapter 5, *this volume*.
4. See for instance *Report of the Ocean/Atmosphere Panel on ARCSS*. National Science Foundation, Directorate for Geosciences, Washington DC, 1989.
5. Polar Research Board, *Study of the Upper Atmosphere and Near-Earth Space in Polar Regions*. National Research Council, Washington DC, 1982.
6. See for instance Polar Research Board, *National Issues and Research Priorities in the Arctic*. National Research Council, Washington DC, 1985; and Washburn, A. L. & Weller, G., Arctic research in the national interest. *Science*, **233** (1986) 633–9.
7. The countries with sovereignty on land and/or waters north of the Arctic Circle: Canada, Denmark/Greenland, Finland, Iceland, Norway, Sweden, the USA and the USSR.
8. Roederer, J. G., International scientific cooperation in the Arctic. In *Proceedings of the Conference on Technology-Based Confidence Building: Energy and Environment*, ed. J. C. Allred, R. C. Eckhardt and A. S. Nichols, Center for National Security Studies, Los Alamos National Laboratory LA-11728-C, Los Alamos N. M., 1989, pp. 309–17. For a detailed analysis, see Young, O. R., *The Arctic in World Affairs*. Washington Sea Grant Program, University of Washington, Seattle WA, 1989.
9. See for instance Möttölä, K. (ed.), *The Arctic Challenge*. Westview Press, Boulder and London, 1988; and Ostreng, W., *The Role of the Arctic for Defense: Implications for Scientific Cooperation*. Preprint, Fritjof Nansen Institute, Lysaker, Norway, 1988.
10. Published by United Nations, General Assembly Security Council document A/42/621 S/19 180, 1987.
11. Roederer, J. G., Glasnost in the Arctic. *EOS*, Transactions of the American Geophysical Union, Vol. 69 May 3, 1988 p. 561. See also Friedheim, R. L., The regime of the Arctic—Distributional or integrative bargaining? *Ocean Development and International Law*, **19** (1988), 493–510.
12. Hickok, D. M., Weller, G., Davis, T. N., Alexander, V. & Elsner, R., *United States Arctic Science Policy*. Alaska Division of the American Association for the Advancement of Science, University of Alaska Fairbanks, Alaska, 1981.
13. *A Study of United States Arctic Research Policy and the Possible Role of the Naval Arctic Research Laboratory*. Report, Departments of the Interior, Defense, and Energy, Washington, DC, 1982.
14. United States arctic research plan, biennial revision 1990–1991. *Arctic Research of the United States*, **3** (1989) 3–49.
15. For a critical discussion of the greenhouse effect and global warming, see Singer, S. F. (ed.), *Global Climate Change*. Paragon House Publishers, New York, 1989.

Chapter 2

The Climatology and Meteorology of Arctic Air Pollution

WOLFGANG E. RAATZ

Deutscher Wetterdienst, Wallneyer Str. 10, D-4300 Essen
Federal Republic of Germany

2.1 INTRODUCTION

Arctic air pollution was first observed as a visible haze over the Alaskan Arctic during the Ptarmigan weather reconnaissance flights.[1] Subsequent research has shown that this visible haze may be found over the entire arctic region and that visibility-reducing particles constitute just one portion of the arctic air pollution burden.[2,3,4] Haze layers consist of aerosols and gases, most of them having pollution-derived components from anthropogenic activities. However, at times natural components like desert dust particles may dominate.[5,6] Recently, it has been proposed that a portion of the visibility-reducing haze in the Arctic may be composed of precipitating ice crystals.[7]

Arctic air pollution, as documented by a number of ground-based sampling sites, constitutes an arctic-wide phenomenon, exhibiting a strong seasonal variation with a winter/spring maximum and a summer minimum. This seasonal variation repeats itself remarkably from year to year, hence the occurrence of arctic air pollution constitutes a distinct feature of the arctic atmosphere. In addition to this seasonal trend, episodic variations on time-scales down to a day or even hours have been noted.[8,9]

Transport and deposition of aerosols and gases are controlled to a large extent by the state of the atmosphere, hence the synoptic and climatological conditions of the arctic atmosphere need to be discussed in terms of their relevance to the characteristic features of arctic air pollution. Overviews on arctic climates have been given by Wilson[10,11] and Vowinckel & Orvig,[12] for example.

In general terms the Arctic is an ice-covered mediterranean sea surrounded by the vast continental areas of North America and Eurasia. Despite the distinctive environment of the Arctic there is no single definition available to delineate its limit. The Arctic Circle forms the southern boundary of the Arctic in an astronomical context, but this criterion bears little relationship to climatic characteristics or vegetation, for example. In many respects, a meteorological definition remains equally unsatisfactory. For example, the limit of arctic air masses, known as 'the polar front', fluctuates from day to day with major longitudinal differences, as well as seasonal and interannual variations. However, for the meteorological/climatological aspects of arctic air pollution this definition of the arctic region as being bounded to the south by the polar front appears the most suitable.

The origins of aerosols and gases in arctic haze may be both natural and pollution-derived. These sources may be within the close vicinity of the observed haze event, or they may be thousands of kilometres away.

Local sources within the inner Arctic are usually point sources. They may include open leads, forest fires,[13] smoke from naturally burning carbonaceous materials and pyrites,[14] volcanoes, local industry like oil fields/wells and smelters.[15] These sources may be of local importance, but they seem to contribute little to the arctic-wide phenomenon of arctic air pollution.[16,17]

Industrial source regions covering large areas of the mid-latitudes appear to be the major contributors to arctic air pollution, especially during times when these regions lie north of the polar front.[18] Major industrial regions are located in the northeastern USA/southeastern Canada, western and eastern Europe, western USSR, the Ukraine, southern Urals, western Siberia, Korea and Japan. It is of interest to note that the main industrial areas of the USSR are located at about 50–55°N, followed by western Europe at 50°N, eastern Europe and Canada at 45°N, and the northeastern USA at 40°N. East Asian industrial source regions are farthest away from the Arctic at 30–35°N.[19] A distant natural source of arctic haze may be the deserts of northern and western China (~40–45°N), and those of the southern USSR (~45°N), as well as the

Saharan desert of Africa (20–30°N). In general, the chances of being north of the polar front are less the farther south the source region is located. The importance of a source contributing to arctic air pollution depends on its strength, proximity to the arctic region (or arctic air mass), and the frequency of synoptic conditions creating a poleward flow.

2.2 FEATURES OF THE MIDDLE AND UPPER TROPOSPHERIC FLOW

2.2.1 The circumpolar vortex

In the middle and upper troposphere together with the lower stratosphere (i.e. the atmosphere between about 3 and 15 km), the arctic region is covered by an extensive cold-cored centre of low pressure.[10,20] The centre of this circumpolar vortex does not coincide with the geographic North Pole, and at times there are two or three centres discernible. During winter, these centres seem to cluster over Ellesmere and Baffin Islands, to the north of Novaya Zemlya, and northwest of the Sea of Okhotsk (Fig. 2.1(a)). During summer, there is usually only one centre of the vortex which is then located in the vicinity of the pole (Fig. 2.1(b)).

The southern boundary of the polar vortex is characterized by a zone of strong pressure and temperature gradients, and consequently by a zone of strong westerly winds. These are the so-called upper westerlies. The upper westerlies usually do not flow parallel to a latitude circle, but instead show wave-like disturbances. A circulation of the upper westerlies with small amplitude wave disturbances is referred to as a zonal situation (Fig. 2.1(a), (b)), large amplitudes characterize a meridional situation like the extended trough over the Atlantic Ocean in Fig. 2.1(c). During a zonal circulation there is little air mass exchange between the middle and northern latitudes, whereas during strong meridional flow air masses may travel from subtropical latitudes to the north.

Waves with small amplitudes move relatively rapidly towards the east (transient waves), whereas waves with larger amplitudes move more slowly, eventually becoming stationary (standing waves), or may even show a slight westward movement.

During winter the temperature difference between pole and equator is largest resulting in a circulation which is about three times more intense than during summer. The magnitude of the winter amplitudes is about twice the magnitude of the summer amplitudes.

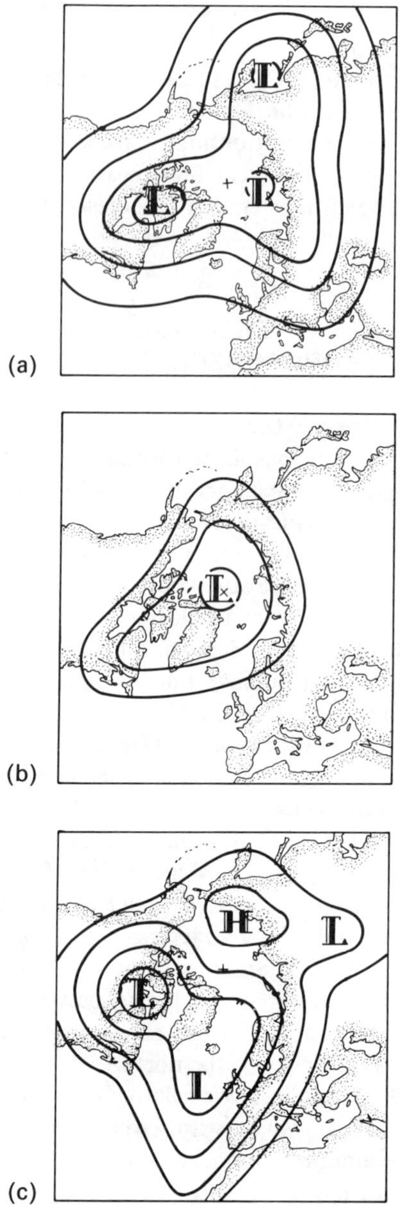

Figure 2.1. Schematic representation of the upper tropospheric flow. (a) Mean winter circulation conditions; (b) mean summer conditions; (c) a meridional circulation pattern during winter with a deep trough over the Atlantic Ocean and blocking over Eurasia.

The mean winter conditions as presented in Fig. 2.1(a) are characterized by deep troughs of low pressure extending southward towards Japan and towards eastern Canada, and a weaker trough towards the Black Sea. During summer, these troughs are much less vigorous, but the one over eastern Canada is still clearly discernible (Fig. 2.1(b)).

The upper air circulation is characterized by a number of large-scale patterns.[21] They represent well-defined modes of the general circulation, which may not be identified when looking at mean synoptic maps. However, some features of the mean trough/ridge pattern can be explained by the permanent influence of mountain barriers to atmospheric circulation as well as by the occurrence of blocking situations (see Section 2.2.2). The degree of expansion/contraction of the circumpolar vortex and the subsequent abnormal displacement of the upper westerlies, as well as the occurrence of standing or transient wave disturbances, has a strong influence on the air exchange between high and low latitudes both at upper levels as well as at low levels, since the movement of near-surface pressure systems is controlled by upper-air flow.

2.2.2 Blocking situations

When the amplitude of the wave disturbances of the upper westerlies becomes large enough, a quasi-stationary meridional flow pattern develops, the so-called blocking situation. Although there are several definitions used to characterize a blocking situation, the main features are: (a) the splitting of the upper westerly current into two branches (as in Fig. 2.1(c) over Eurasia) and (b) the formation of a persistent, vertically well-developed surface anticyclone in mid-latitudes.[22,23,24]

Blocking situations occur with pronounced seasonal and longitudinal variations. In general, blocking develops most frequently during late winter/spring, whereas during summer/autumn a low frequency of blocking is noted. In addition, most blocking ridges and anticyclones tend to be located over the eastern North Atlantic region (roughly 10°E to 20°W) and over the western North Pacific region (roughly 120–180°W). On the average, Atlantic blocks occur twice as often as Pacific blocks, hence there is considerably more air mass exchange along the Atlantic side of the arctic region.

The annual variation of blocking frequencies differs slightly for the two regions. In the Atlantic region there is a maximum of occurrence from February through April. During the winter months most blocking situations form over the Eurasian continent (0–20°E), whereas during spring the position of blocking highs moves westward on to the Atlantic

Ocean (0–25°W).[25] The minimum of blocking situations occurs during the summer months over the Atlantic Ocean. In the Pacific region, most of the blocking situations occur during January, whereas during the following months the frequency is considerably less. There is, however, a secondary maximum in June, which is most pronounced over the central and western parts of the North Pacific.

A blocking situation most often lasts for about four to eight days. It appears that the Atlantic region has a much greater ability to support long-lasting episodes of strong meridional flow than the Pacific region.

It has been noticed that the seasonal variation of blocking frequencies over the Atlantic region roughly follows the seasonal variation of air pollution concentration in the Norwegian Arctic.[26,27] However, a significant correlation between individual pollution episodes and blocking events could not be found. Similarly, no significant relationship could be established between occurrence, position and amplitude of wave disturbances of the upper westerlies and air pollution episodes at Barrow, Alaska.[28]

From a synoptic point of view, it cannot be expected that near-surface pollution concentrations will correlate with upper air flow. Near-surface air flow transporting the pollutants may differ significantly in direction and strength from the upper air flow, especially in the Arctic when air masses are decoupled from each other by strong temperature inversions. On the other hand, as mentioned before, the upper air circulation does, to a large extent, control the near surface pressure patterns so that a general relationship between the state of the upper air circulation and pollution episodes should exist.

It has been noticed that most transport events of pollutants from European sources to the Alaskan Arctic are characterized by initial blocking over Europe and a meridional circulation pattern over the Arctic, especially during late winter/spring. On the other hand, during early/mid-winter the circulation appeared more often zonal in character, thus supporting flow along the rim of the Arctic Basin from northern and central Soviet sources to Alaska.[8]

2.2.3 The polar front

The boundary between tropical and polar air masses is referred to as the polar front. Near the surface, the polar front is recognizable as a zone of frequent cyclonic activity. It varies rapidly in time, location and intensity.[29] At upper levels, the polar front coincides with the southern boundary of the circumpolar vortex. The polar front in winter may be identified at the 500 mb level by the isotherm band between −24°C and −34°C, whereas during summer its position lies between −14°C and −24°C.[30] In winter the polar front lies roughly between 40–50°N, whereas during summer its mean position is shifted northward between 60° and 70°N.

It has been proposed that the seasonal variation of the polar front may be responsible for the seasonal variation of the pollution-derived aerosols observed in the near-surface arctic air.[18] During winter/spring, industrial source regions of the mid-latitudes would be located north of the polar front (i.e. within the arctic air mass), whereas during summer the same source regions would be located to the south of the front, thus being 'decoupled' from the arctic region. This coupling–decoupling hypothesis is supported by the concept of isentropic flow (see Section 2.2.4).

Sometimes, and then quite often on a regional scale, a further distinction between polar air and arctic air masses is made, both being separated by the arctic front. Since the extreme cold air masses of the Arctic are restricted to the lower troposphere, the arctic front has only a small vertical extent, and quite often shows little weather activity associated with it.

2.2.4 The concept of isentropic flow

The adiabatic process is defined as one in which there is no heat added or taken away. This means that, to a first approximation, an air parcel in the free atmosphere carried upwards cools adiabatically (about 1°C per 100 metres) with the decreasing pressure that it experiences and, conversely, a descending air parcel will heat up as the pressure increases. During a dry-adiabatic process (assuming no saturation over water vapour) an air parcel can be traced by a conservative quantity, the potential temperature. Surfaces of equal potential temperature are also called isentropic surfaces: they slope upward towards colder regions.[31]

An air parcel starting at mid-latitudes north of the polar front (point A in Fig. 2.2) has to rise when moving into a region covered by a colder air mass, finally arriving at point A1. However, the rise will be moderate since the air mass at point A has temperature characteristics similar to those of the air mass at point A1, because both air masses are part of the arctic air mass north of the polar front. An air parcel starting at mid-latitudes south of the polar front (point B) initially rises moderately until it reaches the polar front. Within the polar front, a zone of sharp horizontal temperature gradient, the air has to rise strongly, finally followed by a moderate rise over the polar region up to point B1.

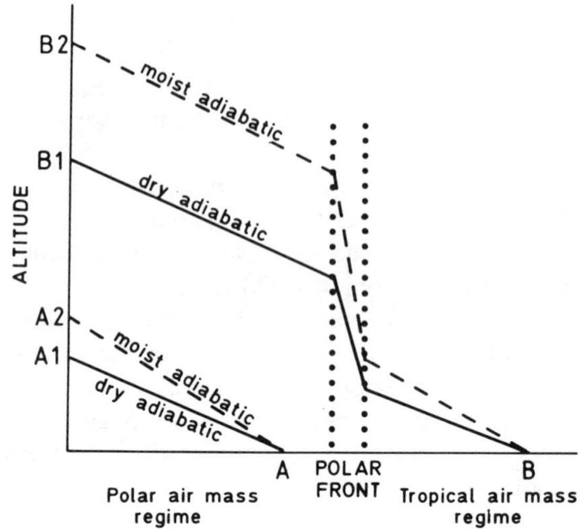

Figure 2.2. Schematic representation of isentropic flow into the Arctic from sources being either located north of the polar front (A) or south of the polar front (B).

The concept of a dry-adiabatic process has to be modified when the air containing moisture eventually becomes saturated, The release of water adds energy to the air parcel, so that during a moist-adiabatic process, the decrease of temperature with decreasing pressure is less than during a dry-adiabatic process. Air that leaves point A in a state of saturation with respect to water vapour will ascend along a moist-adiabatic path, now arriving at point A2. If an air parcel were to leave point A, initially being unsaturated, but achieving saturation before reaching its final position, it would have to ascend to an altitude between points A1 and A2. The same arguments apply to the situation when an air parcel starts at point B. Its final position will be somewhere between B1 and B2. However, a moist air mass will experience enhanced scavenging by clouds and precipitation along the polar front. The amount of moisture an air parcel can contain depends on the parcel's temperature, i.e. moisture content decreases with decreasing temperature. As a result, the difference in height between A1 and A2 is of the order of several hundreds of metres, whereas between B1 and B2 the difference may amount to several kilometres.

The concept of isentropic flow has been applied to the transport of arctic air pollution.[19,32,33] A dry continental air mass from the Eurasian continent having temperatures similar to those over the Arctic Basin would allow for transport within the lower troposphere. Slight variations in the moisture content and subsequent saturation of the continental air mass will slightly lift the level of transport. This might cause some layering of arctic haze as has been observed over the Arctic (see Section 2.6). Very dry and warm air masses originate from the desert areas of the subtropics. Desert dust may enter the Arctic at high altitudes without having experienced much scavenging. In contrast, moist and warm air masses from mid-latitudes south of the polar front would lose most of their pollutants due to scavenging at the polar front.

A simulation of transport of SO_2 and SO_4 into the Arctic along isentropic surfaces suggested that during winter conditions emissions of Soviet sources contribute about 75% of the total near-surface SO_4 burden, whereas European sources contribute only 24%.[34] At heights above 3 km, emissions from European sources become more important than Soviet sources. North American sources may have a slight influence as well.

It should be pointed out that in the atmosphere adiabatic conditions are not always fulfilled. For example, during the dark arctic winter months air masses tend to cool radiatively (i.e. diabatically) which may not be neglected in the case of long travel times.[32] In spring and summer, the haze aerosols may absorb solar radiation and thus raise the temperature of their atmospheric environment.[35]

2.3 FEATURES OF THE NEAR-SURFACE ATMOSPHERIC FLOW

2.3.1 The four semi-permanent pressure systems

The near-surface atmospheric circulation of the Arctic and the mid-latitudes is characterized by four major semi-permanent pressure systems: the Icelandic and Aleutian lows as the two oceanic low pressure systems, and the Asiatic and North American highs as the two continental high pressure systems. The two continental highs are occasionally linked by a ridge across the Arctic Ocean.[20,21,36–38] However, there is no semi-permanent anticyclone over the inner Arctic as was assumed in the early years of arctic meteorology.[39]

During winter (Fig. 2.3), the Icelandic low extends over a large area, stretching from the south of Greenland to Novaya Zemlya. Its influence even reaches the pole. This large low pressure complex includes three

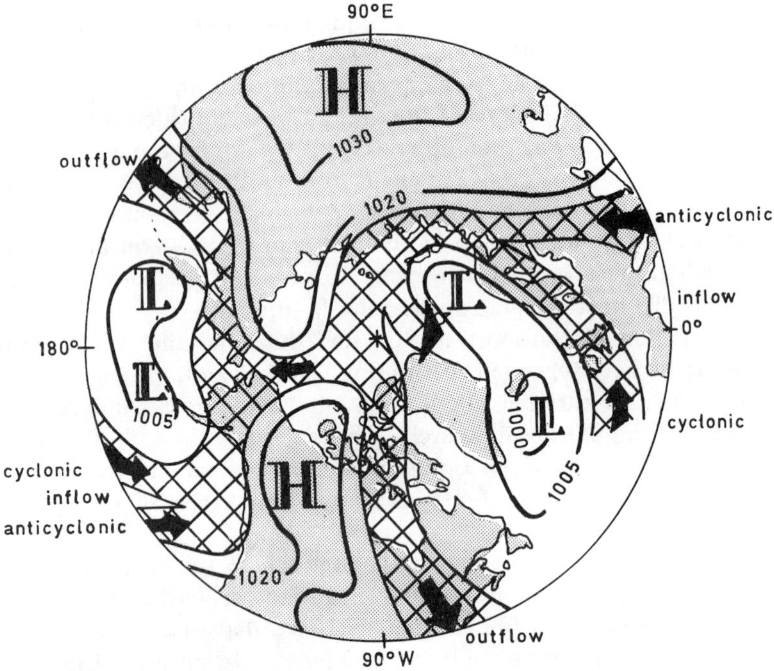

Figure 2.3. Schematic representation of the mean winter near-surface pressure distribution (mb) as well as the main zones of air flowing into/out of the arctic region.

subcentres: one south of Greenland, another north of Norway, and one north of Cape Chelyuskin. The Icelandic low is the result of migratory low pressure systems which tend to decay at these locations. The Icelandic low remains strong during the winter months, October to March,[40] whereas during the summer months it weakens. The subcentres to the east disappear and a rather formless area of low pressure remains over southern Baffin Island.

The Aleutian low is located over the Pacific Ocean occupying a more southerly position than the Icelandic low. During winter, its mean position extends from Kamchatka to the Gulf of Alaska, representing the graveyard of numerous Pacific cyclones. In contrast to the Icelandic low, the Aleutian low has little influence on the inner Arctic, because the mountain barriers of Alaska and northeast Siberia hinder the penetration of cyclones into the North. The Aleutian low weakens rapidly after January[19] and practically disappears during the summer months.

The two high pressure systems are primarily thermally produced. They form as the land mass cools during the beginning of winter and disappear towards summer. Therefore these anticyclones are usually shallow surface features, unless warm air advection in the upper troposphere converts them into a dynamically controlled high pressure system.

Since the Eurasian continent is two to three times wider in mid-latitudes than the North American continent, the Asiatic high is by far the larger and the more persistent of the two high pressure systems.[41,42] During October, the Asiatic anticyclone forms over eastern Siberia. Besides its central position, the Asiatic anticyclone tends to form secondary cells which at times are better developed than the main cell. During early winter an eastward extension becomes of importance, whereas towards spring the Asiatic high expands its influence towards the west, reaching central Europe. During April, the Asiatic high weakens and eventually disappears being replaced by a weak low pressure system.

The North American high is much less developed than the Asiatic high. Even during the winter months its persistence is poor, thus varying rapidly in strength and position. Only at times does it control the surface circulation, e.g. when it is supported by a high pressure ridge in the upper troposphere.

Due to these well-developed pressure systems the winter circulation of the mid-latitudes and the Arctic is intense, whereas during summer, due to the weakening pressure gradients, the near-surface circulation quite often appears sluggish.

Rahn evaluated the importance of these four semi-permanent pressure systems.[19] He concluded that the Icelandic low and the Asiatic high create much better atmospheric conditions for air flow from mid-latitudes into the Arctic than the two other pressure systems. In another study it was suggested that most pathways of pollution transport to Alaska were controlled by the Asiatic high, and the seasonal variation in frequency could also be explained to a large extend by the seasonal variation of the position of the high.[8]

2.3.2 Tracks of cyclones and anticyclones

Most mid-latitude cyclones develop from wave disturbances in the polar front. In general, cyclones will then travel in an easterly direction, steered by the upper westerlies. During their life cycle from a wave disturbance to an occluded system, cyclones cross from the warm side of the polar front to the cold side, then stagnate and fill up in these areas which have been identified as the mean position of the semi-permanent lows. Thus, cyclone tracks spiral northward into the arctic region.[43] Since cyclones are connected to the polar front, the position of their mean track shifts as the position of the polar front changes during the year. Locally, and only for short periods of time, thermal contrasts between land and ocean and their varying snow/ice cover create secondary cyclone tracks which exist only as long as the thermal contrast lasts. In general, cyclones move along marine pathways; when cyclones reach land masses (or ice-covered oceans) they usually fill up rapidly.

In contrast anticyclones, as part of the polar front, have a tendency to move southward. Thus they have little importance in transporting air masses into the arctic region. Only during blocking situations do quasi-stationary anticyclones initiate northward transport along their western flank. During these conditions they may also drift westward. However, mid-latitude anticyclones rarely move into the Arctic.

It has been emphasized that pollution transport along marine (cyclonic) pathways will experience massive scavenging by clouds and precipitation.[8,19] Continental (anticyclonic) pathways, on the contrary, are characterized by less scavenging. However, since anticyclones do not move as rapidly as cyclones, and they rarely move northward, long-range transport may occur during circulation patterns which are controlled by a number of quasi-stationary anticyclones.[8] In this case, polluted air masses travel within so-called transport zones, in zones of relatively strong pressure gradients along the flanks of the anticyclones.

2.3.3 Large-scale circulation patterns

Since the polar vortex and the upper westerlies exert a strong influence on surface pressure systems, it is not surprising to observe that the near-surface circulation also tends to be characterized by a number of quasi-stationary large-scale sea-level pressure patterns. For the European continent the Katalog der Grosswetterlagen Europas (a catalogue of large-scale weather patterns) represents a well-known example.[44] For the Arctic, a number of distinct circulation types has been identified as well.[45] These circulation patterns were most persistent during December through May and changed abruptly from one pattern of stability to another. From June to November the general level of persistence was lower, and the patterns less developed. A set of quasi-persistent circulation patterns could be related to pollution episodes at Barrow, Alaska.[8] The persistence of these circulation patterns provided sufficient time for air masses to cross the Arctic. The seasonal variation of aerosol types at Barrow could be related to the seasonal variation of transport types from different mid-latitude industrial source regions.

The above mentioned studies were based upon subjective identification and classification. An objective approach was performed using a correlation technique, the Kirchhofer classification.[46] Eight patterns were identified. The most frequent pattern looks like the mean surface pressure distribution in winter (as shown in Fig. 2.3). The seasonal variation of this most frequent pattern, however, exhibits a notable feature. Its occurrence is most frequent during winter and shows an abrupt decrease towards April followed by a minimum during the summer months. This result strongly supports the concept that the seasonal variation of transport may be the primary cause for the seasonal variation of arctic air pollution as postulated in several studies.[8,26,47] Unfortunately, objective classification schemes are not as objective as one might hope. There is still a high degree of subjectivity left within each approach, and different classification schemes tend to produce different synoptic types.[48]

2.4 SPECIAL METEOROLOGICAL FEATURES OF THE ARCTIC TROPOSPHERE

2.4.1 Temperature inversions

One of the most characteristic features of the lower arctic troposphere is the persistence of intense temperature inversions.[49] An inversion is present almost continuously over the entire arctic region in winter and over snow and ice-covered areas in summer. The increase of temperature with height within the inversion layer represents a barrier for vertical turbulent mixing and may therefore prevent significant mixing of pollutants injected into the Arctic.

The structure of polar inversions can be complex and is particularly sensitive to local conditions.[11] It may explain some of the layered structures observed in arctic haze (see section 2.6).

There are two main types of inversions. The first type is represented by the surface inversions, when the increase of temperature with height starts at the ground. The second type includes the lifted inversion. In this case, a turbulent mixing layer separates the inversion base from the ground (Fig. 2.4).

During winter, about 85% of all temperature profiles show a surface inversion, whereas only 10% show a lifted inversion. In contrast, during summer lifted inversions reach a frequency of occurrence of about 50%, whereas the frequency of surface inversions drops to 30%. There is an abrupt change from winter to summer conditions occurring in April/May, the corresponding autumn transition period is far less abrupt.

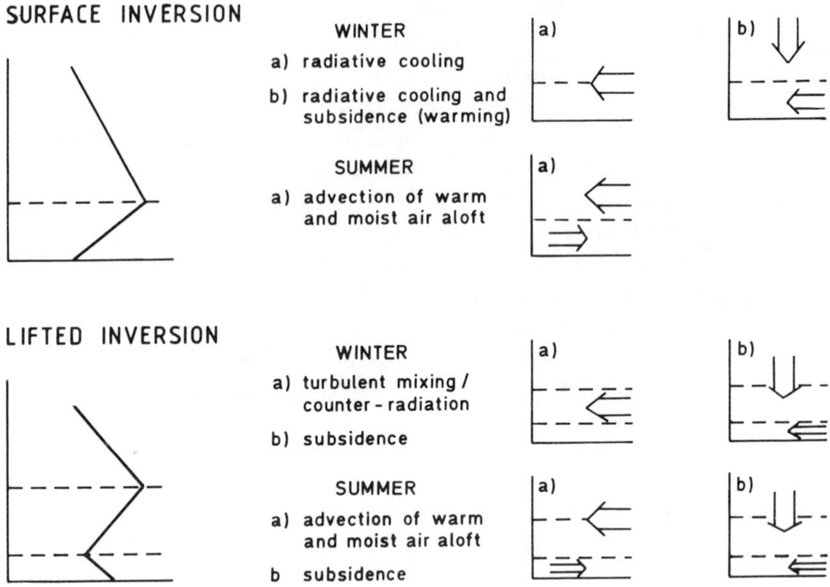

Figure 2.4. Schematic overview on inversion types, main processes of formation, air masses involved.

In winter, surface inversions form during anticyclonic conditions as a result of energy loss from the snow-covered surface in the absence of incoming solar radiation. The inversion may be weakened or even be destroyed by counter-radiation from a sufficiently thick cloud cover or by strong winds. Typically, surface inversions extend up to 2000 m in height accompanied by a temperature increase of about 10°C or more from the ground to the inversion top.

During such synoptic conditions there is only one air mass present in which the temperature inversion develops. This may effect the distribution of pollutants. In the originally homogeneously mixed âir mass the pollutants in the upper layers become decoupled from the pollutants of the same origin in the lower layers as the surface inversion develops. Dry deposition, for example, may then effectively remove pollutants from that lowest layer. On the other hand, pollutants released at very cold temperatures within the inversion layer will be restricted to low-level transport without much vertical mixing. In the first case, pollution concentrations at upper levels will be higher than those near the surface, whereas in the second case the vertical distribution will be reversed.

During summer, surface inversions form due to the advection of warmer (mid-latitude) air masses over a cold snow or ice-covered surface. The surface temperature is kept constant at a temperature near freezing point, since all incoming solar energy is used up for the melting process and is therefore not available to warm up the lowest layers of the troposphere which are in contact with the ground. These surface inversions have only a modest thickness of 200–300 m, and the temperature increase with height remains within a range of 1–4°C. Since there are two air masses involved producing the surface inversion, the concentration and the chemical/physical characteristics of the pollutants in these air masses may differ significantly. Since the summer surface inversions are often capped by a ceiling of stratus clouds, concentrations near the surface generally may be lower than aloft due to scavenging by clouds and precipitation.

During winter, the occurrence of lifted inversions is of little importance. The primary reason for their development is energy loss of the surface, resulting in a surface inversion. Due to subsequent turbulent mixing (strong winds) or increased atmospheric counter-radiation by clouds the lowest layer may be destroyed, thus producing a lifted inversion. The possible distribution of pollutants may be similar to that described for the winter surface inversions. Lifted inversions may also develop due to subsiding air within anticyclonic systems. Quite often the lifted inversion and the surface inversion combine to form one very stable surface inversion. In this case, the upper/middle tropospheric air mass may differ significantly from the air mass in the lower troposphere. In some circumstances, stratospheric air may subside down to levels of the middle and lower troposphere.[50,51]

During summer, lifted inversions form when warm and moist air masses from mid-latitudes enter the Arctic. The warm air cannot penetrate to the surface and glides on top of a shallow near-surface cold air layer. The two air masses mix gradually, and the cooling of the warmer mid-latitudinal air leads to condensation and to the formation of stratus clouds (see Section 2.4.2). Evidently, the two air masses involved will have different chemical/physical characteristics due to their different origin. The cloud cover will increase scavenging and affect the concentration of pollutants in the near-surface air mass. Subsidence in summertime anticyclones also leads to the formation of lifted inversions. In this case, dry air is being transported from upper to lower levels, so that relative humidity rapidly decreases above the inversion, a clear distinction between lifted inversions due to subsidence and those due to advection.

Again, two different air masses are involved determining the vertical distribution of pollutants. Although quite a number of vertical profiles of arctic haze have been obtained, no systematic study of the different air masses involved has been attempted.

2.4.2 Clouds and precipitation

Among the many factors determining the formation and distribution of clouds the main ones include: frontal activity and cyclonic systems, the amount of open water, inversion structure, air temperature relative to the water temperature, and such local effects as onshore winds and orographic uplift. The first of these factors is most important for the formation of the middle and high cloud types, while the other factors have a strong influence on the formation of low cloud types.[11]

There exists a marked seasonal variation of total cloud cover over the Arctic Basin. The total average cloud cover increases from 40–60% in winter to 70–90% in summer, the change occurring quite rapidly during April/May.[12,52] This seasonal variation of total cloud cover is mainly due to the seasonal variation of low-level stratus clouds, being a quasi-permanent summer phenomenon. Further to the south of the rim of the Arctic Ocean the seasonal variation of cloud cover varies considerably with longitude owing to the distribution of land and sea, areas of frequent cyclonic activity and orographic influence.

The geographic variation of total cloud amount in winter is closely related to the distribution of cyclones/anticyclones and of frozen/open waters.[53] The greatest cloud amount is observed over the Norwegian–Barents Seas due to the many cyclonic systems there. Towards the east, over the Polar Basin and western Siberia, the available moisture and consequently the cloud amount decreases. Over North America, the Alaska Range forms a barrier restricting the penetration of moist air from the Pacific, and the high total cloud amount is confined to the coastal areas of southern Alaska and Kamchatka. Over Canada, the increased total cloud amount over the Davis Strait–Baffin Bay area again reflects the frequent passage of winter storms and the presence of ice and open water.

During summer, the arctic region is quite often covered by persistent and extensive layers of stratified clouds. One type (referred to as type I stratus cloud) is represented by a thin low level stratus cloud, which is a local phenomenon and occurs predominantly near the coast. The occurrence near the coast is due to the advection of moisture from the open leads in the sea ice by the summer sea breeze.[54] The second type (type II stratus cloud), observed over the Arctic Ocean and normally referred to as the arctic stratus, is characterized by horizontal homogeneity, with mean horizontal extensions of 400 km, in extreme cases up to 2000 km. This type is formed by the advection of moisture from the south by cyclonic activity.[54,55] Recent studies, however, found significant variability in the type and distribution of clouds throughout the Arctic Basin during the spring and summer months.[55,56] The variability appears to be closely related to synoptic pressure systems.

The highest amount of precipitation is measured in the Subarctic, along the windward side of the coastal mountains bordering the North Pacific and the North Atlantic.[11] Towards the interior of the continents and northward to the Arctic Ocean precipitation amounts decrease rapidly. Interior arctic sites are deserts with annual precipitation of about 100 mm. Precipitation mainly occurs with cyclonic activity, and the amount and seasonal variability reflects the passage of cyclones. Over the more southerly continental areas, the mean monthly values show a marked summer and autumn maximum due to the convective nature of the precipitation there. For the very low temperatures of the high arctic and sub-arctic winter, the amount of precipitable water in the atmosphere is negligible. Owing to the continental nature of much of the cold regions, distance from moisture sources becomes an important factor in controlling precipitation patterns. The existence of open water in summer/autumn is a dominant factor in the seasonal distribution of precipitation at arctic coastal and island sites.

It has been hypothesized that the seasonal variation of arctic air pollution could be controlled by the seasonal variation of clouds and precipitation due to scavenging by increased dry and wet deposition.[57,58] During stratus conditions the total amount of water in the atmospheric boundary layer rains out and becomes resuspended 10–20 times per 24 hours.[59] On the other hand, from weather reconnaissance flights it has been reported that during summer months haze was observed at different altitudes than the cloud layers.[6] This might suggest that during the summer months transport of arctic haze may occur at upper tropospheric layers above the quasi-permanent stratus clouds.

It has been pointed out that during winter conditions cloud amount and precipitation encountered should be low when the transport of pollutants occurs under anticyclonic control.[8] Occasionally, anticyclonic circulation patterns may also be present during summertime.

Hypothetically, the atmosphere may be subdivided into two synoptic-scale systems: (1) regions where there is little or no precipitation (dry regions), and (2) regions where precipitation is occurring (wet regions).[60]

Thus, there is always the statistical possibility that pollutants will travel entirely through dry regions. In addition it may be speculated that clouds existing for only short periods of time may scavenge out pollutants locally from a homogeneous hazy air mass, leaving the typical patchy distribution of arctic haze (see also Section 2.6).[58]

2.5 POLLUTION EPISODES AND TRANSPORT PATHWAYS

2.5.1 Episodic behaviour of arctic air pollution

Besides their seasonal variation, pollution concentrations at all arctic sites exhibit short-term fluctuations on time-scales down to a day or even hours.[8,9] The term 'episodic' may be applied to periods of relatively high concentrations of air pollutants compared to the background level.

By their very nature pollution episodes are associated with specific (favourable) meteorological/synoptic situations, e.g. rapid/direct transport without much scavenging. Ideally, episodes should be dealt with as individual case studies using all combinations of relevant and available meteorological, chemical, and physical information. In such a manner they provide examples to illustrate and identify the characteristic features of long-range transport of air pollutants to the Arctic. After having identified these characteristics a climatological interpretation or the application of transport models may be useful.

Unfortunately, the available information is quite often limited. Transport leading to pollution episodes has often only been identified by isobaric or isentropic back-trajectories as a relatively simple 'means'.[61,62] Besides the problems arising when computing representative trajectories over a region of sparse data coverage, little knowledge is gained about the meteorological processes involved.[63,64] In a different approach, synoptic considerations were emphasized when identifying and classifying transport pathways from synoptic charts.[8,65,66] As an additional piece of information on the origin and fate of the pollutants, the chemical signature of aerosols at the probable source region and at the receptor site was used. Thus, in these studies a combination of meteorological and chemical evidence was employed in order to strengthen the analyses.

2.5.2 Transport pathways to the Canadian Arctic

Rahn has argued that North American sources contribute little to the arctic pollution burden.[19] Even at the Canadian Arctic monitoring sites of Alert and Mould Bay little influence from these sources has been noted.[57] The main reason is that polluted air masses originating from northeastern North America will usually travel east to the Atlantic Ocean along a cyclonic (marine) path, with efficient scavenging en route. Most pollutants will have been washed out before reaching Cape Hatteras.[13] In addition, cyclonic flow into the Arctic will be blocked by the high plateau of Greenland.[19]

On the other hand, trajectories have shown that pollution episodes are most often correlated with flow across the Arctic Basin.[57,67] In these cases transport may be controlled by the Asiatic high, its extension over Kamchatka, and another cell over the Alaska/Yukon area (Fig. 2.5(a)). In some rare circumstances, there may be transport of pollution along a continental/anticyclonic path from the south to the Canadian Arctic.[8] During such a situation (Fig. 2.5(b)), the upper westerlies are blocked over the North American continent, and a surface high pressure cell dominates over the eastern half of the continent.

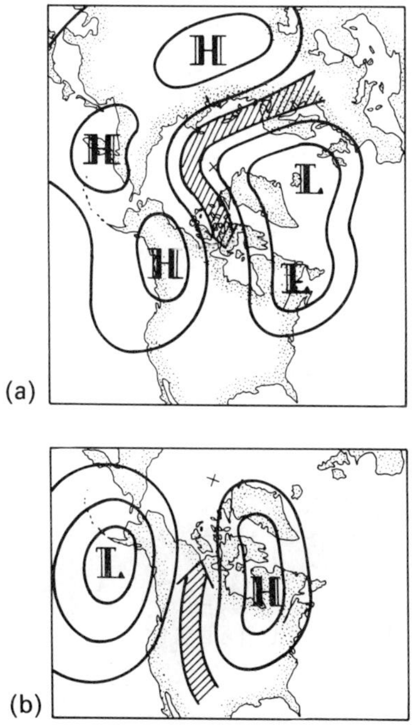

Figure 2.5. Schematic representation of near-surface air flow during air pollution transport to the Canadian Arctic: (a) from Eurasian source regions; (b) from North American source regions.

2.5.3 Transport pathways to the Alaskan Arctic

There have been numerous studies on pollution episodes at Barrow, Alaska, using trajectories,[61] air mass characteristics,[50,68] and synoptic patterns.[8] Pollution episodes occur when an arctic (dry continental) air mass from the north reaches the Alaskan coast. Raatz & Shaw identified different transport pathways from different Eurasian source regions.[8] During winter, zonal circulation types favour transport from Soviet source regions (Fig. 2.6(a)), whereas during spring meridional types support transport from European sources across the Arctic (Fig. 2.6(b)). During the zonal circulation types, the Asiatic anticyclone dominates over the eastern and central regions of the Asian continent and cyclones from the Atlantic Ocean may penetrate into the Arctic Basin. During meridional circulation types, blocking over western Asia/eastern Europe leads to a state of the arctic circulation which is characterized by the

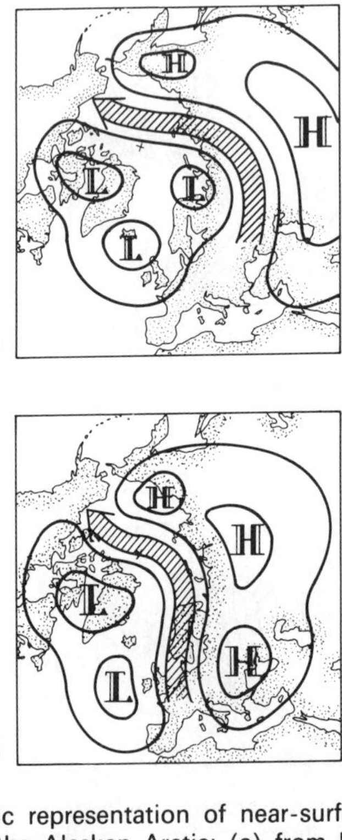

(a)

(b)

Figure 2.6. Schematic representation of near-surface air flow during air pollution transport to the Alaskan Arctic: (a) from Eurasian source regions (zonal flow type); (b) from European source regions (meridional flow type).

presence of just one cold pole over the Baffin Island area (as in Fig. 2.1(c)).

Many of the pollution transport types identified for Barrow pollution episodes may also lead to episodes over the Canadian Arctic when the final segment of the path is slightly modified in such a manner that the polluted air mass reaches Canada.

2.5.4 Transport pathways to the Norwegian Arctic

Pollution transport into the Norwegian Arctic has been studied using trajectories as well as by chemical tracers.[68–72] So-called black and white episodes were identified according to colour and particle composition of air samples.[73] These different episodes were successfully related to distinct atmospheric flow patterns.[65,74] During a 'direct flow' pattern (black episode), a meridional (northward) transport of pollution was established between a cyclone positioned over Iceland/western Europe and a (blocking) anticyclone over eastern Europe (Fig. 2.7(a)). During a 'return-flow' pattern (white episode), the Asiatic anticyclone is positioned over central Asia, in its normal winter location, and the Icelandic cyclonic system extends into the Barents and the Kara Seas.

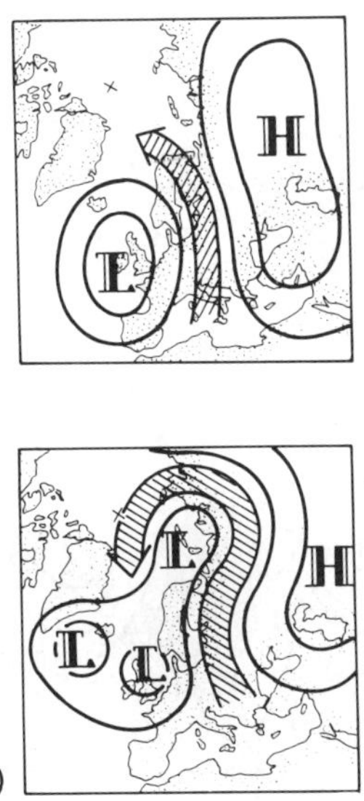

(a)

(b)

Figure 2.7. Schematic representation of near-surface air flow during air pollution transport to the Norwegian Arctic: (a) from European source regions (direct flow); (b) from European/Soviet source regions (return flow).

Transport occurs initially in a northeastward direction, before bending back and 'returning' to the Norwegian Arctic (Fig. 2.7(b)). An example of the meteorological conditions existing during such a return-flow pattern has been studied using detailed meteorological information.[75]

2.5.5 Transport of desert-dust to the Arctic

Occasionally desert dust from eastern Asia may contribute to arctic haze over the Alaskan Arctic.[5] This may be especially true during spring/summer and for haze layers observed at higher altitudes.[6]

During desert dust storms, material is lifted by turbulence associated with a series of cold fronts, usually during dry weather conditions.[76] The dust is then transported on to the Pacific Ocean and may reach the Arctic when blocking of the upper westerlies over the eastern Pacific Ocean/western North America initiates meridional, poleward flow in the middle and upper troposphere (Fig. 2.8(a)). Occasionally, dust from the Saharan desert may also find its way into the Norwegian Arctic (not yet

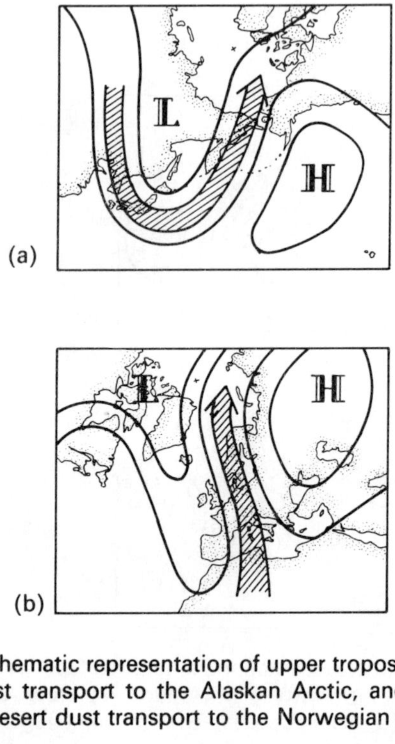

Figure 2.8. Schematic representation of upper tropospheric flow (a) during Asian desert dust transport to the Alaskan Arctic, and (b) during Saharan desert dust transport to the Norwegian Arctic.

documented). Again, strong meridionality is necessary. In those circumstances, a deep trough extends from Iceland to the west coast of north Africa, with cyclonic systems near the surface lifting the material. A corresponding blocking high pressure ridge to the east supports northward flow at altitudes between 3 and 8 km (Fig. 2.8(b)). Dust deposits over central Europe and the British Isles have been reported frequently.[77–79]

2.5.6 Intrusions of stratospheric air into the arctic troposphere

Stratospheric air may enter the arctic troposphere through tropopause folds and descent within anticyclones. Several events of such stratospheric intrusions have been documented.[51,80–82] In some instances, stratospheric air may contain volcanic debris. However, these events appear to occur too sporadically to be considered as a significant source of arctic haze.

2.6 METEOROLOGICAL FACTORS IN THE SPATIAL VARIABILITY OF ARCTIC HAZE

Arctic haze as observed during the Ptarmigan weather reconnaissance flights was characterized by vertical and horizontal variability.[1,6] Surface measurements of air pollutants also exhibit noticeable temporal variability which led to the investigation of pollution episodes due to distinct synoptic patterns causing the injection of pollution plumes into the Arctic. Considering the stable conditions of the arctic atmosphere it was speculated how representative surface measurements could be for the conditions present aloft. Much information on the vertical and horizontal variability was gained when aircraft measurements became available.[83–87]

From these investigations it became apparent that arctic air pollution is characterized by pronounced vertical and horizontal variability, which is most often related to meteorology, but there are cases where a relationship could not be found.

Arctic haze may be observed at all levels throughout the troposphere. In contrast to a polluted mid-latitude air mass, a polluted arctic air mass is characterized by a banded structure,[14] individual layers having diffuse or sharp boundaries.[75,88] The main reason that this banded structure can be maintained over large areas of the Arctic is the extreme stability of the arctic atmosphere.

To a first approximation the banded structure may be explained by the concept of isentropic flow (see Section 2.2.4). Air masses from different source regions and of different meteorological characteristics that conserve their potential temperatures will end up at different altitudes. Sloping of haze layers following isentropic surfaces has been observed occasionally.[14] The banded structure may also be caused or be modified by selective and temporary scavenging of pollutants within an initially homogeneous air mass by clouds and precipitation[58] or by ice crystals[87] leaving 'clean' interstices.

It has been suggested that the layered nature of arctic haze correlates with wind and temperature profiles.[14,89] The haze layer was advected into a region of much cleaner, slower-moving air.

Vertical variability is most pronounced and easily understood when there are two or more distinctly different air masses present, e.g. continental and marine air masses.[50] During anticyclonic conditions the air mass descending in the upper and middle troposphere may even have a stratospheric origin[50] (see Section 2.5.6).

Quite often surface measurements show concentrations lower than those obtained at upper levels. This is especially true when the transport occurs at upper levels, when there is little vertical mixing across the surface inversion, and when scavenging due to dry or wet deposition within the boundary layer is more efficient than aloft. However, different profiles may be observed when the main transport of pollutants occurs within the boundary layer (from sources at extremely cold temperatures),[33] or when either during cyclonic conditions 'clean' marine (cyclonic) air glides on top of the arctic air, or during anticyclonic conditions when 'clean' tropospheric/stratospheric air is subsiding to the top of the surface inversion.[90]

Horizontal variability has been noted to range from 20 km to 200 km or more, with both diffuse and abrupt changes.[6,75,83] Many arguments presented above to explain the vertical variability apply in a similar fashion.

Horizontal variability (patchiness) may also be characteristic of atmospheric diffusion on synoptic scales.[91] Besides the large-scale synoptic control on the transport and the distribution of arctic air pollution, local meteorological phenomena may modify the vertical and horizontal distribution, e.g. land/sea breezes,[92,93] coastal low-level jet streams, downslope winds such as foehn,[87] mountain waves,[14] and influence of the surface (water, ice, etc.) on the inversion structure.

2.7 CONCLUSION

Arctic haze is composed of aerosols and gases from a variety of different sources. The concentration and composition of these haze layers or near-surface pollution episodes may vary significantly in space and time. These variations are controlled, to a large extent, by the arctic atmosphere. The arctic atmosphere cannot be represented by a uniform air mass. Instead, there is constant in and outflow of air masses occurring at all atmospheric levels controlled by a synoptic variability similar to mid-latitudes.[21,36] This synoptic variability is also reflected in the variability of clouds and precipitation, hence scavenging processes. In addition, local effects modify the meteorological conditions as provided by the large-scale synoptic situation. In Fig. 2.9 an attempt has been made to roughly sketch some of the characteristic features of the arctic atmosphere, the various levels of transport from different source regions, as well as the different types of surfaces which have a pronounced influence on the structure of the boundary layer and the air mass involved. Therefore, it should not come as a surprise that the synoptic/meteorological variability as schematically presented in Fig. 2.9 is reflected in a variability of the physical/chemical properties and the distribution of the constituents of arctic air pollution.

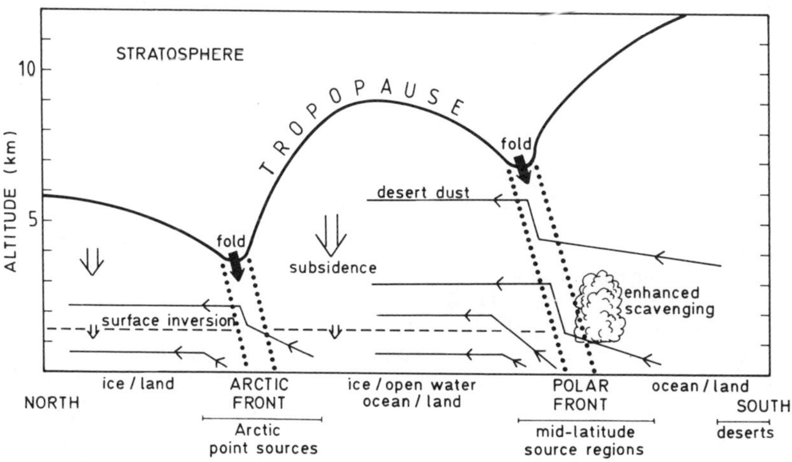

Figure 2.9. Schematic representation of the arctic atmosphere, levels of transport from different source regions, and typical surface conditions.

ACKNOWLEDGEMENTS

Special thanks to all colleagues, especially Glenn Shaw, Kenneth Rahn, Russell Schnell and Barry Bodhaine, who shared with me their ideas and concepts on the meteorological aspects of arctic haze in numerous discussions. My wife Angelika typed the manuscript and drafted the figures.

REFERENCES

1. Mitchell, M., Visual range in the polar regions with particular reference to the Alaskan Arctic. *J. Atmos. Terr. Phys.*, Special supplement (1956) 195–211.
2. Rahn, K. A. (ed.), Arctic air chemistry. *Atmos. Environ.*, **15** (1981) 1345–1516.
3. Rahn, K. A. (ed.), Arctic air chemistry. *Atmos. Environ.*, **19** (1985) 1987–2208.
4. Schnell, R. C. (ed.), Arctic haze. *Geophys. Res. Lett.*, **11** (1984) 359–472.
5. Rahn, K. A., Borys, R. D. & Shaw, G. E., Asian desert dust over Alaska: Anatomy of an arctic haze episode. Geological Society of America, special paper 186 (1981) 37–70.
6. Raatz, W. E., Observations of 'arctic haze' during the 'Ptarmigan' weather reconnaissance flights, 1948–1961. *Tellus*, **36B** (1984) 126–36.
7. Curry, J. A., Arctic cloudiness in spring from satellite imagery: Some comments. *J. Climatology*, **8** (1988) 533–8.
8. Raatz, W. E. & Shaw, G. E., Long-range tropospheric transport of pollution aerosols into the Alaskan Arctic. *J. Clim. Appl. Meteor.*, **23** (1984) 1052–64.
9. Bodhaine, B. A., Dutton, E. G. & DeLuisi, J. J., Surface aerosol measurements at Barrow during AGASP. *Geophys. Res. Lett.*, **11** (1984) 377–80.
10. Wilson, C. V., Climatology of the cold regions. Northern Hemisphere 1. Cold Regions Research & Engineering Laboratory, Report I-A3a, Hanover, New Hampshire, 1967.
11. Wilson, C. V., Climatology of the cold regions. Northern Hemisphere II. Cold Regions Research & Engineering Laboratory, Report I-A3b, Hanover, New Hampshire, 1969.
12. Vowinckel, E. & Orvig, S., The climate of the North Polar Basin. In *Climates of the Polar Regions*, ed. S. Orvig. World Survey of Climatology, 14. Elsevier, Amsterdam, 1970, pp. 129–252.
13. Chung, Y. -S., Air pollution detection by satellites: The transport and deposition of air pollutants over oceans. *Atmos. Environ.*, **20** (1986) 617–30.
14. Radke, L. F., Brock, C. A., Lyons, J. H., Hobbs, P. V. & Schnell, R. C., Aerosol and lidar measurements of hazes in midlatitude and polar air masses. *Atmos. Environ.*, **23** (1989) 2417–30.
15. Shaw, G. E., Evidence for a central Eurasian source area of arctic haze in Alaska. *Nature*, **299** (1982) 815–18.
16. Rahn, K. A. & Heidam, N. Z., Progress in arctic air chemistry, 1977–1980: A comparison of the first and second symposia. *Atmos. Environ.*, **15** (1981) 1345–8.

17. Rahn, K. A., Lewis, N. F., Lowenthal, D. H. & Smith, D. L., Noril'sk, only a minor contributor to arctic haze. *Nature*, **306** (1983) 459–61.
18. Rahn, K. A. & McCaffrey, R. J., On the origin and transport of the winter arctic aerosol. *Ann. N. Y. Acad. Sci.*, **338** (1980) 486–503.
19. Rahn, K. A., Relative importances of North America and Eurasia as sources of arctic aerosol. *Atmos. Environ.*, **15** (1981) 1447–55.
20. O'Connor, J. F., Mean circulation patterns based on 12 years of recent northern hemispheric data. *Mon. Wea. Rev.*, **89** (1961) 211–27.
21. Namias, J., Synoptic and climatological problems associated with the general circulation of the Arctic. *Transactions American Geophysical Union*, **39**, (1958) 40–51.
22. Rex, D. F., Blocking action in the middle troposphere and its effect upon regional climate. *Tellus*, **2** (1950) 275–301.
23. Treidl, R. A., Birch, E. C. & Sajecki, P., Blocking action in the northern hemisphere: A climatological study. *Atmosphere-Ocean*, **19** (1981) 1–23.
24. Lejenäs, H. & Økland, H., Characteristics of northern hemisphere blocking as determined from a long time series of observational data. *Tellus*, **35A** (1983) 350–62.
25. Speth, P. & Meyer, R., The interference of planetary waves for the formation of Atlantic–European blocking: Observational evidence. *Beitr. Phys. Atmosph.*, **57** (1984) 463–76.
26. Iversen, T., On air pollution transport to the Norwegian Arctic. Norwegian Institute for Air Research, OR 59/85, Lillestrøm, 1985.
27. Iversen, T. & Joranger, E., Arctic air pollution and large scale atmospheric flows. *Atmos. Environ.*, **19** (1985) 2099–108.
28. Reiter, E. R., Planetary-wave behavior and arctic air pollution. *Atmos. Environ.*, **15** (1981) 1465–71.
29. Reed, R. J., Principal frontal zones of the northern hemisphere in winter and summer. *Bull. Am. Met. Soc.*, **41** (1960) 591–8.
30. Palmén, E. & Newton, C. W., *Atmospheric Circulation Systems*. Academic Press, New York, 1969.
31. Nagle, F. W., Visualizing a cold air dome. *Bull. Am. Met. Soc.*, **60** (1979) 128.
32. Carlson, T. N., Speculations on the movement of polluted air to the Arctic. *Atmos. Environ.*, **15** (1981) 1473–7.
33. Raatz, W. E., Meteorological conditions over Eurasia and the Arctic contributing to the March 1983 arctic haze episode. *Atmos. Environ.*, **19** (1985) 2121–6.
34. Iversen, T., Simulation of the atmospheric transport of sulphur dioxide and particulate sulphate to the Arctic. Norwegian Institute for Air Research, OR 83/86, Lillestrøm, 1987.
35. Wendling, F., Wendling, R., Renger W., Covert, D. S., Heintzenberg, J. & Moerl, P., Calculated radiative effects of arctic haze during a pollution episode in spring 1983 based on ground-based and airborne measurements. *Atmos. Environ.*, **19** (1985) 2181–93.
36. Keegan, T. J., Arctic synoptic activity in winter. *J. Meteor.*, **15** (1958) 513–21.
37. Reed, R. J. & Kunkel, B. A., The arctic circulation in summer. *J. Meteor.*, **17** (1960) 489–506.

38. Walsh, J. E., Temporal and spatial scales of the arctic circulation. *Mon. Wea. Rev.*, **106** (1978) 1532–44.
39. Jones, P. D., The early twentieth century arctic high — fact or fiction? *Clim. Dynamics*, **1** (1987) 63–75.
40. Glowienka, R., Studies on the variability of Icelandic low and Azores high between 1881 and 1983. *Beitr. Phys. Atmosph.*, **58** (1985) 160–70.
41. Dimitriev, A. A., Situations characterizing various states of the Siberian anticyclone and its northeastern extension in the cold period of the year. In *Problems of the Arctic and the Antarctic*, ed. A. F. Treshnikov. Israel Progr. Sci. Transl., Jerusalem, **29–32**, (1968) 448–55.
42. Lydolph, P. E., Climates of the Soviet Union. *World Survey of Climatology*, 7. Elsevier, Amsterdam, 1977.
43. Klein, W. H., Principal tracks and mean frequencies of cyclones and anticyclones in the northern hemisphere. Res. paper No. 40, US Department of Commerce, Washington DC, 1957.
44. Hess, P. & Brezowsky, H., Katalog der Grosswetterlagen Europas. *Bericht des Deutschen Wetterdienstes*, **13**. Offenbach, 1969.
45. Wilson, C. V., The synoptic regimes in the lower arctic troposphere during 1955. Publication in Meteorology 8, McGill University, Montreal, 1958.
46. Crane, R. G. & Barry, R. G., Comparison of the MSL synoptic pressure patterns of the Arctic as observed and simulated by the GISS General Circulation Model. *Meteorol. Atmos. Phys.*, **39** (1988) 169–83.
47. Raatz, W. E., An anticyclonic point of view on low-level tropospheric long-range transport. *Atmos. Environ.*, **23** (1989) 2501–4.
48. Key, J. & Crane R. G., A comparison of synoptic classification schemes based on 'objective' procedures. *J. Climatology*, **6** (1986) 375–88.
49. Vowinckel, E. & Orvig, S., The inversion over the Polar Ocean. WMO Technical Note 87 (1967) 39–59.
50. Raatz, W. E., Schnell, R. C., Bodhaine, B. A., Oltmans, S. J. & Gammon, R. H., Air mass characteristics in the vicinity of Barrow, Alaska, 9–19 March 1983. *Atmos. Environ.*, **19** (1985) 2127–34.
51. Oltmans, S. J., Raatz, W. E. & Komhyr, W., On the transfer of stratospheric ozone into the troposphere near the North Pole. *J. Atmos. Chemistry*, **9** (1989) 245–53.
52. Huschke, R. E., Arctic cloud statistics from 'calibrated' surface weather observations. Rand Corporation, RM-6173-PR, 1969.
53. Vowinckel, E., Cloud amount and type over the Arctic. Publication in Meteorology 51, McGill University, Montreal, 1962.
54. Jayaweera, K.O.L.F., Meteorological conditions conducive to the formation of stratus clouds over the Beaufort Sea. *Nat. Wea. Dig.*, **7** (1982) 25–32.
55. Barry, R. G., Crane, R. G., Schweiger, A. & Newell, J., Arctic cloudiness in spring from satellite imagery. *J. Climatology*, **7** (1987) 423–51.
56. Kukla, G. J. & Robinson, D. A., Variability of summer cloudiness in the Arctic Basin. *Meteorology & Atmos. Physics*, **39** (1988) 42–50.
57. Barrie, L. A., Hoff, R. M. & Daggupaty, S. M., The influence of mid-latitudinal pollution sources on haze in the Canadian Arctic. *Atmos. Environ.*, **15** (1981) 1407–19.

58. Shaw, G. E., Eddy diffusion transport of arctic pollution from the mid-latitudes: A preliminary model. *Atmos. Environ.*, **15** (1981) 1483–90.
59. Holmgren, B. & Enger, L., Stratus and the atmospheric boundary layer over the Arctic Ocean. Paper presented at the 8th meeting of the European Geophysical Society, Uppsala, 24–29 August, 1981.
60. Smith, F. B., The significance of wet and dry synoptic regions on long-range transport of pollution and its deposition. *Atmos. Environ.*, **15** (1981) 863–73.
61. Harris, J. M., Trajectories during AGASP. *Geophys. Res. Lett.*, **11** (1984) 453–6.
62. Ottar, B. & Pacyna, J. M., Sources of Ni, Pb, and Zn during the arctic episode in March 1983. *Geophys. Res. Lett.*, **11** (1984) 441–4.
63. Moritz, R. E., Accuracy of surface geostrophic wind forecast in the Central Arctic. *Mon. Wea. Rev.*, **111** (1983) 1746–58.
64. Kahl, J. D. & Samson, P. J., Uncertainty in trajectory calculations due to low resolution meteorological data. *J. Clim. Appl. Meteor.*, **25** (1986) 1816–31.
65. Rahn, K. A., The Eurasian sources of arctic aerosol. Norwegian Institute for Air Research, OR 34/79, Lillestrøm, 1979.
66. Rahn, K. A., The Mn/V ratio as a tracer of large-scale sources of pollution aerosol for the Arctic. *Atmos. Environ.*, **15** (1981) 1457–64.
67. Hoff, R. M. & Trivett, N. B. A., Ground-based measurements of arctic haze made at Alert, N.W.T., Canada during the Arctic Gas and Aerosol Sampling Project (AGASP). *Geophys. Res. Lett.*, **11** (1984) 389–92.
68. Raatz, W. E., On the meteorological characteristics of polluted air masses at Barrow, Alaska. PAGEOPH, **120** (1982) 662–72.
69. Heintzenberg, J. & Larssen, S., SO_2 and SO_4 in the Arctic: Interpretation of observations at three Norwegian arctic–subarctic stations. *Tellus*, **35B** (1983) 255–65.
70. Joranger, E. & Ottar, B., Air pollution studies in the Norwegian Arctic. *Geophys. Res. Lett.*, **11** (1984) 365–8.
71. Pacyna, J. M. & Ottar, B., Transport and chemical composition of the summer aerosol in the Norwegian Arctic. *Atmos. Environ.*, **19** (1985) 2109–20.
72. Pacyna, J. M. & Oehme, M., Long-range transport of some organic compounds to the Norwegian Arctic. *Atmos. Environ.*, **22** (1988) 243–57.
73. Brosset, C., Black and white episodes. *Ambio*, **5** (1976) 157–63.
74. Rahn, K. A., Joranger, E., Semb, A. & Conway, T. J., High winter concentrations of SO_2 in the Norwegian Arctic and transport from Eurasia. *Nature*, **287** (1980) 824–6.
75. Raatz, W. E., Schnell, R. C. & Bodhaine, B. A., The distribution and transport of pollution aerosols over the Norwegian Arctic on 31 March and 4 April 1983. *Atmos. Environ.*, **19** (1985) 2135–45.
76. Uematsu, M., Duce, R. A., Prospero, J. M., Chen, L., Merrill, J. T. & McDonald, R. L., Transport of mineral aerosol from Asia over the North Pacific Ocean. *J. Geophys. Res.*, **88** (C9) (1983) 5343–52.
77. Haeberli, W., Sahara dust in the Alps—A short review. *Zeitschrift f. Gletscherkunde & Glazialgeologie*,**13** (1977) 206–8.
78. File, R. F., Dust deposits in England on 9 November 1984. *Weather*, **41** (1986) 191–5.

79. Reiff, J., Forbes, G. S., Spieksma, F. Th. M. & Reynders, J. J., African dust reaching northwestern Europe: A case study to verify trajectory calculations. *J. Clim. Appl. Meteor.*, **25**, (1986) 1543–67.
80. Shapiro, M. A., Schnell, R. C., Parungo, F. P., Oltmans, S. J. & Bodhaine, B. A., El Chichon volcanic debris in an arctic tropopause fold. *Geophys. Res. Lett.*, **11** (1984) 421–4.
81. Raatz, W. E., Schnell, R. C., Shapiro, M. A., Oltmans, S. J. & Bodhaine, B. A., Intrusions of stratospheric air into Alaska's troposphere, March 1983. *Atmos. Environ.*, **19** (1985) 2153–8.
82. Shapiro, M. A., Hampel, T. & Krueger, A. J., The arctic tropopause fold. *Mon. Wea. Rev.*, **115** (1987) 444–54.
83. Radke, L. F., Lyons, J. H., Hegg, D. A. & Hobbs, P. V., Airborne observations of arctic aerosols. I: Characteristics of arctic haze. *Geophys. Res. Lett.*, **11** (1984) 393–6.
84. Raatz, W. E., Schnell, R. C. & Bodhaine, B. A., Atmospheric cross sections for the Arctic Gas and Aerosol Sampling Program, March–April, 1983. NOAA Tech Memo. ERL ARL-134, Silver Spring, Maryland, 1985.
85. Pacyna, J. M., Ottar, B., Vitols, V. & Arnesen, K., Aircraft measurements of air pollution in the Norwegian Arctic. Appendices A, B, C, and D. Norwegian Institute for Air Research, OR 16/86, Lillestrøm, 1986.
86. Herbert, G. A., Bridgman, H. A., Schnell, R. C., Bodhaine, B. A. & Oltmans, S. J., The analysis of meteorological conditions and haze distribution for the second Arctic Gas and Aerosol Sampling Program (AGASP II), March–April 1986. NOAA Tech. Memo. ERL ARL-158, Silver Spring, Maryland, 1987.
87. Hoff, R. M., Vertical structure of arctic haze observed by lidar. *J. Appl. Meteor.*, **27** (1988) 125–39.
88. Pacyna, J. M. & Ottar, B., Vertical distribution of aerosols in the Norwegian Arctic. *Atmos. Environ.*, **22** (1988) 2213–22.
89. Radke, L. F., Hobbs, P. V. & Bailey, J. H., Airborne observations of arctic aerosols. III: Origins and effects of air masses. *Geophys. Res. Lett.*, **11** (1984) 401–4.
90. Schnell, R. C. & Raatz, W. E., Vertical and horizontal characteristics of arctic haze during AGASP: Alaskan Arctic. *Geophys. Res. Lett.*, **11** (1984) 369–72.
91. Gifford, F. A., Barr, S., Malone, R. C. & Mroz, E. J., Tropospheric relative diffusion to hemispheric scales. *Atmos. Environ.*, **22** (1988) 1871–9.
92. Moritz, R. E., On a possible sea-breeze circulation near Barrow, Alaska. *Arctic & Alpine Res.*, **9** (1977) 427–31.
93. Kozo, T. L., Mountain barrier baroclinity effects on surface winds along the Alaskan arctic coast. *Geophys. Res. Lett.*, **7** (1980) 377–80.

Chapter 3

Arctic Air Pollution as Reflected in Snowpits and Ice Cores

CLIFF I. DAVIDSON[a], JEAN-LUC JAFFREZO[b] and PAUL A. MAYEWSKI[c]

[a]*Departments of Civil Engineering and Engineering & Public Policy, Carnegie Mellin University, Pittsburgh, Pennsylvania 15213, USA*
[b]*Department of Civil Engineering, Carnegie Mellon University, Pittsburgh, Pennsylvania 15213, USA*
[c]*Glacier Research Group, Institute for Study of Earth, Oceans, and Space, University of New Hampshire, Durham, New Hampshire 03824, USA*

3.1 INTRODUCTION

The atmospheric concentration data presented elsewhere in this volume show that the Arctic receives considerable pollution from the mid-latitudes. Anthropogenically produced sulfates, nitrates, trace metals, and carbon-containing compounds are among the species influenced by long-range transport. The arctic atmosphere also contains a variety of natural constituents, such as soil dust and seaspray, which result from local and distant sources.

Many studies have involved direct measurement of the airborne concentrations of these species. While producing a wealth of data, this approach has provided information only over the past decade or so, when arctic air monitoring programs began. Furthermore, most of the studies have included sampling only during spring or summer due to logistical constraints.

In contrast to these direct air monitoring programs, the glacial record provides a valuable tool for assessing atmospheric constituents in the Arctic during all seasons and over long time periods. Much of the glacial snow and ice is perpetually frozen; gases and particles reaching the surface are likely to remain in the snowpack for indefinite periods. Successive snowfalls, each containing traces of material from the atmosphere, form a time series record that can be used to infer the history of atmospheric constituents.

The glacial record can be studied using snowpits as well as ice cores. Snowpits involve excavation of an area that is typically 2–3 m on a side and extends downward as much as several meters below the surface. Samples of snow and ice are removed manually from the wall in layers <1 cm to several centimeters in depth. The samples may be removed continuously down the wall so that all layers of snow are represented, or they may be removed only from discrete layers of particular interest. Physical properties such as snow crystal grain size, opacity, and density also may be determined for each layer. To minimize contamination, individuals working in the snowpits wear special clean suits that do not produce dust, as well as face masks; all sampling tools and sample containers are pre-cleaned. Snowpit samples obtained in this way can be used to provide detailed time series records, with relatively large samples for each layer (as much as several kg of snow), for time periods up to several decades if accumulation rates are small. They are particularly useful for studying atmospheric transport and deposition, and also for identifying changes that occur within the first few years after deposition. In addition, snowpit data can provide insights on the extent to which data from a single deep ice core are representative of the entire glacier, since spatial inhomogeneities in concentration can be readily determined through a series of shallow snowpits.

Although snowpits are relatively easy to excavate and sample, they can be used only to obtain information over the recent past. Ice cores allow retrieval of glacial record samples from the surface all the way down to bedrock, spanning time periods well over 100 000 years, depending on site. A wide range of drilling equipment is now available for the recovery of shallow (<200 m), intermediate (200–1000 m), and deep (>1000 m) ice cores. The diameters of the cores range in size up to $\sim$ 16 cm. Drilling systems are most commonly electric, powered by solar panels (to minimize on-site contamination) or by generators. Once brought to the surface, the core is cleaned using procedures that are often tedious. The upper part of the core is generally composed of firn, where the snow has densified but not yet formed ice; these samples are scraped mechanically to remove the contaminated outer layers. Ice from deeper sections is rinsed with ultra-clean water. A number of physical measurements as well as electrical conductivity tests are sometimes conducted on-site as the core sections are retrieved. As with the snowpit work, all individuals working with core samples wear clean suits and face masks, and sampling equipment is pre-cleaned. These steps are generally conducted in specially prepared trenches excavated in the snow.

The time series records provided by snowpits and ice cores have been used for a variety of purposes. Of primary importance are studies of greenhouse gases that affect the Earth's energy balance and hence influence global climate. These studies include detailed examination of recent changes in concentration due to anthropogenic activities, as well as changes occurring over very long time periods influenced by natural variations. For example, increasing concentrations resulting from anthropogenic activities have been documented in ice cores for CO_2,[1,2] CH_4,[2,3] and N_2O.[2,4] Natural changes in concentrations occurring over time scales of millenia have also been reported (CO_2,[5,6] CH_4,[7,8] and N_2O[4,9]). In addition, there is evidence that depletion of O_3 in the atmosphere over Antarctica may be reflected through proxy indicators in glaciers, such as NO_3^- or Cl^-.[10]

Second, the glacial record has been used to document changes in concentrations of natural particulate matter emitted from the continents and the oceans. Such concentration changes often indicate variations in climatic conditions. For example, data on microparticles have provided information on changes in atmospheric dust loadings related to global aridity.[11,12] Concentrations of marine constituents have been used as indicators of sea-ice cover.[13]

Third, snowpits and ice cores have provided information on global cycling of chemical species that affect biological activity such as sulfur, nitrogen, and ammonium. Biological productivity is known to affect climate; for example, biogenic nuclei have been identified in clouds over Antarctica.[14] Recent work suggests that dimethyl sulfate (DMS) emitted from microorganisms in the oceans may be oxidized to methane sulfonic acid (MSA) aerosol which may serve as condensation nuclei.[15]

Fourth, snowpits and ice cores have been valuable for identifying changes in the global distribution of toxic species, particularly those emitted by anthropogenic activities. As an example, glacial record data have documented increased atmospheric loadings of industrial Pb in both northern and southern hemispheres.[16,17] Increases in airborne SO_4^{2-} and NO_3^- from industrial sources have been similarly documented in the northern hemisphere.[18–20]

Fifth, glacial record data have been used to identify specific point source releases of anthropogenic and natural chemical species. This in turn has provided information on global atmospheric circulation patterns and major geological events. Chernobyl radioactivity released during April and May 1986 is one example;[21,22] debris from volcanic eruptions is another.[23–26] Layers containing material from these sources may also serve as definitive date markers.

Finally, snowpits and ice cores have been used to explore solar phenomena, particularly sunspot cycles. Radionuclides produced by cosmic radiation have been related to cyclic variations in sunspots,[27] and there have been attempts to relate nitrate concentrations to sunspot cycles.[28,29]

This chapter reviews glacial record data as an indicator of atmospheric constitutents in the Arctic, focusing on anthropogenically derived material. The chapter is divided into two main sections. First, factors influencing chemical constituents in glacial snow and ice are discussed. Macroscale phenomena such as long-range transport and microscale phenomena such as deposition processes are included. Second, several examples of glacial record data for the Arctic are reviewed. The data are categorized by chemical constituent, with separate subsections for anions and cations, trace metals, radionuclides, and trace gases.

Because this chapter focuses on anthropogenically derived material, the categories of natural constituents summarized above have not been addressed here. Furthermore, many anthropogenic chemical species are known to be important components of arctic haze,[30] but have not been studied in detail in the glacial record. Examples include soot carbon aerosol[31–33] and several organic vapors.[34–37] These studies are considered outside the scope of this manuscript.

3.2 FACTORS INFLUENCING CHEMICAL CONSTITUENTS IN THE GLACIAL RECORD

Proper interpretation of glacial record data requires an understanding of processes by which constituents are transported from source regions to their ultimate sink in the snow or ice. There are essentially three major steps involved in this transport. First, the constituents are transported from the point of emission to the atmosphere over the glacier. The constituents are then brought from the atmosphere to the surface of the glacier by some deposition process. Finally, physical and chemical changes occur as the constituents become buried by successive snowstorms and as the snow changes to firn and ice. Although our understanding of the first two steps is far from complete, we are able to begin applying this knowledge to help interpret the glacial record. In contrast, our understanding of the third step is much more limited; we have little information about chemical changes occurring within glaciers, particularly over long time scales, that can help us decipher chemical species data.

In this section, we explore processes influencing chemical constituents in glaciers by considering each of these three steps. For the first step, we examine source regions and pathways of long-range atmospheric transport. We discuss current knowledge of these topics as well as problems in using meteorological data to derive air mass trajectories. Note that these issues are considered for the transport of constituents that are of interest in glaciers; for the more general case of long-range transport of arctic haze, the reader is referred to Chapter 2. For the second step, we consider separately the mechanisms of dry deposition and wet deposition. Finally, we examine briefly the phenomena believed to influence chemical constituents in glaciers after deposition. Processes that cannot be easily categorized as dry or wet deposition, such as the settling of ice fog crystals or direct interception of water droplets with surfaces, may be important under certain conditions but have not been well studied in the Arctic. Therefore they are considered outside the scope of this manuscript.

3.2.1 Sources regions and long-range transport pathways

Several detailed studies have examined source regions responsible for arctic haze, as discussed in detail in Chapter 2. Focusing on the time of annual maximum airborne concentrations in February and March, most of these studies have reported Eurasia as the dominant source region for chemical constituents in the atmosphere in the northern latitudes.[38–41] However, these studies have not in general considered receptor locations on glaciers. Therefore the source regions and transport pathways influencing the glacial record may be somewhat different from those reported for arctic haze.

A number of problems exist when attempting to conduct trajectory analyses for glacial sites in the Arctic. Of primary importance is that most glaciers are located at high elevations above sea-level. The Greenland Ice Sheet, where the majority of arctic glacial record data have been obtained, attains an elevation of ~3000 m as shown in Fig. 3.1. Air reaching the Ice Sheet is sometimes transported to coastal Greenland at low elevations, and then rises sharply to 1000–2000 m following the

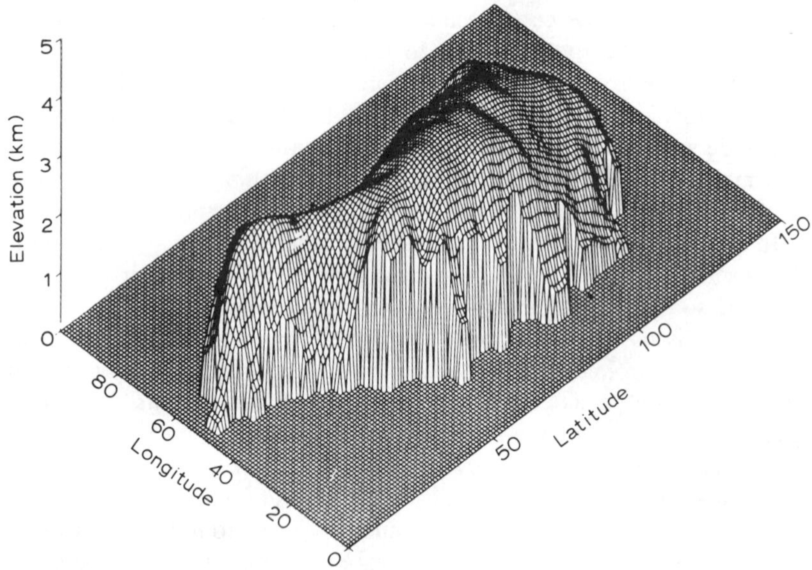

Figure 3.1. Topography of Greenland with a view from the southeast, reproduced from McInnes and Radok.[109] Each grid square represents 20 km × 20 km. (From McInnes, B. and Radok, U. (1985). A steady-state prediction of Dye 3 core features. In *Greenland Ice Core: Geophysics, Geochemistry and the Environment*, ed. C. C. Langway, Jr., H. Oeschger, and W. Dansgaard, Geophysical Monograph 33, 111–17. Copyright by the American Geophysical Union).

terrain while moving inland. At other times, airflow patterns bring material to the Ice Sheet by transport at relatively high elevations over long distances. Vertical air motions are difficult to account for in trajectory analysis, and hence there is considerable uncertainty in trajectories reported for Greenland.

Another problem is that trajectory computations rely on meteorological data from networks of weather stations positioned over large geographical areas. There are few such stations in the Arctic, and therefore considerable interpolation between stations is necessary.

An additional difficulty concerns the lack of airborne chemical constituent data for locations on arctic glaciers. Trajectory computations at mid-latitude receptor sites have been checked on occasion by airborne chemical species concentrations; consistencies between the airborne data and source regions predicted by trajectory analyses have been used to argue for the validity of the trajectories. Such supporting information is not yet available for arctic glaciers.

Despite these problems, a limited number of trajectory computations have been carried out for the Greenland Ice Sheet at Dye 3 (65°11′N, 43°50′W, elevation 2479 m).[42] The trajectories have been computed for the year August 1988–July 1989, using a model developed by Olson *et al.*[43] The model is three-dimensional and incorporates vertical air motions which are not necessarily adiabatic. Three sets of trajectories backward from Dye 3 have been computed, for air reaching Dye 3 at 775 mb (near the surface of the Ice Sheet), 700 mb, and 500 mb.

The results of these computations for 775 mb are shown in Fig. 3.2. Each point represents the location of the air mass five days before arriving at Dye 3, with one trajectory computed for each day. The plots show pronounced seasonal differences. In winter, most of the points are generally within the Arctic, although some of these involve travel over great distances. Nearly half of the trajectories show long-range transport with travel distances greater than 2000 km; most of these are to the west and north. The predominant source region is the Canadian Arctic, with many points around Hudson's Bay. There are also scattered points in all directions surrounding Dye 3. Of particular interest are the points over the Arctic Ocean to the north, indicating possible transport from Asia.

The spring and summer maps show more transport from the south and east, although transport over the pole from Asia is also possible. Many of the points are over the Atlantic Ocean south of Greenland. As in winter, there is considerable scatter in all directions.

During fall, the sources have shifted to the west. The predominant source regions are the western Atlantic Ocean and eastern North America, with long-range transport indicated by 40% of the points. Again, there is considerable scatter.

Overall, these plots suggest North America as an important source region during fall and winter; during spring and summer, more air masses arrive from the central Atlantic. Transport from Europe and Asia over the pole appears to be possible during all seasons, according to this limited dataset. Air masses from Europe may reach Dye 3 on winds from the south and east during all seasons.

The trajectories for 700 and 500 mb indicate considerably more long-range transport. Although there are some differences, the 700 mb points generally show rough agreement with those in Fig. 3.2. North America is a dominant source region during fall and winter, while the Atlantic Ocean is important during spring, summer, and fall. Transport over the

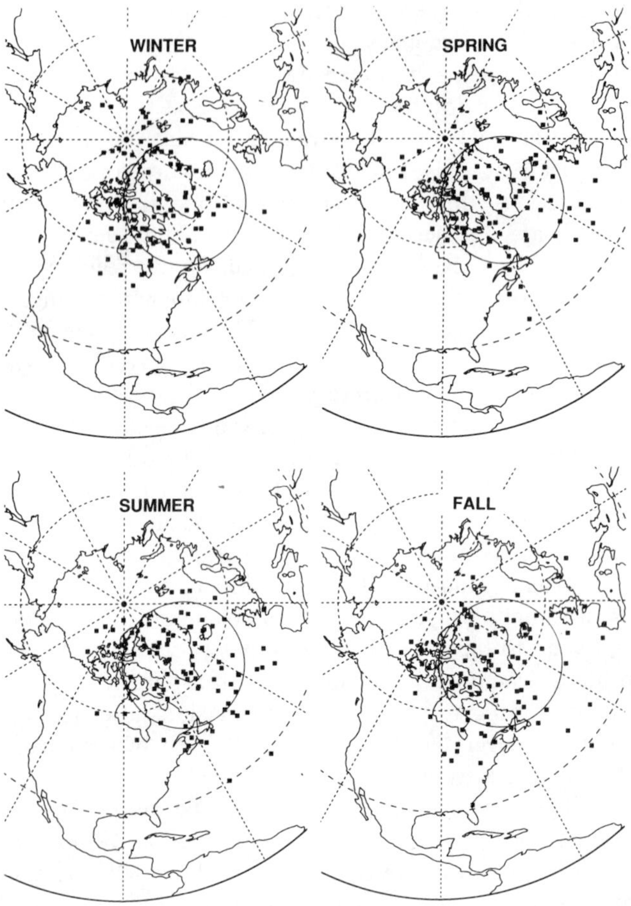

Figure 3.2. Locations of air masses five days prior to reaching Dye 3 at the 775 mb level, taken from Jaffrezo *et al.*[42] Each point represents one day during the year August 1988–July 1989. The seasons correspond to three month periods, beginning with winter represented by January 1–March 31. The circle on each map indicates 2000 km distance from Dye 3.

pole may also be important. In contrast, the 500 mb plots suggest North America as the most important source region during all seasons, with some points occurring over the Pacific.

The different results at these three elevations imply that it is difficult to determine *a priori* the dominant source regions influencing Dye 3. Winds near the surface may be important on some days, while air motion aloft may be the determining factor on other days. Our understanding of long-range transport may be improved by considering other meteorological information and airborne chemical species data in conjunction with the trajectories, although this has not yet been attempted in detail.

As a further limitation, it must be emphasized that the points in Fig. 3.2 may not reflect long-term average conditions. For example, Miller[38] has obtained five-day backward trajectories for the years 1975–1979 for Barrow, Alaska, using a 200–3000 m averaging layer. During 1975, transport from Europe or Asia occurred for 70% of the trajectories with travel distances greater than 2000 km; during 1979, only 10% of the long-range trajectories showed this pattern. Figure 3.2 shows that roughly 20% of the long-range trajectories for Dye 3 are associated with transport from Europe or Asia, based on those points located in the eastern hemisphere and within 15° west of Greenwich. It is apparent that additional trajectory work over several years is needed to better define source regions influencing Greenland as well as other glacial sites in the Arctic.

3.2.2 Dry deposition

In the absence of precipitation, atmospheric gases and particles can reach the surface of glaciers by dry deposition. In this section, we consider the mechanisms of dry deposition, and also the mathematical models developed to predict dry deposition onto snow.

3.2.2.1 Mechanisms of dry deposition

The overall process of dry deposition onto snow may be conveniently divided into three steps. During the first step, particles and gases are carried from the free atmosphere down to the region of nearly quiescent air just above the snow surface. This step, termed aerodynamic transport, is dependent upon windspeed and atmospheric stability. The roughness of the surface also affects aerodynamic transport, although this is not likely to be of major importance for a relatively smooth horizontal snowfield. Aerodynamic transport affects both gases and particles. The principal mechanism is eddy diffusion, where net motion is from regions of high concentration to regions of low concentration. If the surface is a perfect sink, the concentration near the surface will be small and hence there will be continual flow of the constituent from the atmosphere towards the surface. Turbulent wind eddies provide the driving force for this motion. For particles that are sufficiently large (diameters greater than a few tenths of a micron), sedimentation by gravity will also affect transport from the free atmosphere towards the surface.

In the second step, particles and gases are carried across the nearly quiescent air just above the snow. Termed the quasi-laminar sublayer, this region may be as small as a fraction of a millimeter over a relatively smooth snow surface. In many cases, the sublayer is not a simple horizontally oriented region but rather follows the contour of each ice needle or other irregularity on the surface.

Transport of gases across the sublayer is by Brownian diffusion. For particles, the transport is by Brownian diffusion, interception, impaction, and turbulent inertial deposition. If the particles are sufficiently large, sedimentation by gravity across the sublayer may also be important. These mechanisms are illustrated in Fig. 3.3.

Brownian diffusion refers to motion of gases and particles from regions of high concentration to regions of low concentration similar to the mechanism of eddy diffusion. However, the driving force in this case is the random thermal energy of the molecules of air and of the constituent species. Thus Brownian diffusion occurs even in quiescent air. Interception occurs when a particle of diameter d_p is on an air streamline that passes within a distance $d_p/2$ of a roughness element on the surface, so

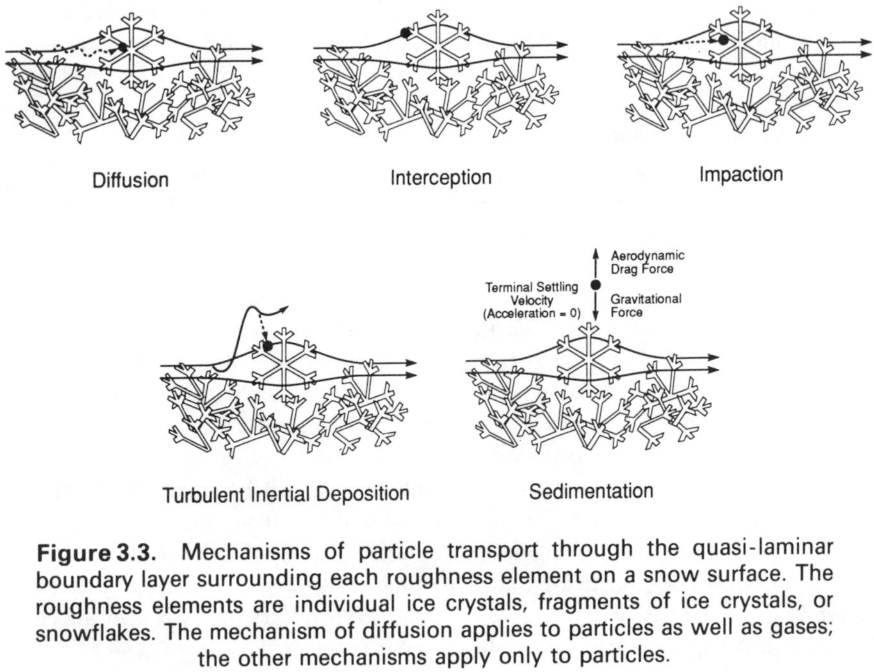

Figure 3.3. Mechanisms of particle transport through the quasi-laminar boundary layer surrounding each roughness element on a snow surface. The roughness elements are individual ice crystals, fragments of ice crystals, or snowflakes. The mechanism of diffusion applies to particles as well as gases; the other mechanisms apply only to particles.

that the particle effectively collides with the roughness element. Impaction occurs when a particle cannot follow an air streamline as it turns to go around roughness elements: its inertia carries it in a straight line through the sublayer and onto the surface. Turbulent inertial deposition refers to transport across the sublayer by turbulent air eddy motion. This mechanism differs from impaction in that energy is derived from the component of turbulent flow normal to the direction of mean flow, rather than from the mean flow itself. Thus impaction can occur only if the surface is sufficiently rough, while turbulent inertial deposition can occur over even the smoothest snow surface. Note that Brownian diffusion affects particles smaller than about 0·1 μm, while interception, impaction, and turbulent inertial deposition apply mainly to particles larger than this size. Sedimentation through the sublayer applies primarily to particles greater than a few tenths of a micron. Electrostatic attraction and forces such as thermophoresis (motion along a temperature gradient) and diffusiophoresis (motion along a humidity gradient) also may affect transport through the sublayer, although the relative importance of these mechanisms for a snow surface is not well understood.

In the third and final step, the particles and gases interact to a varying extent with the surface. Gases may be *absorbed* irreversibly or *adsorbed* reversibly by the snow; unreactive gases may not deposit effectively. Similarly, particles may bounce off the surface or may stick to it, in some cases reacting chemically and/or physically with the snow.

For many particulate and gaseous species, interaction with the surface is likely to be affected by the presence of liquid water associated with the snow or ice.[44] Such a layer has been found to enhance SO_2 capture by ice depositing from the vapor phase,[45] and it is likely to enhance the deposition of highly soluble gases such as HNO_3.[46] The porosity of snow may also be significant in allowing air to penetrate beneath the surface, increasing the surface area available for deposition.[47]

3.2.2.2 Mathematical relations

Dry deposition is usually defined by noting that the deposition flux F_d ($g\,cm^{-2}\,sec^{-1}$) is linearly proportional to the airborne concentration C_a($g\,cm^{-3}$), thus defining the dry deposition velocity v_{ddry} ($cm\,sec^{-1}$) at height z:

$$v_{ddry}(z) = -F_d/C_a(z) \tag{3.1}$$

The minus sign is needed because the flux downward to the surface is a negative quantity, yet the deposition velocity is defined as positive.

Because C_a is a function of height z above the surface, v_{ddry} is also a function of height

The deposition velocity may be considered the inverse of a resistance to transport r (sec cm^{-1}), having three components representing the steps in the deposition process:

$$v_{\mathrm{ddry}}(z)=1/r_{\mathrm{total}}=(1/[r_a(z)+r_b+r_c])+(1/r_g) \tag{3.2}$$

where r_a, r_b, r_c, and r_g represent resistances to aerodynamic transport, boundary layer transport, surface transport, and gravitational transport, respectively. The gravitational resistance is a separate term since r_g is considered to be in parallel with the sum of the series resistances $(r_a+r_b+r_c)$. The term r_a is derived by analogy with heat transport in a turbulent atmosphere and is a function of eddy diffusivity which is proportional to height. Corrections for buoyant forces in a nonadiabatic atmosphere are often necessary. The term r_b is described by eddy and Brownian diffusion, interception, impaction, turbulent inertial deposition, and sedimentation. Finally, r_c is usually assumed to be zero for particles, implying perfect retention by the surface, although this resistance may be significant in many circumstances. For gases, r_c is often large enough to be rate-limiting.

Most deposition models have been developed for vegetation. However, Ibrahim *et al.*[48] discussed a dry deposition model for snow which includes expressions for the various steps in the process, accounting for the mechanisms described above. In addition, hygroscopic growth in the humid layer just above the snow is included. The model was applied to ammonium sulfate aerosol, with results shown in Fig. 3.4. The model has also been applied to conditions at Dye 3,[49] with an example of the results shown in Fig. 3.5.

Verification of the results of Fig. 3.4 and 3.5 is not possible at present, mostly because of a lack of input data on particle size and a lack of dry deposition data. These limitations in our ability to model dry deposition are discussed further by Davidson and Wu.[50]

3.2.3 Wet deposition

Precipitation scavenging, or wet deposition, may occur by nucleation during formation of rain or snow, or by transport to existing raindrops and snowflakes. Each of these two cases is discussed for conditions in the Arctic, focusing on the Greenland Ice Sheet. Mathematical relations for wet deposition are also briefly discussed.

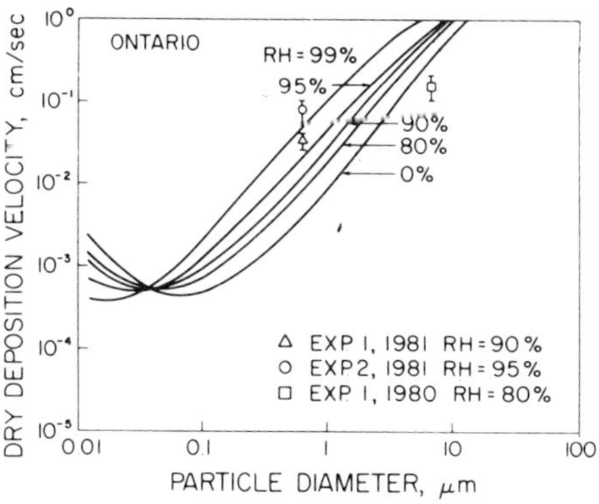

Figure 3.4. Predicted dry deposition velocity as a function of particle diameter, based on a 10 cm reference height and conditions in Ontario (redrawn from Ibrahim *et al.*[48]). Experimental data of Ibrahim *et al.* obtained with monodisperse ammonium sulfate aerosol are also shown. (From Ibrahim, M., Barrie, L. A., and Fanaki, F. (1983). An experimental and theoretical investigation of the dry deposition of particles to snow, pine trees and artificial collectors. *Atmos. Environ.*, **17**, 781–88).

3.2.3.1 Mechanisms of wet deposition

Figures 3.6 and 3.7 illustrate the two categories of wet deposition, namely nucleation and transport to exisiting raindrops and snowflakes. In order to understand nucleation, it is necessary to identify the steps involved in the formation of snow. Figure 3.6 shows a rough illustration of these steps for the Greenland Ice Sheet. In the first step, air near the coast becomes saturated with water vapor. This vapor condenses out on airborne particles which serve as cloud condensation nuclei, forming cloud droplets. Particles which contain significant amounts of soluble salts, such as SO_4^{2-} emitted from the oceans, generally make good condensation nuclei. Other hygroscopic particles including those emitted from anthropogenic combustion activities may also serve in this capacity.

In the second step, water vapor continues to diffuse to the growing cloud droplets. Coalescence of the droplets also enhances the growth process. The droplets reach diameters that are typically 5–20 μm in diameter.

Some of the newly formed clouds then move inland, rising in elevation as they follow the surface of the Ice Sheet. At some point in their travels, the temperature becomes cold enough to activate ice nuclei, shown as the

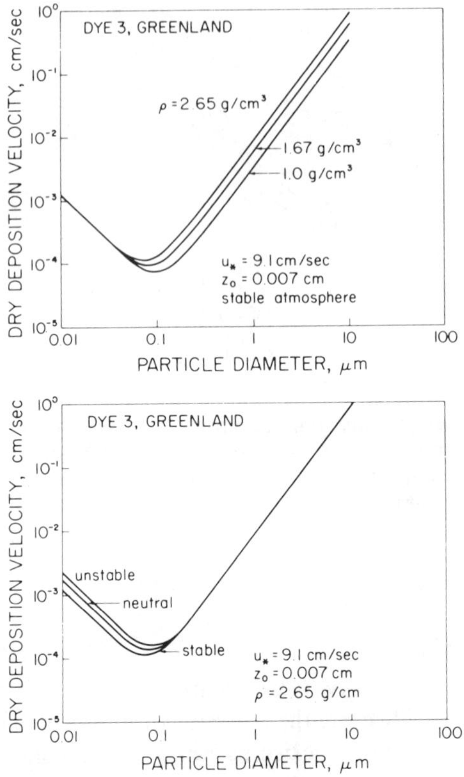

Figure 3.5. Predicted dry deposition velocity as a function of particle size using the model of Ibrahim *et al.*[48] applied to measured meteorological conditions at Dye 3 (taken from Davidson[49]). The density of 2·65 g/cm^3 is typical of soil dust, while 1·67 g/cm^3 corresponds to ammonium sulfate aerosol. The atmospheric stabilities are assumed to have Monin-Obukhov Length equal to -100 m (unstable), infinity (neutral), and $+100$ m (stable). The relative humidity is assumed to be 0%.

third step. Airborne particles that have a crystal lattice structure resembling that of ice, such as some particles derived from the Earth's crust, make good ice nuclei.[51-53] Upon initiation of the ice phase, much of the remaining water vapor in the air condenses onto the ice nuclei. Furthermore, since the supersaturation with respect to ice is always greater than that with respect to liquid water, the cloud droplets formed earlier on condensation nuclei will tend to evaporate, producing water vapor that is consumed by the growing ice crystals. Some of the cloud droplets will

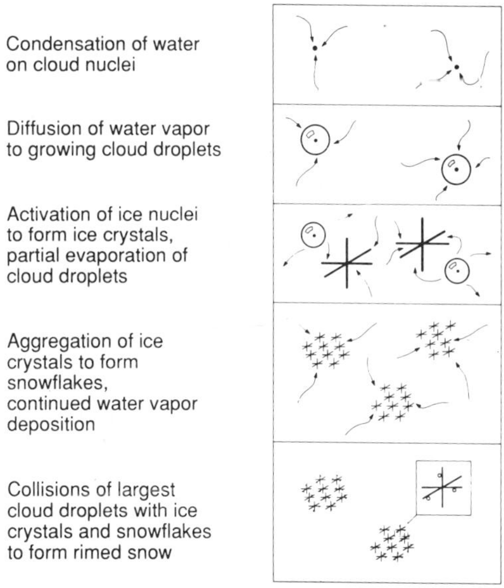

Figure 3.6. Various steps in the scavenging of particles by nucleation during formation of snow as applied to the Greenland Ice Sheet.

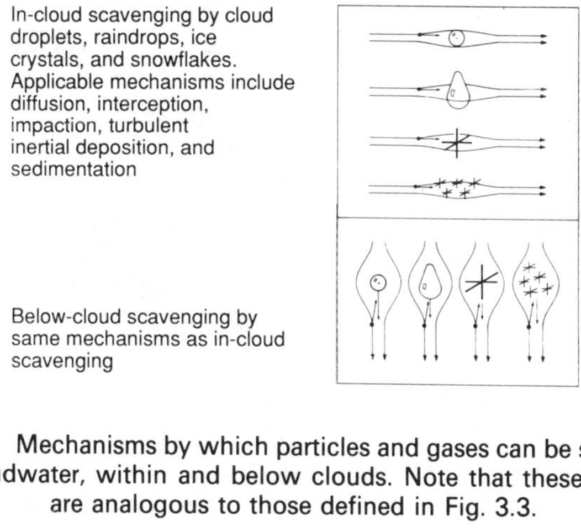

Figure 3.7. Mechanisms by which particles and gases can be scavenged by existing cloudwater, within and below clouds. Note that these mechanisms are analogous to those defined in Fig. 3.3.

evaporate completely, particularly those that are swept out of the cloud into regions of low humidity. Other cloud droplets may only partially evaporate. The particles which served as condensation nuclei for these evaporated droplets will be released back into the atmosphere; many of these particles may be re-collected subsequently by ice crystals or cloud droplets. As a result of these processes, the cloud at this point contains water vapor, supercooled liquid water droplets and ice crystals, where the latter are typically 50–100 μm in diameter. The water droplets and ice crystals are collectively known as cloudwater.

The fourth step involves aggregation of ice crystals to form snowflakes. Technically, a snowflake consists of one or more ice crystals; in the case of wet snow formed near 0°C, snowflakes may contain dozens or even hundreds of crystals. The aggregation of ice crystals is in marked contrast to the formation of raindrops in warm clouds: typical raindrops are formed by coalescence of hundreds or thousands of cloud droplets. It is unlikely that the cloud droplets in a mixed-phase cloud will form raindrops, due to the lengthy time period needed for the coalescence of a large number of droplets (several minutes or tens of minutes). Collisions of the droplets with ice crystals and snowflakes will occur before this can happen. Note that raindrops and snowflakes are collectively know as hydrometeors, and they are distinguished from the original cloudwater on the basis of size: hydrometeors are greater than the critical fall size, which typically lies between 100 and 2000 μm in diameter. The fall size is dependent on the speed of updrafts in the cloud which counteract gravitational settling.

In the fifth step, which occurs simultaneously with step 4, the cloud droplets collide with the ice crystals and snowflakes at rates which depend on their respective sizes. The droplets freeze on contact with the ice phase, producing small frozen spheres and other irregularities on the surface of the ice. This is known as rimed snow; the extent of riming depends on the amount of liquid water present in the cloud and on the sizes of the droplets and ice crystals, and hence on the temperature and humidity.

The amount of riming has important implications for the scavenging of atmospheric pollutants. Riming can lead to efficient removal of particles which are hygroscopic or otherwise make good condensation nuclei. Unrimed snow, on the other hand, is inefficient for removing these types of particles.[54–57] In contrast, the amount of riming may be less important in affecting the scavenging of crustal particles, for which removal is dependent on the fact that these particles make good ice nuclei. Because riming is dependent on temperature, the types of particles in the ice and snow of a glacier are likely to depend on the time of year that the snow was originally deposited. This aspect is discussed further in Section 3.3.

Note that variations in the steps of Fig. 3.6 are likely for a particular event. For example, condensation at the coast may not occur if temperatures are sufficiently cold; the cloud may form solely by activation of ice nuclei. In that case, the snow will be essentially unrimed.

It is significant that the formation of snow can begin by condensation even at temperatures well below 0°C. For example, a salt particle (e.g. ammonium sulfate) may serve as a condensation nucleus, resulting in a droplet of highly concentrated salt solution. The droplet remains a liquid even if the air temperature is below 0°C, since the freezing point of the solution is depressed below that of pure water. As the droplet grows by water vapor diffusion, the solution becomes more dilute. Eventually the droplet achieves a dilution sufficient for it to freeze, from which point the frozen droplet grows like an ice crystal. This process may occur for air temperatures down to −40°C: below this temperature, significant amounts of liquid water cannot exist in clouds.

Temperature gradients in the atmosphere may also affect the formation of precipitation. Ice crystals and snowflakes may form at elevations well above the surface where cold temperatures prevail, but the falling snow may melt and become rain when it encounters warmer temperatures near the ground. Rain formed in this manner may have a chemical composition resembling that of snow.

Figure 3.7 illustrates the second category of wet deposition, transport of constituents to existing cloudwater and hydrometeors. The mechanisms apply both within clouds and below the cloud base. The process of diffusion mentioned in Fig. 3.7 can remove gases as well as particles; the other mechanisms mentioned in the figure apply only to particles. Note that nucleation scavenging illustrated in Fig. 3.6 pertains only to particles.

In general, the transport mechanisms within and below clouds are similar to those discussed for dry deposition to glaciers. However, the flow patterns surrounding the various components of a cloud differ from those over a stationary snow surface. Each cloud droplet, ice crystal, raindrop, and snowflake is surrounded by a quasi-laminar sublayer through which transport must occur if scavenging is to take place. Factors such as the sizes and shapes of the cloudwater and hydrometeors, as well as the sizes and shapes of the particles being collected, are therefore important.[58,59] Solubility and chemical reactivity also affect collection of both gases and particles; meteorological parameters such as the temperature, windspeed, and amount of water vapor within and below the clouds are similarly important.

It is significant that the humidity in clouds is above the deliquescent point for the bulk of the aerosol. Therefore most of the particles will have become activated, taking on at least modest amounts of water. Within-cloud scavenging pertains mainly these activated particles Below-cloud scavenging, on the other hand, pertains mainly to dry aerosol.

3.2.3.2 Mathematical relations

Analogous to the flux due to dry deposition, the wet deposition flux F_w is defined as:

$$F_w = -\rho_w C_s P_o = -\int_0^\infty \psi(z) C_a(z)\, dz \tag{3.3}$$

where P_o is the precipitation rate at the ground (cm sec^{-1}), C_s is the concentration of chemical species in the snow (g constituent per g snow), C_a is the concentration of chemical species in the air (g constituent per cm^3 air), and ρ_w is the density of snow meltwater (g cm^{-3}). The second part of eqn (3.3) expresses F_w in terms of the airborne concentration and the scavenging rate ψ (sec^{-1}) defined as a first order loss process:

$$dC_a/dt = -\psi C_a \tag{3.4}$$

Slinn[59] among others suggested expressions for ψ based on the efficiency of collisions between atmospheric particles and hydrometeors. However, use of these fundamental expressions is difficult because of a lack of input data on characteristics of atmospheric constituents and characteristics of clouds. Therefore an empirical approach is often used to relate wet deposition to airborne concentration data, assuming a linear relation between concentrations in snow and in air. This relationship is defined through the scavenging ratio W:

$$W = \rho_a C_s / C_a \tag{3.5}$$

where ρ_a is the air density (g cm^{-3}). When defined in this manner, the scavenging ratio is thus dimensionless on a mass basis {(g constituent per g snow)/(g constituent per g air)}. The ratio may also be defined as dimensionless on a volume basis {(g constituent per cm^3 snow)/(g constituent per cm^3 air)}, leading to values which are roughly 1000 times greater. By measuring W under various conditions in the field, and assuming the processes affecting W for a species are unchanging, it is possible to estimate C_a based on measurements of C_s. More importantly, the scavenging ratios can provide insight into the relative importance of the various mechanisms involved,[60,61] so that more sophisticated models of wet deposition can be developed. Examples of values of W for snow are presented in Section 3.3.

It is clear that for each chemical species, the time series data recorded in the snow are a combination of inputs due to both dry and wet deposition. The relative importance of each process is a function of the characteristics of the constituents as well as of meteorological features; the latter category includes temperatures that determine the relevant scavenging mechanisms, the snow accumulation rate at different times of the year, and winds that control delivery to the surface. The overall deposition rate thus varies greatly among different constituents in both space and time.

3.2.4 Post-deposition processes

The above discussion has not considered processes occurring after deposition, as the chemical constituents on the surface become buried by successive layers of snow. Examples include redistribution of the uppermost layers of snow by the wind, formation of depth hoar, meltwater percolation downward from the surface, and physical and chemical changes which occur as the snow densifies. In addition, diffusion of the chemical constituents may occur within the snow or ice matrix, in some cases leading to loss of chemical species to the atmosphere through the top of the snowpack.

Some of these processes are beginning to be explored. For example, Alley[62] has studied the formation of wind crusts (surface snow that has formed a hard surface due to exposure to wind) and depth hoar (low density firn or snow containing large crystals) which are likely to impact upon the transport of water vapor and heat in the uppermost layers of snow. The resulting changes in physical properties such as density and porosity are likely to affect chemical constituents. Wind crusts may decrease rates of diffusion within the snowpack of certain constituents, particularly gases; depth hoar is likely to alter the location of constituents on individual ice grains, leading to changes in relationships between electroconductivity and chemical species concentrations. In other studies, Bales *et al.*[63] have shown that a small amount of percolating meltwater carries a disproportionate fraction of salts within the snowpack. The presence of ice lenses and presumably other discontinuities within the snow also can affect meltwater transport.[64]

It is difficult to apply this limited knowledge of post-deposition processes to arctic glaciers. However, we expect that many of these processes will have significant impact on glacial record data, particularly in areas of low snow accumulation rate where several years of record may be disturbed. Meteorological conditions can provide insight into the specific processes of concern. For example, areas of high windspeed are likely to be affected by surface redistribution problems; regions where temperatures approach or exceed 0°C are likely to be affected by meltwater percolation.

We can gain insight into these processes by studying changes over time in the record of chemical constituents in the snowpack, and by examining spatial variations in snow chemistry to assess horizontal homogeneity. Work of this type has begun at a number of locations; one example is a cooperative effort known as the Dye 3 Gas and Aerosol Sampling Program (DGASP). Conducted during August 1988 – July 1989, the field work in this program involved near-continuous gas and aerosol sampling, collection of fresh precipitation, and subsequent excavation of snowpits. Results of the study will provide information on all three steps in the overall transport from source to ultimate sink in the Greenland Ice Sheet. With the exception of the trajectory data discussed earlier, results of the program are not yet available.

3.3 EXAMPLES OF ANTHROPOGENIC POLLUTANTS IN THE GLACIAL RECORD

The combined effect of the mechanisms described in the previous section result in a continuous time series record of chemical constituents in glaciers. Several research programs have studied these constituents in glaciers in the Arctic, mostly involving the Greenland Ice Sheet. Figure 3.8 indicates the locations of many of these studies. The sites shown, with the exception of Camp Century, represent those established by the first Greenland Ice Sheet Program (GISP 1). The first continuous deep core to bedrock was obtained at Camp Century in 1966; intermediate cores were drilled as part of GISP 1 at Dye 3 (1971), Milcent (1973), and Crete (1974), all roughly 400 m deep. A second Greenland core to bedrock was obtained in 1981 at Dye 3. Numerous other shallow cores ranging up to 100 m depth were also obtained as part of this program.[65]

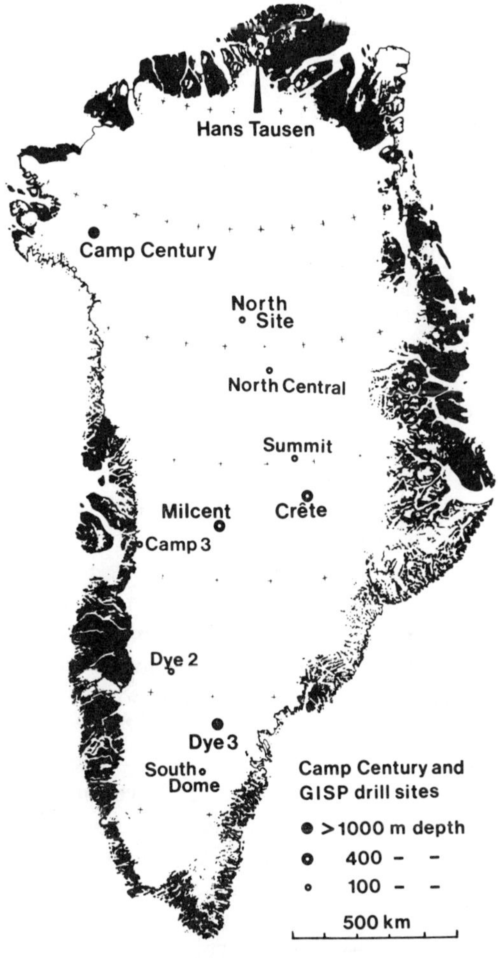

Figure 3.8. Sites of ice coring in Greenland, after Langway *et al.*[65] With the exception of Camp Century, all sites shown were established as part of the first Greenland Ice Sheet Program (GISP 1). (From Langway, C. C., Oeschger, H., and Dansgaard, W. (1985). The Greenland Ice Sheet Program in perspective. In *Greenland Ice Core: Geophysics, Geochemistry and the Environment,* eds. C. C. Langway, Jr., H. Oeschger, and W. Dansgaard, Geophysical Monograph 33, 1–8. Copyright by the American Geophysical Union).

Since GISP 1, several other independent snowpit and ice coring projects have taken place in the Arctic. Most of these have been on the Greenland Ice Sheet, near the original sites shown in Fig. 3.8. Within the last two to three years, a new major program to obtain glacial record data from Summit using snowpits, shallow cores, and two deep cores has commenced. This program includes two components: GISP 2, run by American scientists, and the Greenland Ice Program (GRIP), run by European scientists. Together the two components will provide a method for eliminating the effects of ice deformation from the resultant record. A considerable amount of new information is expected on historical trends in anthropogenic pollutants, as well as on climate change and geological events dating back perhaps 200 000 years.

In this section, we summarize examples of arctic glacial record data with a focus on anthropogenic pollutants. The records include snowpit and ice core data for anthropogenic constituents obtained from GISP 1, as well as similar data for these constituents from a number of independent studies. Results for GISP 2 are not yet available.

The section is divided according to chemical constituent. First, we discuss anions and cations with particular attention to SO_4^{2-} and NO_3^-. Next we examine trace metals, focusing on the problem of obtaining reliable data for metals at the low concentrations encountered in the Arctic. We then consider radionuclides, summarizing information on nuclear bomb explosions and on the Chernobyl accident. Finally, we consider the problem of trace gases in arctic snow.

3.3.1 Anions and cations

A number of snowpit and ice core studies have focused on analysis of anions and cations for the purpose of studying anthropogenic material in the Arctic. The snowpit studies have in general attempted to identify some of the most important factors influencing these constituents in the glacial record at different times of the year, as well as the spatial homogeneity of the chemical signals. The ice core studies have focused on broader issues such as determination of the extent to which man's activities have impacted the arctic atmosphere over time.

Interpretation of anion and cation data is complicated by the fact that atmospheric gases as well as particles may be responsible for the measured concentrations in snow and ice. At present, there are limited airborne concentration data for particulate anion and cation species; attempts have been made to relate these particulate concentrations to levels in snow and ice. Very little is known of the gaseous fractions of these species in the atmosphere, such as SO_2, NH_3, HCl, and HNO_3, and hence they are omitted from the discussion that follows.

3.3.1.1 Snowpits

Perhaps the most detailed snowpit studies involving anions and cations in the Arctic have been conducted in the vicinity of Dye 3, as well as on the central Greenland Ice Sheet. One such study is that of Mayewski *et al.*[66] These investigators characterized the spatial distribution of chemical species by excavating several snowpits within a 150 km × 150 km grid centered on the GISP 2 drill site at Summit. Total ionic balance measurements reveal that the snow contains the following major anions and cations, in order of decreasing abundance within each category: anions NO_3^-, SO_4^{2-}, and Cl^-; cations H^+, NH_4^+, Ca^{2+}, Na^+, Mg^{2+}, and K^+. The major sources of these species at Summit include mid-latitude anthropogenic activity, marine aerosols, and soil dust. The time of maximum annual concentration of these species is highly consistent, with mid-winter/early spring peaks for total SO_4^{2-}, marine SO_4^{2-}, total Cl^-, marine Cl^-, Ca^{2+}, Na^+, Mg^{2+}, and marine K^+. The time of maximum concentration is later spring or summer for NO_3^- and NH_4^+. Examples of the time series records from one of the snowpits (6 m deep) covering ten years is shown in Fig. 3.9.

A recent study conducted by Davidson *et al.*[67,68] has involved two snowpits near Dye 3, excavated during 1986 and 1987. The resulting data have been combined with concentrations reported by Mayewski *et al.*[69] for a 1984 snowpit near Dye 3 to represent a continuous record over 1982–1987. Species of interest include SO_4^{2-}, NO_3^-, and Cl^-. The layers have been dated by comparing the variations in $\delta^{18}O$ values, which are functions of the temperature during the formation of the snow, with temperature and precipitation records at Dye 3. This has enabled roughly one-month resolution. The snowpit data are then compared with airborne concentrations from routine monitoring stations throughout the Arctic.

Monthly average concentrations in the snow are plotted in Fig. 3.10. SO_4^{2-} shows a broad annual maximum during January, February, and March. This peak is coincident with the annual maximum in airborne concentrations observed at sea-level monitoring stations in the Arctic, suggesting that the glacial record is a reasonable indicator of seasonal variations in airborne concentration of this species. However, it is noteworthy that the marked seasonal variation in airborne levels is somewhat muted in the snow: the difference between minimum and

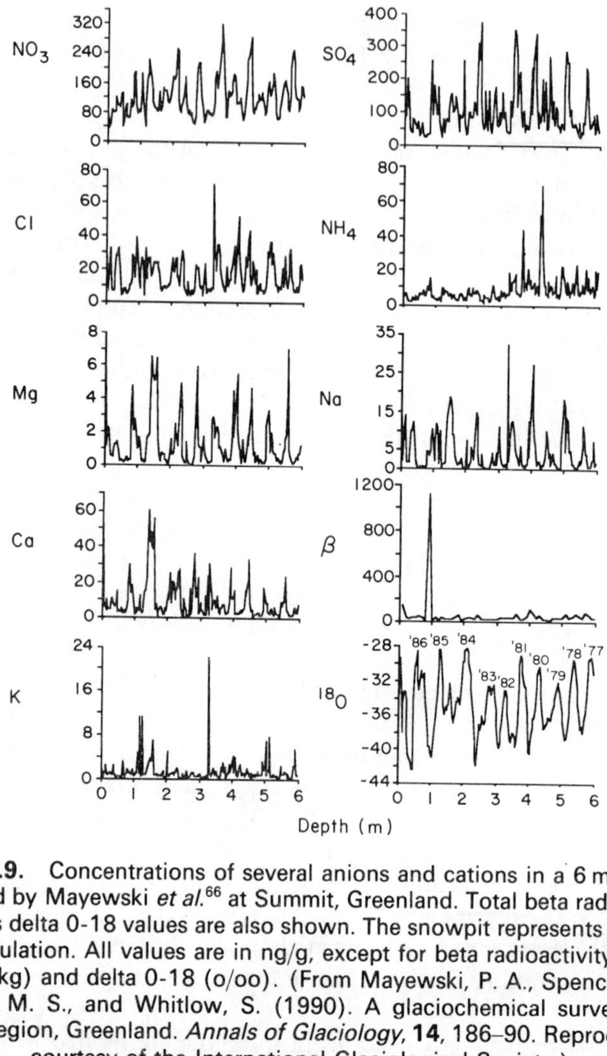

Figure 3.9. Concentrations of several anions and cations in a 6 m snowpit excavated by Mayewski *et al.*[66] at Summit, Greenland. Total beta radioactivity as well as delta 0-18 values are also shown. The snowpit represents ten years of accumulation. All values are in ng/g, except for beta radioactivity (counts per hour/kg) and delta 0-18 (o/oo). (From Mayewski, P. A., Spencer, M. J., Twickler, M. S., and Whitlow, S. (1990). A glaciochemical survey of the Summit region, Greenland. *Annals of Glaciology*, **14**, 186–90. Reproduced by courtesy of the International Glaciological Society.)

maximum monthly average concentrations in Fig. 3.10 is only a factor of three, rather than a factor of 20 or greater as observed in the sea-level airborne data. Smaller variability in concentrations in precipitation relative to variability in airborne levels has been observed for other chemical species in the Arctic[70,71] as well as in mid-latitude precipita-

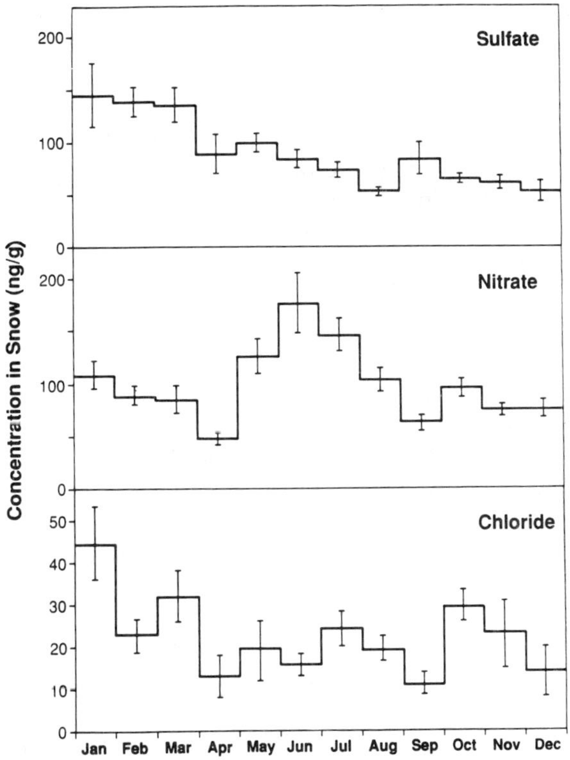

Figure 3.10. Concentrations from three snowpits near Dye 3 after dating the layers and averaging the data by month (taken from Davidson *et al.*[68]). A total of 823 samples from 227 snowpit layers are represented; the number of layers averaged within each month ranges from 7 to 40. Error bars show one standard deviation about the arithmetic mean. (From Davidson, C. I., Harrington, J. R., Stephenson, M. J., Small, M. J., Boscoe, F. P., and Gandley, R. E. (1989). Seasonal variations in sulfate, nitrate and chloride in the Greenland Ice Sheet: relation to atmospheric concentrations. *Atmos. Environ.*, **23**, 2483–93. Reprinted with permission from Pergamon Press.)

tion.[60] This may partly represent a damping effect as time varying airborne concentrations are integrated into a single precipitation event. The difference may also reflect more efficient nucleation scavenging in summer in the Arctic when airborne concentrations are low and when riming is prevalent.

Figure 3.10 shows that NO_3^- in the snow peaks in mid-summer. In contrast, airborne NO_3^- at sea-level monitoring stations peaks in late February, similar to the pattern for airborne SO_4^{2-}. For NO_3^-, differences between seasonal variations in the air and those in the snow are attributed to two possible factors. First, scavenging of NO_3^- may be much more efficient in summer, due to the occurrence of rimed snow. Second, the seasonal variation in airborne NO_3^- at the elevation of Dye 3 may be quite different from that at the sea-level monitoring sites due to the complexities of NO_x sources and transport. Legrand and Kirchner[72] suggest that stratospheric NO_x resulting from lightning, N_2O oxidation, and possibly cosmic rays, as well as tropospheric sources, may influence global nitrogen cycling.

Cl^- concentrations in the snow peak in January, with relatively high values during October through to March and a small peak in July. The winter maxima probably represent long-range transport of seaspray from storms in the North Atlantic. The small summer peak is attributed to seaspray from the nearby coast, when the sea around Greenland is unfrozen. As with SO_4^{2-} and NO_3^-, the concentrations in the snow are not likely to be directly proportional to airborne concentrations due to possible greater scavenging efficiencies in summer.

The limited anion and cation data available suggest that concentrations in the snow are dependent upon airborne concentrations in complex ways: the scavenging ratio defined by eqn (3.5) is likely to vary from storm to storm. In addition, the relative importance of dry deposition may differ among anions and cations and at various times of the year. For example, the SO_4^{2-} inputs reflected in Fig. 3.10 are mostly the result of wet deposition when considered on an annual basis; during the winter, however, dry deposition may be relatively important due to low snowfall rates, inefficient precipitation scavenging, and high atmospheric concentrations. The snowpit data also suggest that concentrations can vary greatly over small spatial scales, mostly due to post-deposition processes such as redistribution of surface snow by the wind. This is likely to affect short-term signals. On the other hand, comparison of data from different snowpit studies on the Ice Sheet suggests that the entire inland portion of Greenland may receive inputs fairly synchronously on the average.

Despite the complexities of air–snow transfer and post-deposition changes, data for anions and cations have shown certain consistencies that have enabled us to identify anthropogenic influence in the Arctic. We now explore examples of these ice core data from several studies.

3.3.1.2 Ice cores

One of the earliest ice core projects which included anion data was the GISP 1 study of Herron *et al.*[73] who sampled for SO_4^{2-} as well as trace metals in the 398-m core from Milcent (Fig. 3.8). Results show SO_4^{2-} concentrations of 136 and 276 ng g^{-1} for two near-surface snow samples, compared with an average concentration of 81 ± 22 ng g^{-1} for several core samples representing the years 1170–1886. Herron *et al.* attribute the three-fold increase between surface snow and core concentrations to anthropogenic activities.

Other work by Herron[18] shows higher concentrations of both SO_4^{2-} and NO_3^- in recent Greenland snow compared with older snow from cores. Concentrations of SO_4^{2-} in pre-1900 ice taken from Dye 3, Camp Century, and North Central sites (Fig. 3.8) average 40 ng g^{-1}, compared with 110 ng g^{-1} in near-surface snow. Concentrations of NO_3^- in these two categories of samples average 58 ng g^{-1} and 130 ng g^{-1}, respectively. It is of interest that comparisons of recent and older snow from Antarctica show no significant difference for either anion.[18,74] The lack of anthropogenic influence for these anions in antarctic snow reflects the small proportion of industrial activity in the southern hemisphere: over 90% of the man-made global emissions occur north of the equator.[75]

Estimates of the increases in SO_4^{2-} and NO_3^- concentration over time in Greenland snow have been refined in several additional ice core studies. Three of these studies involve separate drilling efforts in the vicinity of Dye 3, reported by Neftel *et al.*,[19] Mayewski *et al.*,[20] and Finkel & Langway.[76] A fourth study conducted by Steffenson[77] has involved coring at Summit.

Neftel *et al.*[19] report concentrations of SO_4^{2-} and NO_3^- in an ice core covering the years 1895–1978 (annual resolution). This information has been used with global emissions data to develop a simple model for the fraction of SO_4^{2-} and NO_3^- in the core resulting from anthropogenic emissions. Results of this work are shown in Fig. 3.11. Assuming no anthropogenic influence in the ice core samples before 1900, the graph shows that anthropogenic NO_3^- surpassed natural NO_3^- in the ice in the late 1950s; anthropogenic SO_4^{2-} surpassed natural SO_4^{2-} a couple of years later. Neftel *et al.* also estimate that each increase of 1 Tg (10^{12} g) of land-based nitrogen emissions results in an increase of 1·4 ng g^{-1} NO_3^- in the core. Similarly, each increase of 1 Tg of land-based sulfur emissions results in an increase of 0·6 ng g^{-1} SO_4^{2-} in the core. These values are based on total northern hemisphere emissions in 1975 of 43 Tg year^{-1} for nitrogen and 133 Tg year^{-1} for sulfur.

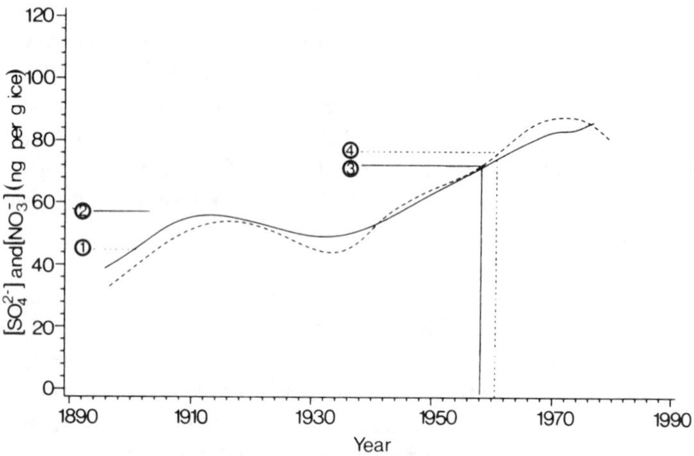

Figure 3.11. Concentrations of SO_4^{2-} (broken line) and NO_3^- (solid line) over time in a core obtained near Dye 3 by Neftel *et al.*[19] Point 1 corresponds to the level of sulfate in the core in 1900, assumed to be entirely from natural sources, while point 2 shows the level of nitrate in the core in 1900, also assumed to be entirely natural. Point 3 shows the level of nitrate when anthropogenic inputs had increased to the point of just equaling natural inputs; point 4 shows the level of sulfate when anthropogenic inputs just equaled natural inputs. (From Neftel, A., Beer, J., Oeschger, H., Furcher, F., and Finkel, R. C. (1985). Sulphate and nitrate concentrations in snow from South Greenland 1895–1978. *Nature*, **314**, 611–13. Reprinted by permission from *Nature*, Copyright 1985 Macmillan Magazines Limited.)

Mayewski *et al.*[20] report concentrations of SO_4^{2-}, NO_3^-, and Cl^- in a core covering the years 1869–1984 (seasonal resolution). The core was obtained 40 km up the katabatic gradient from Dye 3, at a site relatively free of local contamination from the station.[69] Ten-year running means for excess (non sea-salt) SO_4^{2-} and for NO_3^- are pictured in Fig. 3.12. Distinct increases in SO_4^{2-} are seen during the years 1900–1910, 1940–1950, and 1960–1970. Mayewski *et al.* suggest that the sharp increase during 1900–1910 is mostly volcanic input. The other increases are attributed to anthropogenic sulfur emissions in North America and Eurasia. Overall, the excess SO_4^{2-} concentrations for 1968–1984 average 84 ng g^{-1}, nearly a three-fold increase over the 1870–1900 average of 30 ng g^{-1}.

Concentrations of NO_3^- in Fig. 3.12 show a different trend. Levels are relatively constant until 1955, after which there are steady increases to

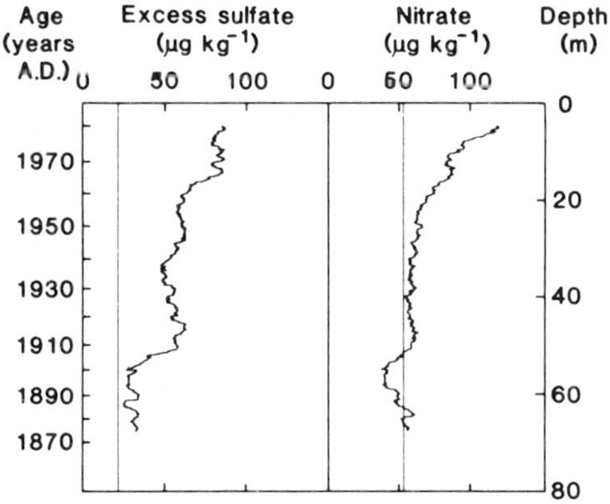

Figure 3.12. Ten-year running mean concentrations in excess (non-sea salt) SO_4^{2-} and in NO_3^- from a core obtained near Dye 3 by Mayewski *et al.*[20] (From Mayewski, P. A., Lyons, W. B., Spencer, M. J., Twickler, M., Dansgaard, W., Koci, B., Davidson, C. I., and Honrath, R. E. (1986). Sulfate and nitrate concentrations from a south Greenland ice core. *Science*, **232**, 975–77. Copyright 1986 by the American Association for the Advancement of Science.)

the end of the record in 1984. The average NO_3^- concentration for 1869–1955 is 55 ng g^{-1}; the average for 1975–1984 is 115 ng g^{-1}, suggesting a doubling in concentration since the 1950s. It is significant that Mayewski *et al.*[78] find a close parallel between concentrations in the core and North American emissions for SO_4^{2-}, but not for NO_3^-. This suggests different sources and/or transport mechanisms for the two species.

Concentrations of Cl^- are relatively constant over the period 1869–1984 (seasonal resolution). Most of the Cl^- in the core is attributed to seaspray, implying that distances and transport pathways from the ocean to the drill site have not changed significantly over the past century.

Finkel & Langway[76] report concentrations of SO_4^{2-}, NO_3^-, Cl^-, and H^+ in a core dating back two centuries. Four 10-m depth intervals (seasonal resolution) have been used to compare concentrations in different sections of the core, representing the years 1764–1775, 1823–1832, 1920–1930, and 1944–1954.

Results show similar levels during the first two intervals, corresponding to pre-industrial concentrations. These values for SO_4^{2-} and NO_3^- are 24 ng g^{-1} and 52 ng g^{-1}, respectively, in rough agreement with the Dye 3 values discussed above. The data show that SO_4^{2-} concentrations in the ice doubled by 1920–1930, and are three times as high as pre-industrial values by 1944–1954. Concentrations of NO_3^- are roughly the same for the first three intervals; a 30% increase is observed for the final time period. The later increase in NO_3^- compared with SO_4^{2-} is consistent with the findings of Mayewski *et al.*[20] No appreciable difference in Cl^- concentrations is observed for any of the time intervals.

For estimation of H^+ levels, Finkel and Langway[76] have used the electroconductivity method of Hammer[23] and have compared these measurements with calculations from an ion balance model incorporating the SO_4^{2-}, NO_3^-, and Cl^- data. The model assumes that all of the Cl^- is of marine origin and that the Cl^- and marine-derived SO_4^{2-} anions are associated with cations other than H^+. Results show good agreement between measured and calculated values for the first two time intervals. For the latter two intervals, the calculated H^+ concentrations are often greater than the measured values, suggesting that recent increases in SO_4^{2-} and NO_3^- are accompanied by cations other than H^+.

The ice core sections analyzed by Finkel and Langway[76] include ten samples per year, allowing assessment of seasonal variations. Some of the results are shown in Fig. 3.13. The seasonal patterns in the 1920s and 1950s agree with the results of Davidson *et al.*[68] shown in Fig. 3.10. However, it is interesting that concentrations of SO_4^{2-}, NO_3^-, and H^+ in recent ice show greatest increases above the pre-industrial concentrations during the winter and spring, when long-range transport from the mid-latitudes is most efficient. The effect is seen for the latter two time intervals in the case of SO_4^{2-} but only for the last interval in the case of NO_3^-, again consistent with the finding that increasing concentrations are observed later for NO_3^-. The effect is less pronounced for H^+.

Finally, Steffensen[77] has examined SO_4^{2-}, NO_3^-, and Cl^- concentrations in two cores from Summit for the years 1891–1910. The cores were collected on the east and west sides of the ice divide in central Greenland. SO_4^{2-} on both sides of the divide show a distinct increase after 1903; concentrations prior to this date resemble the pre-industrial levels reported by Herron[18] for central Greenland. The trend after 1903 agrees with the sharp increase during 1900–1910 reported by Mayewski *et al.*[20] for Dye 3. Steffensen attributes the trend to mid-latitude anthropogenic emissions as well as pronounced volcanic activity during 1902. In approximate agreement with previous studies, SO_4^{2-} peaks in the spring or early summer.

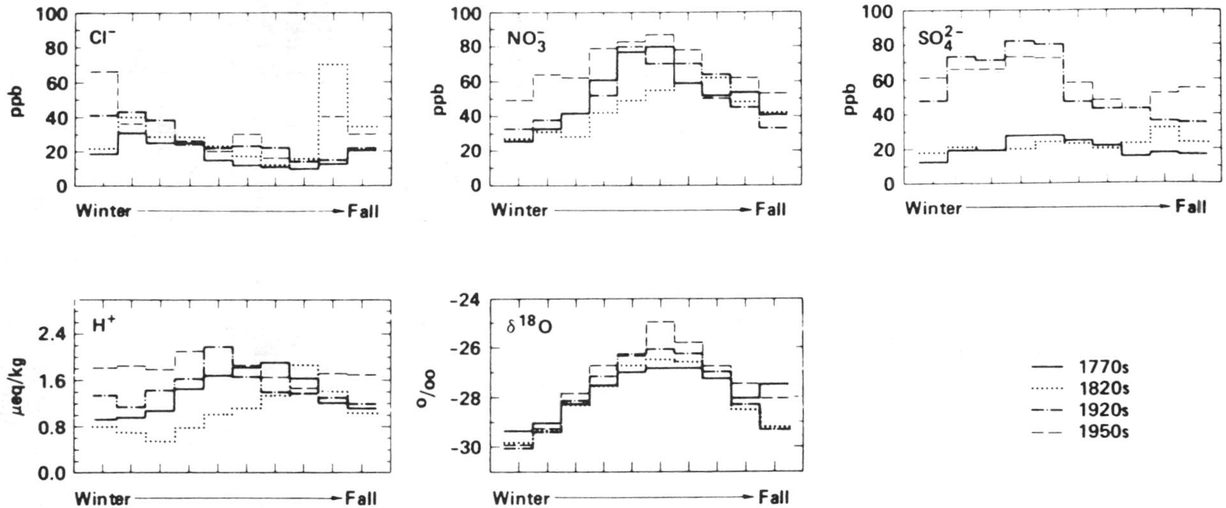

Figure 3.13. Seasonal average concentrations from sections of a Dye 3 core representing four different time periods, taken from Finkel and Langway.[76] (From Finkel, R. C., Langway, C. C., Jr., and Clausen, H. B. (1986). Changes in precipitation chemistry at Dye 3, Greenland *J. Geophys. Res.*, **91**, 9849–55. Copyright by the American Geophysical Union.)

NO_3^- shows no change in concentration over the twenty-year interval, with strong summer peaks as reported in many of the studies cited above. Similar levels occur on both sides of the ice divide. Cl^- also shows no trend over this time period; concentrations peak during winter–early spring. For some years, the Cl^- concentration is higher on the eastern side of the divide, attributed to a greater marine influence.

Steffensen[77] explains the seasonal differences in anion concentration in terms of atmospheric circulation patterns, the influence of photochemistry, and seasonal variations in precipitation based partly on the work of Heidam[79] and Barrie.[80] Of particular interest is the importance of vertical mixing in transporting atmospheric constituents from the lower troposphere to the elevation of the Greenland Ice Sheet: the strong winter–early spring peak in Cl^- in the ice suggests that variations in long-range transport and vertical atmospheric mixing have more impact than seasonal variations in concentration in the lower troposphere over the Greenland coast, where aerosol Cl^- peaks in summer.

All of the above studies involve snow and ice in Greenland. Very few glaciers in the Arctic outside Greenland have been sampled; one such program has been conducted on Agassiz Ice Cap, located on Ellesmere Island in the Canadian Arctic.[71,81] This has involved a series of snowpits and ice cores covering the years 1950–1983 that have been sampled for H^+ and conductivity, as well as two cores sampled for conductivity only, covering 1912–1980 and 1945–1978, respectively.

The conductivity data in the longest core show a general decrease of about 30% during 1912–1950, followed by a rapid increase of 50–70% during 1950–1980.[71] The shorter cores show similar rapid increases after 1950. These increases are in general agreement with trends in the Greenland data discussed above. However, the effects of pollution are much more evident on Agassiz: by inferring H^+ concentrations based on the conductivity data, Barrie *et al.* note that H^+ levels are three to four times higher than the sum of SO_4^{2-} and NO_3^- ion equivalents in the Dye 3 core of Neftel *et al.*[19] This difference is attributed to the relatively low elevation of Agassiz (1·6 km) compared with Dye 3 (2·5 km), implying that Agassiz is influenced by more polluted air in the lower troposphere. The hypothesis of Finkel & Langway[76] that cations other than H^+ may be associated with SO_4^{2-} and NO_3^- implies that the true difference between the Agassiz and Dye 3 records may be even greater than suggested. However, differences in source regions influencing the two sites as well as differences in elevation may be partly responsible.

Barrie *et al.*[71] also compared acidity levels in one of the snowpits with concentrations of airborne acidity measured at Alert, 240 km to the north. The time series of concentrations in the air and concentrations in the snowpit are shown in Fig. 3.14. Both curves show maxima in late winter. However, as observed for SO_4^{2-} at Dye 3,[68] the variation in acidity in the air is far greater than that in the snow. The points in Fig. 3.14 correspond to an overall average scavenging ratio of 240, which is within the range of values reported above for SO_4^{2-}.

Other studies in the Canadian Arctic have been reported by Holdsworth and Peake[82] and Holdsworth *et al.*[83] In the first study, an ice core record from Mt. Logan, Yukon Territory (elevation 5630 m) provided an opportunity to sample mid-troposphere air masses that emanate from the eastern Pacific Ocean. The complete record from Mt. Logan is not yet available, but it appears that there is little anthropogenically derived SO_4^{2-} or NO_3^- at this site. In contrast, Holdsworth *et al.*[83] have recovered a series of shallow cores from sites in Northwest Territories,

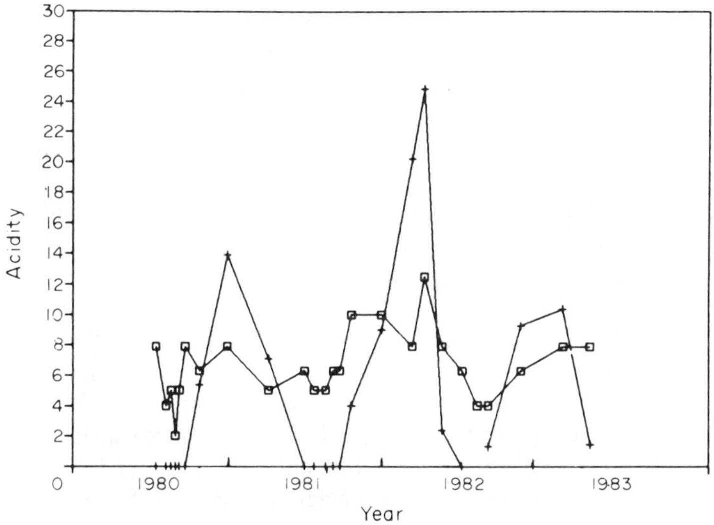

Figure 3.14. Concentrations of acidity in snow samples taken from Agassiz Ice Cap, and in aerosol samples collected at nearby Alert corresponding to the same time periods, taken from Barrie et al.[71] □, acidity ice (μm litre^{-1}) and +, acidity air (nm metre^{-3}). (From Barrie, L. A., Fisher, D., and Koerner, R. M. (1985). Twentieth century trends in arctic air pollution revealed by conductivity and acidity observations in snow and ice in the Canadian High Arctic. *Atmos. Environ.*, **19**, 2055–2063. Reprinted with permission from Pergamon Press.)

Yukon Territory, and Alberta; the records show increases in SO_4^{2-}, NO_3^-, and total acidity in recent decades. The increases are attributed to anthropogenic activity, although the increase in NO_3^- may be partially due to forest fires.

Overall, the ice core data summarized in this section generally agree that SO_4^{2-} and NO_3^- concentrations have increased by factors of two to three since the turn of the century, with smaller but still measureable increases in H^+ over this time period. Cl^- levels have remained roughly constant. The ratio of anthropogenic/natural constituents levels in the ice is maximum during late winter–early spring each year, when transport from the mid-latitudes is most efficient. Although this information can help us to understand the role of anthropogenic activities on global cycling of these species, far more information on atmospheric transport pathways, deposition, and post-depositional changes is needed to make maximum use of the glacial record for anions and cations.

3.3.2 Trace metals

A number of studies have considered trace metals in polar snow. As with anions and cations, most of the work concerning surface snow and snowpits has studied air–snow deposition processes, and establishment of concentrations in recent snow to which levels in deeper ice can be compared. Ice core studies, on the other hand, have considered long-term trends in concentration.

It is convenient to categorize data on trace metals on the basis of the crustal enrichment factor EF_{crust}:

$$EF_{crust} = (X_{air}/Al_{air})/(X_{crust}/Al_{crust}) \qquad (3.6)$$

where X_{air} and Al_{air} represent the airborne concentrations of any element X and of aluminum, respectively, and X_{crust} and Al_{crust} are the respective average concentrations in the Earth's crust. Aluminum is often used as the reference element in eqn (3.6), since it is derived primarily from crustal erosion at most locations. Several investigators have hypothesized that values of EF_{crust} near unity imply crustal erosion as a dominant source. Elements with airborne concentrations that are enriched relative to crustal composition ($EF_{crust} \gg 1$) are likely to have predominantly non-crustal sources, such as anthropogenic activities, natural combustion (volcanism, forest fires), biological emissions, and seaspray.[84,85] Although enrichment factors may help us to identify airborne metals resulting mainly from crustal erosion, the results cannot be considered quantitative. This is because some noncrustal processes (e.g. generation of seaspray) emit certain metals in a ratio similar to that of crustal erosion. Furthermore, emissions of metals to the air during erosion may be different from their relative abundances in the crust, resulting in chemical fractionation.[86] Nonuniformities in crustal composition also confound interpretation of enrichment factors.

Equation (3.6) may be applied to metals in snow as well as in air. However, interpretation of enrichment factor data for snow is more complex, since the metal of interest may be incorporated into snow at a rate different from that of the reference element. This problem of chemical fractionation has been discussed by several authors.[87–89] Examples of metals reported in polar snow that have large EF_{crust} include Ag, As, Cd, Cu, Pb, Se, and Zn. Examples of metals with EF_{crust} near unity include Ca, Fe, K, Mg, Mn, and Na.

Emissions inventories have been estimated for many of these metals. The inventories for some of the metals with high enrichment factors show that emissions from anthropogenic sources exceed those from natural sources on a world-wide basis. For example, Pacyna[90] reports that man-made emissions of Cd, Cu, Pb, and Zn exceed natural emissions by factors of 8, 3, 24, and 7, respectively. Glacial record data for metals should therefore be particularly useful for identifying the extent of long-range transport of industrial emissions to the polar regions.

Although accurate trace metal data would be of great interest, recent studies suggest that the majority of published trace metal concentrations in arctic snow may be erroneous due to contamination problems.[91–93] As an example, Table 3.1 shows data for Cd, Cu, Pb, and Zn in arctic snow taken from the summary of Wolff & Peel.[94] The authors point out discrepancies in the table: concentrations from their own investigation are much smaller than those reported in the other studies. Inadequate contamination control is cited as a posible reason for the differences in concentration. However, Wolff & Peel mention that there may be genuine variations in concentration due to different time periods and locations of snow collection among the studies. Additional data on metals in arctic snow are needed to resolve these discrepancies.

Despite potential contamination problems, results of some surface snow and snowpit studies show certain consistencies in attempting to assess wet and dry deposition. For example, Davidson *et al.*[88] report scavenging ratios based on eqn (3.5) for fresh snow at Dye 3. The crustal elements Al, Fe, K, Mg, Mn, and Na have average values in the range 1200–2000. In contrast, Pb and SO_4^{2-} have average values below 200. The difference is attributed to efficient scavenging of crustal particles by

Table 3.1 Concentrations of Cd, Cu, Pb and Zn in recent arctic snow, taken from Wolff and Peel.[94] Values given in parentheses indicate snow representing only part of a year. (From Wolff, E. W. and Peel, D. A. (1988). Concentrations of cadmium, copper, lead and zinc in snow from near Dye 3 in south Greenland. *Annals of Glaciology*, **10**, 193–7. Reproduced by courtesy of the International Glaciological Society.)

References	*Location*	*pH of analysis*	*Comments*	*Cd*($pg\ g^{-1}$)	*Cu*($pg\ g^{-1}$)	*Pb*($pg\ g^{-1}$)	*Zn*($pg\ g^{-1}$)
Murozumi *et al.*[16]	Camp Century, Greenland	1–2	1964–65 snowfall			200	
Herron *et al.*[73]	Milcent, Greenland	1	1971–73 snowfall	8		144	224
Mart[110]	Eastern Arctic Ocean (from ice floes)	2–3	1979–80 mean (Aug–Sept 1980)	5 (0·4)	97 (13)	190 (13)	
Davidson *et al.*[88]	Dye 3, Greenland	HF digestion	(Mean of 8 days' sampling, mainly in April–May 1983)	(8·5)	(92)	(129)	
Wolff & Peel[94]	36 km from Dye 3, Greenland	3	1983–84 mean (late summer 1983 only)	0·74 (0·74)	6·2 (3·2)	28 (7·2)	27 (17)

ice nucleation; Pb and SO_4^{2-} make poor ice nuclei. This same study also reported that concentrations of crustal elements in surface snow increase over time between snowstorms, presumably due to dry deposition. The enriched elements do not show comparable increases. This observation is consistent with the smaller dry deposition velocities of enriched elements due to smaller particle sizes.[95,96]

Very few studies have attempted to analyze trace metals from deep firn and ice in the Arctic. The most reliable information to date are values reported in the pioneering work of Murozumi *et al.*[16] Shown in Fig. 3.15, the results indicate a 200-fold increase in Pb concentrations in modern Greenland snow compared to concentrations in ice dating back several thousand years. For Antarctica, a few recent studies have produced reliable data for trace metals in modern as well as pre-industrial ice, as summarized by Boutron;[97] these studies indicate an approximate five-fold increase in Pb due to anthropogenic activities. However, reliable data for a variety of metals in deep arctic ice will have to wait for results of the GISP 2 and GRIP programs, where a major effort to obtain such data is underway.

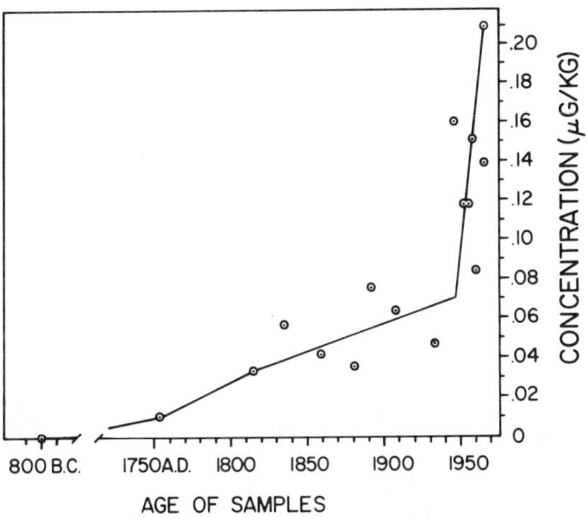

Figure 3.15. Concentrations of Pb in samples obtained from deep ice trenches at Camp Tuto and Camp Century, Greenland, taken from Murozumi *et al.*[16] Concentrations are given in μg Pb/kg snow. (From Murozumi, M., Chow, T. J., and Patterson, C. (1969). Chemical concentrations of pollutant lead aerosols, terrestrial dusts and sea salts in Greenland and antarctic snow strata. *Geochimica et Cosmochimica Acta*, **33**, 1247–94. Reprinted with permission from Pergamon Press.)

3.3.3 Radionuclides

Work on anthropogenic radionuclides in arctic glaciers has encompassed two topic areas. First, ice cores obtained from the Greenland Ice Sheet have been analyzed for species emitted from nuclear bomb testing in the 1950s–1970s. Second, surface snow and snowpits in Greenland have been analyzed for debris from the Chernobyl accident in April–May 1986.

In the first topic area, Fig. 3.16 shows the nuclear detonation yield from bomb tests world-wide as a function of year, taken from Carter & Moghissi.[98] The tests emitted ^{3}H, ^{134}Cs, ^{137}Cs, ^{90}Sr, several isotopes of plutonium, and other radionuclides which have been measured in the Greenland Ice Sheet. For example, data for a number of isotopes in firn at Dye 3 are shown in Fig. 3.17.[99] The patterns in these graphs resemble

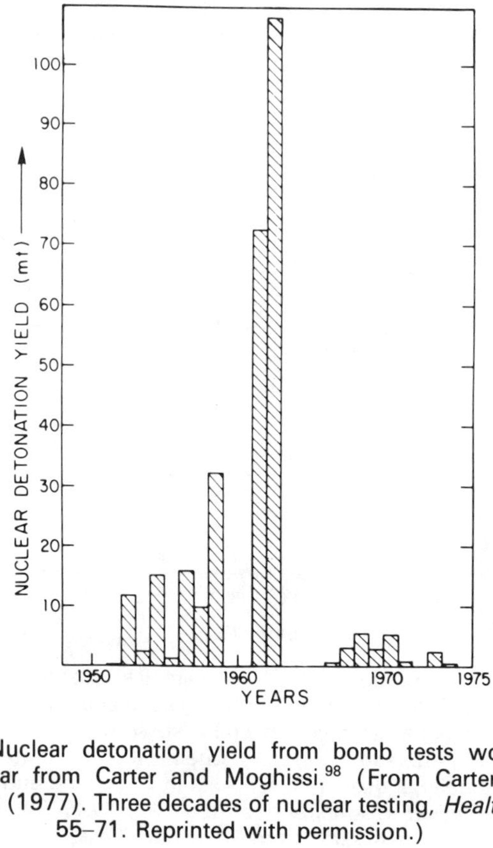

Figure 3.16. Nuclear detonation yield from bomb tests worldwide as a function of year from Carter and Moghissi.[98] (From Carter, M. W. and Moghissi, A. A. (1977). Three decades of nuclear testing, *Health Physics*, **33**, 55–71. Reprinted with permission.)

the detonation yield data in Fig. 3.16; note that there is a lag between the bomb tests and the year of peak concentrations of some radionuclides. It is also significant that the concentrations persist for years after the tests. These observations are explained on the basis of atmospheric residence times: some of the material was injected into the troposphere and thus had a relatively short residence time, while some reached the stratosphere where material may remain for years.[100]

There are also distinct differences in residence times among the radionuclides that reached the stratosphere.[99-101] Tritium is a gaseous species which may persist for long periods of time, while plutonium isotopes are partly associated with particulate matter and likely to be removed from the stratosphere relatively quickly. Cesium and strontium have residence times that are intermediate, possibly affected by their ability to form associations with water. Differences in the characteristics of the weapons also played a role in the glacial record data. For example, much more tritium was produced in the later bomb tests compared with those in the 1950s, as evidenced from Fig. 3.17. Additional glacial record data are available for Antarctica, allowing studies of atmospheric circulation patterns in both hemispheres.[100]

The second topic area, radioactivity from Chernobyl, is important because the release of material provides an unambiguous tracer of air masses from a known source area over a relatively short period of time. Three independent studies have focused on Chernobyl radioactivity in arctic snow. The first study by Pourchet *et al.*[102] included sampling of surface snow from two sites in northern Greenland and two sites on Svalbard in the Norwegian Arctic. These samples were collected during May 1986, just after the release, and analyzed for ^{134}Cs and ^{137}Cs. The second study conducted by Davidson *et al.*[21] involved several layers of the 1986 and 1987 Dye 3 snowpits discussed earlier,[68] with analysis for the same radioactive cesium isotopes. Finally, Dibb[22] sampled snow in seven pits from the vicinity of Dye 3, Summit, and Camp Century in Greenland, with analysis for the same species.

Dibb[22] has summarized results of the three studies, and concludes that the Chernobyl plume has resulted in detectable levels of radioactivity in the snow at all of these widely separated sites. However, both concentrations and inventories of ^{134}Cs and ^{137}Cs vary greatly among the sites. These variations partly reflect nonuniformities in atmospheric transport. Most of the Chernobyl radioactivity was confined to the troposphere, unlike some of the bomb debris that reached the stratosphere. One would expect atmospheric mixing to be less complete in the former case,

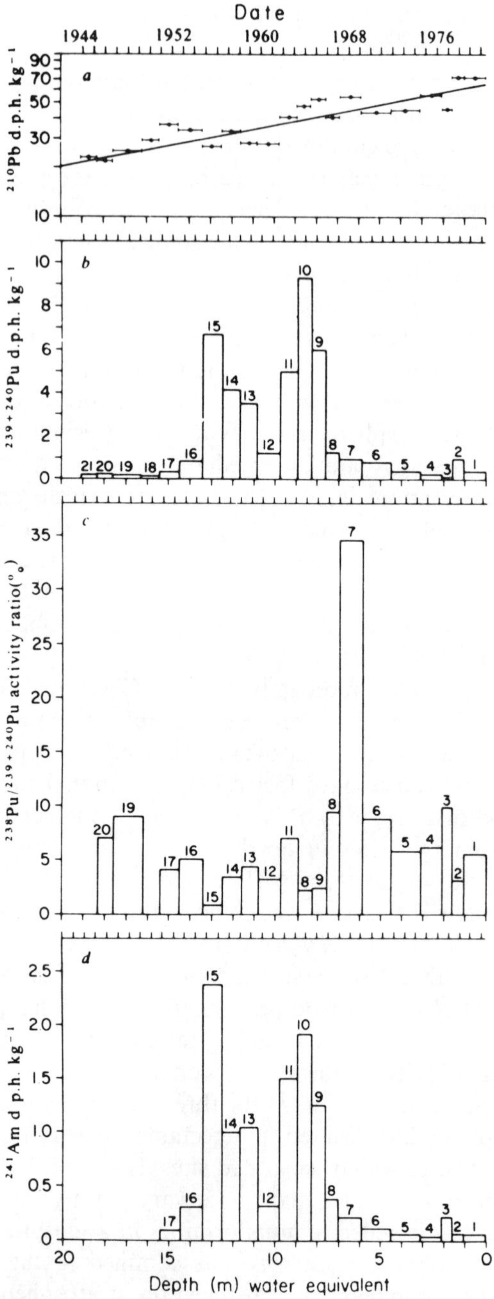

Figure 3.17. Radionuclides from Dye 3 ice cores from Koide *et al.*[99]. The time scale is derived from the ^{210}Pb profile. Data have been corrected to the time of deposition. (From Koide, M., Goldberg, E. D., Herron, M. M., and

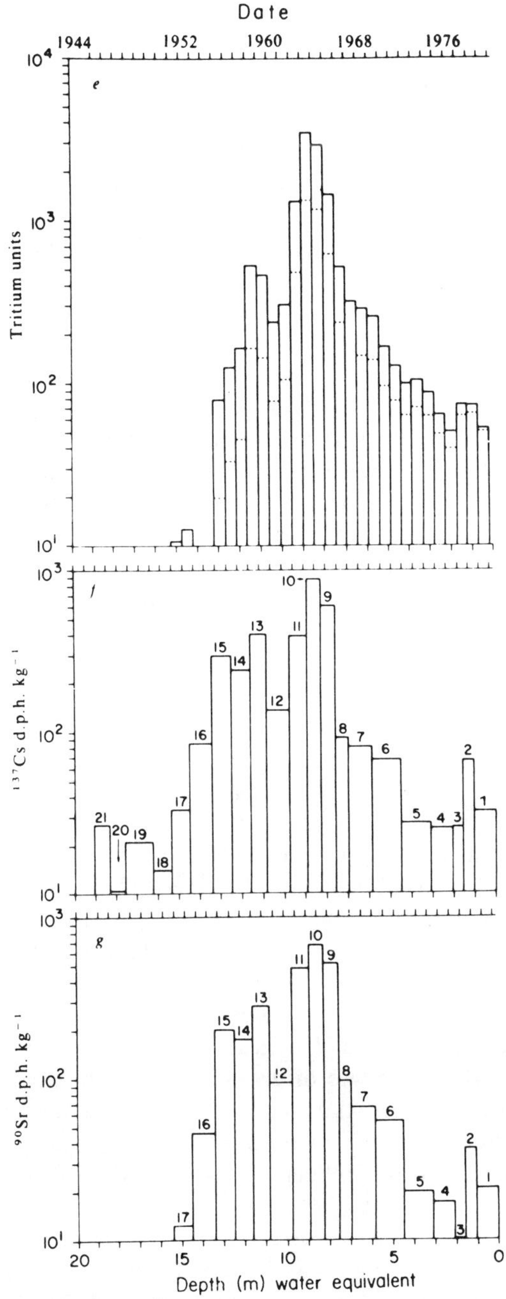

Langway, C. C., Jr. (1977). Transuranic depositional history in south Greenland firn layers. *Nature*, **269**, 137–39. Reprinted by permission from *Nature*, Copyright 1977 Macmillan Magazines Limited.)

resulting in variable airborne concentrations and differences in wet and dry deposition throughout the Arctic. Indeed, a wealth of information on variations in airborne radioactivity from Chernobyl in the middle and high latitudes has been obtained from meteorological models (e.g. Pudykiewicz[103]). Dibb[22] emphasizes that post-depositional processes probably also affected activity levels in the snow: the relatively quick passage of the Chernobyl plume across the Arctic deposited a thin layer of radioactivity onto the snowfields, which later became redistributed by mechanisms which included disturbance of the surface snow by wind, meltwater percolation, and diffusion within the snowpack. The implication is that glacial record data in general must be interpreted with caution because of such spatial inhomogeneities in the snow.

In addition to these anthropogenic radionuclides which serve as tracers of opportunity, several natural radionuclides may be indicators of air masses with particular transport characteristics. For example, ^{222}Rn is emitted primarily from the continents. Significant concentrations of its daughter product ^{210}Pb in glaciers indicate transport from land. The radionuclides ^{7}Be, ^{10}Be, and ^{36}Cl are associated with air masses from the stratosphere. By correlating concentrations of other chemical constituents with those of such radionuclides, we can thus obtain information on transport pathways of the constituents.

3.3.4 Trace gases

There has been a growing concern in recent years about increases in concentrations of greenhouse gases that may affect climate, such as CO_2, CH_4, N_2O, and fluorocarbons. Most of the direct airborne concentration data showing these increases have been obtained in the mid-latitudes, although some measurements have taken place in the Arctic (e.g. Khalil & Rasmussen[34]). As with the other chemical species of interest, such direct measurements have provided information only over the past decade or so. It is therefore of great interest to obtain greenhouse gas data from ice cores, where time series records may extend back thousands of years. In this way, natural variations occurring over scales of millenia as well as anthropogenic influences over decades can be used to study climate change.

The behavior of gases adsorbed on snow and ice is only poorly understood; concentrations may be far out of equilibrium with atmospheric concentrations.[104,105] If the gases are soluble, their behavior is dependent on the presence of a liquid layer on the surface.[45] It is probable that measurements of gases adsorbed onto snow or ice can be used only for studying short-term phenomena.

In contrast, the behavior of gases trapped in air bubbles in ice is more stable and better understood. The air in the bubbles has essentially the same composition as that of the atmosphere at the time of bubble formation. However, the age of the entrapped air is not the same as that of the surrounding ice, because bubbles become isolated from the atmosphere only during the transformation from firn to ice. The age of the ice at pore close-off has been well-studied and is generally between ~100 and 3000 years.[106,107] This age depends mainly on temperature and accumulation rate. In addition, it is important to estimate the time over which pore close-off is occurring, since this determines the width of the age distribution, i.e. the range of ages represented in a gas sample. Schwander & Stauffer[108] have estimated that this range may vary from ~22 years to ~590 years. These values are for ice without traces of summer melting. If melting occurs, the process of pore close-off is much more complicated: ice layers in the firn may act as barriers against homogeneous diffusion, but may also serve to concentrate some of the gases far out of atmospheric equilibrium.[106]

Few studies concern cores that have been obtained at sites where there is high accumulation rate and low summer temperatures, needed to enable precise dating and good time resolution over the last centuries. Two such studies in northern Greenland are reported by Rasmussen & Khalil[3] for CH_4 and by Khalil & Rasmussen[9] for N_2O, respectively. The concentration of CH_4 is observed to increase gradually from ~700 ppbv to ~1600 ppbv over the last 250 years. The concentration of N_2O is roughly constant at 285 ppbv until about 1900, and then increases to 307 ppbv. These results are in good agreement with the antarctic ice core studies of Zardini *et al.*[4] for N_2O and Pearman *et al.*[2] for N_2O and CH_4. Pearman *et al.* also measured CO_2, and found a 20% increase over the past 200 years.

Neftel *et al.*[1] present results for CO_2 from another antarctic core, using precise time resolution. The concentration of CO_2 in air bubbles in the core increases from 280 ppmv in the year 1750 to 345 ppmv in 1984, similar to the results of Pearman *et al.*[2] Increases in concentrations of all three gases are attributed to human activities: CH_4 is linked to food production, while N_2O and CO_2 are related to combustion processes. Biomass burning is a particularly important source category for CO_2. The measured concentrations in air bubbles in the core are shown in Fig. 3.18. Shown in the same graph are calculated concentrations extrapolated

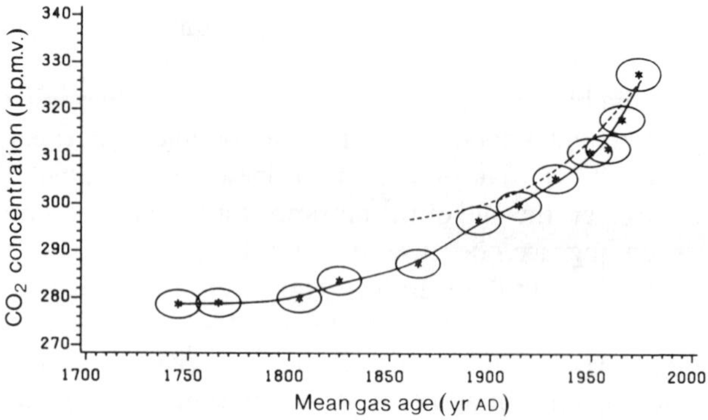

Figure 3.18. Concentrations of CO_2 in air bubbles in an antarctic ice core, taken from Neftel *et al.*[1] Ellipses have horizontal axes which indicate a close-off time interval of 22 years. The dotted line shows calculated concentrations assuming CO_2 emissions are only from fossil fuel combustion. (From Neftel, A., Moor, E., Oeschger, H., and Stauffer, B. (1985). Evidence from polar ice cores for the increase in atmospheric CO_2 in the past two centuries. *Nature*, **315**, 45–7. Reprinted by permission from *Nature*, Copyright 1985 Macmillan Magazines Limited.)

backwards based on a model that assumes all CO_2 emissions are from fossil fuel combustion. The difference between modeled and measured values is attributed to biomass emissions.[1]

There are also a number of studies concerning natural variations in trace gas concentrations occurring over several thousand years. For example, Stauffer *et al.*[5] and Barnola *et al.*[6] have reported wide variations in CO_2, while Raynaud *et al.*[7] and Stauffer *et al.*[8] have found varying concentrations of CH_4. Natural variations in N_2O have been studied by Khalil & Rasmussen[9] and Zardini *et al.*[4] Such long-term changes are also linked to changes in climate, and thus are important to help understand recent changes in concentrations caused by anthropogenic activities.

Additional information concerning anthropogenic influences on greenhouse gas concentrations may result from the GISP 2 project. However, more definitive information is likely to be derived from future ice coring efforts in Antarctica, where conditions are better for defining the ages of gas bubbles in the ice.

3.4 SUMMARY AND CONCLUSIONS

The glacial record has provided valuable data for assessing anthropogenic pollutants reaching the Arctic. Most of the available information has been obtained in Greenland, where surface snow, snowpit, and ice core samples have been analyzed for various chemical species. This chapter has reviewed the topic of pollutants in the glacial record, discussing the processes by which chemical constituents are transported to the glaciers as well as examples of glacial record data.

The processes responsible for chemical constituent levels in arctic glaciers can be divided into three steps. In the first step, the constituents are transported from source regions to the atmosphere over the glaciers. With the exception of a few local sources, most pollution in the arctic atmosphere is the result of long-range transport from the mid-latitudes. North America, Europe, and western/central Asia are important source regions for these pollutants, especially during the late winter and early spring when transport is most efficient. In the second step, a fraction of these pollutants will reach the surface of glaciers by dry or wet deposition. Dry deposition involves the transport of gases and particles from the free atmosphere down to the region of viscous air just above each snowflake or ice crystal, and transport across this viscous sublayer onto the ice. In many cases, air-carrying constituents can penetrate the surface snow, depositing gases and particles onto ice within the snowpack. Wet deposition refers to scavenging by precipitation. This includes the capture of particles during the formation of snow, and collection of gases and particles by existing ice crystals and snowflakes. Finally, the third step involves changes that occur as the constituents on the surface become buried by later snowfall, and as the snow densifies to become firn and ice. Examples of processes occurring in this step include redistribution of surface snow by wind, meltwater percolation, and diffusion within the snowpack.

Examples of glacial record data have been presented for chemical species in four categories. First, information on anions and cations has been presented, focusing on data for SO_4^{2-}, NO_3^-, Cl^-, and H^+ as well as electrical conductivity resulting from ionic species. Limited information is then presented for Ca^{2+}, Na^+, Mg^{2+}, and K^+. Results indicate that time series records of SO_4^{2-} and NO_3^- concentrations in arctic glaciers are consistent with trends in estimated global emissions of these species: SO_4^{2-} levels have increased by a factor of three since 1900, while NO_3^- levels have been roughly constant until the 1950s and have doubled since then. Concentrations of H^+ and measurements of conductivity also show increases beginning in the 1950s. In contrast to these species which are affected by anthropogenic emissions, Cl^-, Ca^{2+}, Na^+, Mg^{2+} and K^+ are mostly derived from soil dust or seaspray. All of these data show distinct seasonal variations which are related to the seasonal cycles in airborne concentrations and to seasonal differences in scavenging.

The second category involves trace metals in arctic glaciers; all available data are for Greenland. Contamination problems have plagued most of these studies. The limited available data indicate that concentrations of Pb have increased by orders of magnitude since pre-industrial times, in general agreement with estimates of global anthropogenic emissions. Data for other trace metals that have dominant anthropogenic sources, with a few exceptions from recent surface snow studies, are of questionable validity.

The third category involves radionuclides produced by anthropogenic activities. These include species released during nuclear bomb tests, and species emitted from the Chernobyl accident. The data show that time series records of plutonium, tritium, and other radoisotopes in Greenland ice cores are similar to the known nuclear detonation yields from bomb tests in the 1950s–1970s. These data have been used to obtain information on atmospheric circulation patterns and residence times of the radionuclides. Data on radioactive cesium from the Chernobyl accident have provided information on the lack of homogeneity in deposition across arctic glaciers; this lack of homogeneity is due to differences in airborne concentration, wet and dry deposition rates, and post-deposition processes such as redistribution of surface snow by winds. Besides these anthropogenic radionuclides, some natural radionuclides can be tracers of particular source categories that can assist interpretation of other chemical constituent data. Examples include ^{210}Pb from continents and several radioisotopes from the stratosphere.

The final category considered is trace gases. Of primary importance are greenhouse gases that affect the Earth's energy balance, and whose atmospheric concentrations have been increasing in recent years due to anthropogenic emissions. The limited available arctic data, all from Greenland, show that CH_4 in the ice has doubled over the past 250 years, while N_2O has increased by about 8% since 1900. Additional data from Antarctica support these results. The antarctic data further show that CO_2 has increased roughly 20% over the past 200 years.

Overall, the data suggest that valuable information can be obtained from the glacial record concerning the influence of anthropogenic activities on global cycling of chemical species. However, our ability to relate chemical constituent concentrations in glaciers to their origins and transport is extremely limited; we are not yet able to provide more than a cursory interpretation of snowpit and ice core data. To improve on this will require investigations of specific aspects of each step in the overall transport process. These must include controlled laboratory investigations as well as field programs, linked with detailed mathematical modeling efforts. Only through such studies can we hope to make progress in interpreting the wealth of information in glaciers.

ACKNOWLEDGMENTS

Randy Borys provided a number of comments on the manuscript as well as detailed advice on Figures 3.6 and 3.7, and Jack Dibb provided additional valuable comments on the manuscript. Parts of the paper were prepared by Shirley Knapp. This work was funded by National Science Foundation Projects DPP-8618223, DPP-8821018 and ATM-8922034.

REFERENCES

1. Neftel, A., Moor, E., Oeschger, H. and Stauffer, B., Evidence from polar ice cores for the increase in atmospheric CO_2 in the past two centuries. *Nature*, **315** (1985) 45–7.
2. Pearman, G. I., Etheridge, D., de Silva, F. and Fraser, P. J., Evidence of changing concentrations of atmospheric CO_2, N_2O, and CH_4 from air bubbles in antarctic ice. *Nature*, **320** (1986) 248–50.
3. Rasmussen, R. A. and Khalil, M. A. K., Atmospheric methane in the recent and ancient atmospheres: Concentrations, trends, and interhemispheric gradients. *J. Geophys. Res.*, **89** (1984) 11599–608.
4. Zardini, D., Raynaud, D., Scharffe, D. and Seiler, W., N_2O measurements of air extracted from antarctic ice cores: Implication on atmospheric N_2O back to the last glacial–interglacial transition. *J. Atmos. Chem.*, **8** (1989) 189–201.
5. Stauffer, B., Neftel, A., Oeschger, H. and Schwander, J., CO_2 concentration in air extracted from Greenland ice samples. In *Greenland Ice Core: Geophysics, Geochemistry and the Environment*, ed. C. C. Langway, Jr., H. Oeschger, and W. Dansgaard. AGU Geophysical Monograph 33, 1985, pp. 85–9.
6. Barnola, J. M., Raynaud, D., Korotkevich, Y. S. and Lorius, C., Vostok ice core provides 160,000-year record of atmospheric CO_2. *Nature*, **329** (1987) 408–14.
7. Raynaud, D., Chappellaz, J., Barnola, J. M., Korotkevich, Y. S. and Lorius, C., Climatic and CH_4 cycle implications of glacial–interglacial CH_4 change in the Vostok ice core. *Nature*, **333** (1988) 655–7.
8. Stauffer, B., Lochbronner, E., Oeschger, H. and Schwander, J., Methane concentration in the glacial atmosphere was only half that of preindustrial Holocene. *Nature*, **332** (1988) 812–14.
9. Khalil, M. A. K. and Rasmussen, R. A., Nitrous oxide: Trends and global mass balance over the last 3000 years. *Annals of Glaciology*, **10** (1988) 73–9.

10. Mayewski, P. A., Spencer, M. J., Lyons, W. B., Twickler, M. S. and Dibb, J., Ice core record and ozone depletion—potential for a proxy ozone record. *Antarctic Journal of the U.S.*, **23** (1990) 64–8.
11. Dansgaard, W., Clausen, N., Gundestrup, N., Johnsen, S. J. and Rygner, C., Dating and climatic interpretation of two deep Greenland ice cores. In *Greenland Ice Core: Geophysics, Geochemistry and the Environment*, ed. C. C. Langway, Jr., H. Oeschger, and W. Dansgaard. AGU, Geophysical Monograph 33, 1985, pp. 71–6.
12. Thompson, L. G., Variations in microparticle concentration, size distribution and elemental composition found in Camp Century, Greenland and Byrd station, Antarctica deep ice cores. In *Proceedings of the Symposium on Isotopes and Impurities in Snow and Ice*, Grenoble, France, Aug–Sep. 1975. IAHS Publication 118, 1975, pp. 351–64.
13. Mayewski, P. A. and Lyons, W. B., Source and climatic implication of the reactive iron and reactive silicate concentrations found in a core from Meserve Glacier, Antarctica. *Geophys. Res. Lett.*, **9** (1982) 190–92.
14. Saxena, V. K., Evidence of the biogenic nuclei involvement in antarctic coastal clouds. *J. Phys. Chem.*, **87** (1983) 4130–34.
15. Legrand, M. R., Delmas, R. J. and Charlson, R. J., Climate forming implications from Vostok in ice core sulfate data. *Nature*, **354** (1988) 418–20.
16. Murozumi, M, Chow, T. J. and Patterson, C., Chemical concentrations of pollutant lead aerosols, terrestrial dusts and sea salts in Greenland and antarctic snow strata. *Geochimica et Cosmochimica Acta*, **33** (1969) 1247–94.
17. Boutron, C. and Patterson, C. C., Relative levels of natural and anthropogenic lead in recent antarctic snow. *J. Geophys. Res.*, **92** (1987) 8454–64.
18. Herron, M. M., Impurity sources of F^-, Cl^-, NO_3^-, and SO_4^{2-} in Greenland and antarctic precipitation. *J. Geophys. Res.*, **87** (1982) 3052–60.
19. Neftel, A., Beer, J., Oeschger, H., Furcher, F. and Finkel, R. C., Sulphate and nitrate concentrations in snow from South Greenland 1895–1978. *Nature*, **314** (1985) 611–3.
20. Mayewski, P. A., Lyons, W. B., Spencer, M. J., Twickler, M., Dansgaard, W., Koci, B., Davidson, C. I. and Honrath, R. E., Sulfate and nitrate concentrations from a south Greenland ice core. *Science*, **232** (1986) 975–7.
21. Davidson, C. I., Harrington, J. R., Stephenson, M. J., Monaghan, M. C., Pudykiewicz, J. and Schell, W. R., Radioactive cesium from the Chernobyl accident in the Greenland Ice Sheet. *Science*, **237** (1987) 633–4.
22. Dibb, J., The Chernobyl reference horizon (?) in the Greenland Ice Sheet. *Geophys. Res. Lett.*, **16** (1989) 987–90.
23. Hammer, C. U., Acidity of polar ice in relation to absolute dating, past volcanism, and radioechoes. *J. Glaciol.*, **25** (1980) 359–72.
24. Rampino, M. R. and Self, S., Historic eruption of Tambora (1815), Krakatoa (1883), and Agung (1963), their stratospheric aerosols and climatic impact. *Quat. Res.*, **18** (1982) 127–43.
25. Herron, M. M. and Langway, C. C., Jr., Chloride, nitrate, and sulfate in the Dye 3 and Camp Century ice cores. In *Greenland Ice Core: Geophysics, Geochemistry and the Environment*, ed. C. C. Langway, Jr., H. Oeschger, and W. Dansgaard. AGU, Geophysical Monograph 33, 1985, pp. 77–84.

26. Delmas, R. J., Legrand, M. R., Aristarain, A. A. J. and Zanolini, F., Volcanic deposits in antarctic snow and ice. *J. Geophys. Res.*, **90** (1985) 12901–20.
27. Beer, J., Andree, M., Oeschger, H., Stauffer, B., Balzer, R., Bonani, G., Stoller, Ch., Suter, M., Wolfli, W. and Finkel, R. C., ^{10}Be variations in polar ice cores. In *Greenland Ice Core: Geophysics, Geochemistry and the Environment*, ed. C. C. Langway, Jr., H. Oeschger, and W. Dansgaard. AGU, Geophysical Monograph 33, 1985, pp. 66–70.
28. Zeller, E. J. and Parker, B. C., Nitrate ion in antarctic firn as a marker for solar activity. *Geophys. Res Lett.*, **8** (1981) 895–8.
29. Zanolini, F., Delmas, R. and Legrand, M., Sulfuric acid and nitric acid concentrations and spikes along a 200 m deep ice core at D57 (Terre Adelie, Antarctica). *Annals of Glaciology*, **7** (1985) 70–75.
30. Shaw, G. E. and Khalil, M. A. K., Arctic haze. In *The Handbook of Environmental Chemistry*, Volume 4, Part B, ed. O. Hutzinger. Springer-Verlag, Berlin, 1989, pp. 69–111.
31. Hansen, A. D. A. and Rosen, H., Vertical distributions of particulate carbon, sulfur, and bromine in the arctic haze and comparison with ground-level measurements at Barrow, Alaska. *Geophys. Res. Lett.*, **11** (1984) 381–4.
32. Hansen. A. D. A. and Rosen, H., Horizontal inhomogeneities in the particulate carbon component of the arctic haze. *Atmos. Environ.*, **19** (1985) 2175–80.
33. Clarke, A. D. and Noone, K. J., Soot in the arctic snowpack: A cause for perturbations in radiative transfer. *Atmos. Environ.*, **19** (1985) 2045–53.
34. Khalil, M. A. K. and Rasmussen, R. A., Gaseous tracers of arctic haze. *Environ. Sci. and Tech*, **17** 1983 157–64.
35. Rasmussen, R. A. and Khalil, M. A. K., Gaseous bromine in the Arctic and arctic haze. *Geophys. Res. Lett.*, **11** (1984) 433–6.
36. Hov, O., Penkett, S. A., Isaksen, I. S. A. and Semb, A., Organic gases in the Norwegian Arctic. *Geophys. Res. Lett.*, **11** (1984) 425–8.
37. Berg, W. W., Heidt, L. E., Pollock, W., Sperry, P. D., Cicerone, R. J. and Gladney, E. S., Brominated organic species in the arctic atmosphere. *Geophys. Res. Lett.*, **11** (1984) 429–32.
38. Miller, J. M., A five-year climatology of five-day back trajectories from Barrow, Alaska. *Atmos. Environ.*, **15** (1981) 1401–5.
39. Rahn, K. A. and Shaw, G. E., Sources and transport of arctic pollution aerosol: A chronicle of six years of ONR research. *Naval Res. Rev.*, **34** (1982) 3–26.
40. Harris, J. M., Trajectories during AGASP. *Geophys. Res. Lett.*, **11** (1984) 453–6.
41. Radke, L. F., Hobbs, P. V. and Bailey, I. H., Airborne observations of arctic aerosols. III: Origins and effects of air masses. *Geophys. Res. Lett.*, **11** (1984) 401–4.
42. Jaffrezo, J. L., Boscoe, F. P., Small, M. J., Davidson, C. I. and Olson, M. P., Source regions and atmospheric pathways for aerosols and gases reaching southern Greenland. American Association for Aerosol Research, Annual Meeting, Philadelphia, PA, June 18–22, 1990.
43. Olson, M. P., Oikawa, K. K. and Macaffee, A. W., A trajectory model applied to the long-range transport of air pollutants. Report LRTAP 78–4,

Atmospheric Environment Service, Downsview, Ontario, November, 1978.
44. Nenow, D., Surface premelting. In *Progress in Crystal Growth and Character*, Volume 9, ed. B. R. Pamplin and D. Elwell. Pergamon Press, New York, 1984, pp. 185–225.
45. Valdez, M. P., Dawson, G. A. and Bales, R. C., Sulphur dioxide incorporation into ice depositing from the vapor. *J. Geophys. Res.*, **94** (1989) 1095–1103.
46. Huebert, B. J., Fehsenfeld, F. C., Norton, R. B. and Albritton, D., The scavenging of nitric acid vapor by snow. In *Precipitation Scavenging, Dry Deposition, and Resuspension*, ed. H. R. Pruppacher, R. G. Semonin, and W. G. N. Slinn. Elsevier, New York, 1983, pp. 293–302.
47. Gjessing, Y. T., The filtering effect of snow. In *Proceedings of the Symposium on Isotopes and Impurities in Snow and Ice*, Grenoble, France, Aug–Sep 1975. IAHS Publication 118, 1977, pp. 200–203.
48. Ibrahim, M., Barrie, L. A. and Fanaki, F., An experimental and theoretical investigation of the dry deposition of particles to snow, pine trees and artificial collectors. *Atmos. Environ.*, **17** (1983) 781–8.
49. Davidson, C. I., Mechanisms of wet and dry deposition of atmospheric contaminants to snow surfaces. In *The Enviromental Record in Glaciers and Ice Sheets*, ed. H. Oeschger and C. C. Langway, Jr. John Wiley and Sons, New York, 1989, pp. 29–51.
50. Davidson, C. I. and Wu, Y. L., Dry deposition of particles and vapors. In *Acid Precipitation*, Volume 3, ed. D. C. Adriano, Springer-Verlag, New York, 1990, pp. 103–216.
51. Kumai, M., Fog modification on the Greenland Ice Cap. In *Proceedings of the First National Conference on Weather Modification*, Albany, NY, 1967, pp. 414–22.
52. Mason, B. J., *Clouds, Rain, and Rainmaking*. Cambridge University Press, Cambridge, UK, 1975.
53. Pruppacher, H. R., and Klett, J. D., *Microphysics of Clouds and Precipitation*. D. Reidel, Dordrecht, Holland, 1980.
54. Scott, B. C., Sulfate washout ratios in winter storms. *J. Applied Meteorology*, **20** (1981) 619–25.
55. Borys, R. D., Demott, P. J., Hindman, E. D. and Feng, D., The significance of snow crystal and mountain surface riming to the removal of atmospheric trace constituents from cold clouds. In *Precipitation Scavenging, Dry Deposition, and Resuspension*, ed. H. R. Pruppacher, R. G. Semonin, and W. G. N. Slinn. Elsevier, New York, 1983, pp. 181–9.
56. Barrie, L. A., Atmospheric particles: Their physical/chemical characteristics and deposition processes relevant to the chemical composition of glaciers. *Annals of Glaciology*, **7** (1985) 100–108.
57. Mitchell, D. L. and Lamb, D., Influence of riming on the chemical composition of snow in winter orographic storms. *J. Geophys. Res.*, **94** (1989) 14831–40.
58. Scott, B. C., Theoretical estimates of scavenging coefficients for soluble aerosol particles as a function of precipitation type, rate, and altitude. *Atmos Environ.*, **16** (1982) 1753–62.
59. Slinn, W. G. N., Precipitation scavenging. In *Atmospheric Sciences and*

Power Production, ed. D. Randerson. Technical Information Center, U.S. Department of Energy, Oak Ridge, TN, 1984.

60. Jaffrezo, J. L. and Colin, J. L., Rain–aerosol coupling in urban area: Scavenging ratio measurements and identification of some transfer processes. *Atmos. Environ.*, **22** (1988) 929–35.
61. Jaffrezo, J.L., Colin, J.L. and Gros, J.M., Some physical factors influencing scavenging ratios. (Submitted to *Atmos. Environ.*).
62. Alley, R. B., Concerning the deposition and diagenesis of strata in polar firn. *Journal of Glaciology*, **34** (1988) 283–90.
63. Bales, R. C., Davis, R. E. and Stanley, D. A., Ion elution through shallow homogeneous snow. *Water Resources Research* (in press).
64. Bales, R. C, Somerfeld, R. and Kebler, D. G., Ionic tracer movement through a Wyoming snowpack. (Manuscript in review).
65. Langway, C. C., Oeschger, H. and Dansgaard, W., The Greenland Ice Sheet Program in perspective. In *Greenland Ice Core: Geophysics, Geochemistry and the Environment*, ed. C. C. Langway, Jr., H. Oeschger, and W. Dansgaard. AGU, Geophysical Monograph 33, 1985, pp. 1–8.
66. Mayewski, P. A., Spencer, M. J., Twickler, M. S. and Whitlow, S., A glaciochemical survey of the Summit region, Greenland. *Annals of Glaciology* (in press).
67. Davidson, C. I., Honrath, R. E., Kadane, J. B., Tsay, R. S., Mayewski, P. A., Lyons, W. B. and Heidam, N. Z., The scavenging of atmospheric sulfate by arctic snow. *Atmos. Environ.*, **21** (1987) 871–82.
68. Davidson, C. I., Harrington, J. R., Stephenson, M. J., Small, M. J., Boscoe, F. P. and Gandley, R. E., Seasonal variations in sulfate, nitrate and chloride in the Greenland Ice Sheet: Relation to atmospheric concentrations. *Atmos. Environ.*, **23** (1989) 2483–93.
69. Mayewski, P. A., Spencer, M. J., Lyons, W. B. and Twickler, M. S., Seasonal and spatial trends in south Greenland snow chemistry. *Atmos. Environ.*, **21** (1987) 863–9.
70. Rahn, K. A., Atmospheric, riverine and oceanic sources of seven trace constituents to the Arctic Ocean. *Atmos. Environ.*, **15** (1981) 1507–16.
71. Barrie, L. A., Fisher, D. and Koerner, R. M., Twentieth century trends in arctic air pollution revealed by conductivity and acidity observations in snow and ice in the Canadian High Arctic. *Atmos. Environ.*, **19** (1985) 2055–63.
72. Legrand, M. R. and Kirchner, S., Origins and variations of nitrate in south polar precipitation. *J. Geophys. Res.* (in press).
73. Herron, M M., Langway, C C., Jr., Weiss, H. V. and Cragin, J. H., Atmospheric trace metals and sulfate in the Greenland Ice Sheet. *Geochimica et Cosmochimica Acta*, **41** (1977) 915–20.
74. Kirchner, S. and Delmas, R. J., A 1000 year glaciochemical study at the South Pole. *Annals of Glaciology*, **10** (1988) 1–5.
75. Cullis, C. F. and Hirschler, M. M., Atmospheric sulphur: Natural and man-made sources. *Atmos. Environ.*, **14** (1980) 1263–78.
76. Finkel, R. C. and Langway, C. C., Jr., Changes in precipitation chemistry at Dye 3, Greenland. *J. Geophys. Res.*, **91** (1986) 9849–55.
77. Steffensen, J. P., Analysis of the seasonal variation in dust, Cl^-, NO_3^-, and

SO_4^{2-} in two central Greenland firn cores. *Annals of Glaciology*, **10** (1988) 171–7.
78. Mayewski, P. A., Lyons, W. B., Spencer, M. J., Twickler, M. S., Buck, C. J. and Whitlow, S., Just how sensitive is the remote atmosphere to anthropogenically produced sulfate and nitrate? (Manuscript in review).
79. Heidam, N. Z., The components of the arctic aerosol. *Atmos. Environ.*, **18**, (1984) 329–43.
80. Barrie, L. A., Arctic air pollution: An overview of current knowledge. *Atmos. Environ.*, **20** (1986) 643–63.
81. Koerner, R. M. and Fisher, D., Acid snow in the Canadian High Arctic. *Nature*, **295** (1982) 137–40.
82. Holdsworth, G. and Peake, E., Acid content of snow from a mid-tropospheric sampling site on Mt. Logan, Yukon Territory, Canada. *Annals of Glaciology*, **7** (1985) 153–60.
83. Holdsworth, G., Krouse, H. R. and Peake, E., Trace acid content of shallow snow and ice cores from mountain sites in Western Canada. *Annals of Glaciology*, **10** (1988) 57–62.
84. Duce, R. A., Hoffman, G. L. and Zoller, W. H., Atmospheric trace metals at remote northern and southern hemisphere sites: Pollution or natural? *Science*, **187** (1975) 59–61.
85. Rahn, K. A., The Chemical Composition of the Atmospheric Aerosol.Technical Report, University of Rhode Island, 1976, 265 pp.
86. Rahn, K. A., Silicon and aluminum in atmospheric aerosols: Crust–air fractionation? *Atmos. Environ.*, **10** (1976) 597–601.
87. Rahn, K. A. and McCaffrey, R. J., Composition differences between arctic aerosol and snow. *Nature*, **280** (1979) 479–80.
88. Davidson, C. I., Santhanam, S., Fortmann, R. C. and Olson, M. P., Atmospheric transport and deposition of trace elements onto the Greenland ice sheet. *Atmos. Environ.*, **19** (1985) 2065–81.
89. Colin, J. L. and Jaffrezo, J. L., Rain–aerosol coupling in coastal, urban, and rural area: Scavenging ratios in relation with transport. In *Proceedings of the Tenth International Cloud Physics Conference*, Bad Hamburg, Germany, 1988, pp. 463–5.
90. Pacyna, J., Atmospheric trace elements from natural and anthropogenic sources. In *Toxic Metals in the Atmosphere*, ed. J. O. Nriagu and C. I. Davidson. John Wiley and Sons, New York, 1986, pp. 33–52.
91. Wolff, E. W. and Peel, D. A., Closer to a true value for heavy metal concentrations in recent antarctic snow by improved contamination control. *Annals of Glaciology*, **7** (1985) 61–9.
92. Boutron, C. and Batifol, F., Assessing laboratory procedures for the decontamination of polar snow and ice samples for toxic metals and metalloids analysis. *Annals of Glaciology*, **7** (1985) 7–11.
93. Boutron, C., Atmospheric toxic metals and metalloids in the snow and ice layers deposited in Greenland and Antarctica from prehistoric times to present. In *Toxic Metals in the Atmosphere*, ed. J. O. Nriagu and C. I. Davidson. John Wiley and Sons, New York, 1986, pp. 467–505.
94. Wolff, E. W. and Peel, D. A., Concentrations of cadmium, copper, lead, and

zinc in snow from near Dye 3 in south Greenland. *Annals of Glaciology*, **10** (1988) 193–7.
95. Milford, J. B. and Davidson, C. I., The sizes of particulate trace elements in the atmosphere—a review. *J. Air Poll. Control Assoc.*, **35** (1985) 1249–60.
96. Milford, J. B. and Davidson, C. I., The sizes of particulate sulfate and nitrate in the atmosphere—a review. *J. Air Poll. Control Assoc.*, **37** (1987) 125–34.
97. Boutron, C. F., Atmospheric lead, cadmium, mercury, and arsenic in Antarctic and Greenland recent snow and ancient ice. In *Proceedings of the SCOPE/UNEP Metals Cycling Workshop*, New Delhi, India, Feb. 16–20, 1987.
98. Carter, M. W. and Moghissi, A. A., Three decades of nuclear testing. *Health Physics*, **33** (1977) 55–71.
99. Koide, M., Michel, R., Goldberg, E. D., Herron, M. M. and Langway, C. C., Jr., Characterization of radioactive fallout from pre- and post-moratorium tests to polar ice caps. *Nature*, **296** (1982) 544–7.
100. Koide, M. and Goldberg, E. D., The historical record of artificial radioactive fallout from the atmosphere in polar glaciers. In *Greenland Ice Core: Geophysics, Geochemistry and the Environment*, ed. C. C. Langway, Jr., H. Oeschger, and W. Dansgaard. AGU Geophysical Monograph 33, 1985, pp. 95–100.
101. Koide, M., Goldberg, E. D., Herron, M. M. and Langway, C. C., Jr., Transuranic depositional history in south Greenland firn layers. *Nature*, **269** (1977) 137–9.
102. Pourchet, M, Pinglot, J. F. and Gascard, J. C., The northerly extent of Chernobyl contamination. *Nature*, **323** (1986) 676.
103. Pudykiewicz, J., Numerical simulation of the transport of radioactive cloud from the Chernobyl nuclear accident. *Tellus*, **40B** (1989) 241–59.
104. Rasmussen, R. A., Khalil, M. A. K. and Hoyt, S. D., Methane and carbon monoxide in snow. *J. Air Poll. Control Assoc.*, **32** (1982) 176–8.
105. Rasmussen, R. A., Khalil, M. A. K. and Hoyt, S.D., Trace gases in snow and rain. In *Precipitation Scavenging, Dry Deposition, and Resuspension*, ed. H. R. Pruppacher, R. G. Semonin, and W. G. N. Slinn. Elsevier, New York, 1983, pp. 1301–14.
106. Stauffer, B., Schwander, J. and Oeschger, H., Enclosure of air during metamorphosis of dry firn to ice. *Annals of Glaciology*, **6** (1985) 108–12.
107. Schwander, J., Stauffer, B. and Sigg, A., Air mixing in firn and the age of air at pore close-off. *Annals of Glaciology*, **10** (1988) 141–5.
108. Schwander, J. and Stauffer, B., Age difference between polar ice and the air trapped in its bubbles. *Nature*, **311** (1984) 45–7.
109. McInnes, B. and Radok, U., A steady-state prediction of Dye 3 core features. In *Greenland Ice Core: Geophysics, Geochemistry and the Environment*, ed. C. C. Langway, Jr., H. Oeschger and W. Dansgaard. AGU, Geophysical Monograph 33, 1985, pp. 111–17.
110. Mart, L, Seasonal variations of Cd, Pb, Cu, and Ni levels in snow from the eastern Arctic Ocean. *Tellus*, **35B** (1983) 131–41.

Chapter 4

Chemical Tracers of the Origins of Arctic Air Pollution

JOZEF M. PACYNA

Norwegian Institute for Air Research, P.O. Box 64, N-2001 Lillestrøm, Norway

4.1 INTRODUCTION

The origin of arctic air pollution has been an intriguing question since the 1950s when Mitchell and his colleagues observed bands of particles in the air over Alaska.[1] These bands were difficult to explain at that time. The Arctic does not have many major pollution sources and these are located as far as a few thousand kilometres from the Arctic. Long-range transport of pollutants within air masses had not been recognized at that time. Mitchell[1] was able to deduce through observations of light scattering that the bands consisted primarily of submicron particles.

Almost two decades later, Shaw carried out measurements of the optical transparency of the atmosphere at Barrow, Alaska, to determine the spectral optical thickness of the arctic atmosphere. From unexpectedly high values of total atmospheric turbidity measured in the Alaskan Arctic it was concluded that the arctic atmosphere was polluted by blowing dust and emissions from nearby sources,[2] at least during the spring months. Subsequent aircraft measurements of the vertical profile of turbidity[3] revealed that about 40% of the total turbidity was above 4 km, and that anomalously high turbidity was measured at various levels up to a few kilometres altitude. However, no obvious explanation of the origin of the arctic air contamination was apparent.

In order to explain the origin of arctic air pollution, Shaw, Rahn, and Borys in North America and Ottar, Larssen, and Hanssen in Europe began in the mid 1970s to determine the chemical composition of the arctic aerosol in Alaska and the Norwegian Arctic. Based on the measurements in Alaska, Rahn *et al.*[4] concluded that although particulate matter in air between distinct haze layers was pollution derived, the haze layers themselves were predominantly of crustal composition, originating presumably from Asian deserts between 40°N and 50°N. Thus, this study concluded that arctic haze consisted mainly of wind blown dust. Larssen & Hanssen[5] indicated on the basis of measurements in the Norwegian Arctic that polluted air masses travelled over continental Europe and Scandinavia by Ny Ålesund, Spitsbergen, and Bear Island. Another probable route, according to the authors, was across the Siberian Arctic border.

During the last decade several ground measurement efforts were carried out in the Arctic to assess the origin of arctic haze. Many of these studies were conducted as part of the Arctic Air-Sampling Network[6], a co-ordinated, widespread programme designed to systematically perform arctic air chemistry observations. These measurements resulted in extensive data sets of air concentrations in Greenland,[7] the Norwegian Arctic,[8,9] northern Canada[10] and Alaska.[11,12] Vertical profiles of air concentrations were then obtained during several aircraft measurement programmes in the Arctic.[8,13,14] As a result of all these research efforts the following explanation of arctic haze can be offered. Very long-range, episodic transport of air masses over several thousand kilometres clearly affects the quality of arctic air during all seasons. Polluted air masses, carrying a mixture of natural and anthropogenic constituents from a variety of sources in different geographical areas, have been identified in the arctic atmosphere at altitudes over 2 km. The altitude of polluted air layers is higher in winter than in summer. The layers of polluted air at altitudes below 2·5 km can be traced to episodic transport of air masses from sources situated in areas with potential air temperatures similar to those in the Arctic. The transport of polluted air below 2·5 km is a dominating feature in winter/spring but less so at other times.

Several efforts have been made during the above mentioned measurements to assess sources and source regions for pollutants in arctic air, as well as to describe their major pathways. These efforts include the application of various statistical methods to the data on the chemical composition of arctic air, and modelling. Both the statistical methods and modelling are used together with meteorological data. The aim of this chapter is to present the application of the above methods to the study of the origin of arctic haze.

4.2 ORIGIN OF ARCTIC AIR POLLUTION AS STUDIED FROM THE CHEMICAL COMPOSITION OF AEROSOLS

Major progress in studying the origin of arctic air pollution was made when information on the chemical composition of the arctic aerosol became available. Historically, this information began to be collected in the mid 1970s and already in 1976 the first use of chemistry as a tracer was made in work at Barrow, Alaska.[4] It concluded that arctic haze consisted mainly of wind blown dust. However, discussion had already begun as to what extent the anthropogenic sources may contribute to arctic haze. In this context Rahn[15] proposed a tracer system to distinguish between contributions from various source regions on a geographical basis. The tracer system of Rahn consisted of the Mn/V ratio. It was expected that the concentration ratios between these trace metals in the arctic aerosol should correspond to their emission ratios. Both Mn and V are emitted from anthropogenic sources, mostly combustion of fossil fuels: coal combustion for Mn and oil combustion for V. In addition, substantial amounts of Mn are emitted during steel and iron production. Thus, the Mn/V ratio was thought to be a good tracer to distinguish between pollution arriving in the Arctic from regions where coal and oil are burned to produce heat and electricity. However, both Mn and V are also emitted from natural sources, mainly wind blown dust. Emission data suggest that the crustal Mn exceeds the anthropogenic Mn by one order of magnitude. The natural and anthropogenic emissions of V on a global scale are comparable. To avoid a bias of Mn and V from natural sources, Rahn[15] assumed non-crustal portions of the metals when applying his tracer system. Another assumption was that there are no significant chemical fractionation processes during the long-range transport to the Arctic.

The application of the Mn/V ratio to data from Barrow, was used by Rahn[15] to conclude that emission sources in Europe and the USSR are the major contributors to the arctic aerosol. This conclusion was supported by Raatz & Shaw[16] and arguments were made in the early 1980s that the USSR must be a major source region of the haze.

Further progress in studying the origin of arctic aerosol was made when Lowenthal & Rahn[12] introduced a seven element tracer scheme, using elements that are primarily found on submicron aerosols. They chose the following seven elements: As, Sb, Se, non-crustal V, Zn, non-crustal Mn, and In, from the over 40 that they measured routinely, because these elements best satisfied the criteria of being pollution-derived, concentrated in fine particle aerosol (capable of long-range transport), and capable of being measured precisely. According to the authors, nearly 20 additional elements could probably be used as regional tracers. With the use of this seven-element tracer system, Lowenthal & Rahn[12] deduced that Eurasia was the dominant source of aerosol at Barrow during the winter 1979-1980 with only 5–10% of the aerosol arriving from North American sources. Within Eurasia, central USSR contributed an estimated 75% of the tracer elements, with the remainder coming from Europe. Europe, however, seemed to be a major source of sulphates. The authors warned, however, that these results may not be generally valid for the Arctic, because the relative importance of the European and Soviet sources vary from year to year.

Sheridan & Zoller[17] analysed aerosol samples from the Arctic Gas and Aerosol Sampling Program (AGASP-I) mission for elemental chemistry and reported similar ratio values to those of Lowenthal & Rahn.[12] Additionally, crustal and marine aerosol concentrations were calculated for various locations in the Arctic Basin.

On the other side of the Arctic, Heidam[18] carried out investigations of the chemical composition of the arctic aerosol in Greenland. Already Mitchell[1] had observed that although haze was widespread over Alaska, it was seldom found over Greenland. Strong temperature inversions build up in the Arctic because of the strongly negative heat balance in winter.[19] Greenland has a high frequency of inversion gradients, stronger than those measured in other parts of the Arctic. During such inversions the air stagnates and turbulence is highly suppressed, although some turbulence may originate due to the propagation of gravity waves that develop in regions of strong wind shear, particularly just above the low-lying inversions.[20] However, the Danish measurement programmes in Greenland already launched in 1978[21] had indicated enhanced concentrations of submicron Aitken particles and SO_2. Further measurements of the Greenland aerosol extended this to several trace metals.[18] An application of statistical methods to data on element concentrations or enrichment factors, and particularly factor analysis, has shown that the Greenland aerosol may be composed of three to five statistically independent components,[21] both natural and anthropogenic. Two of these components consist of a crustal component originating from erosion processes (with prominent elements being Al, Si, Ti, Fe) and a marine component produced by sea spray (Cl). These accounted for 40–60% of the total variance. The remaining components are: (1) a combustion component from very distant industrial source areas, transported only rarely straight towards Greenland (prominent elements: S and Br); (2) an anthropogenic component consistent with direct transport from primary source regions in the Urals (Zn and Cu); and (3) an engine exhaust component (Pb).

Although the investigation of the chemical composition of aerosols in the Norwegian Arctic began in the 1970s, the major measurement campaigns were launched in 1982.[8] During the period August 1982–July 1984, five measurement campaigns were carried out at four ground stations in the Arctic and at five stations in mainland Norway. The chemical composition of the arctic aerosol aloft was obtained on the basis of 35 sampling flights in 1983 and 1984. Recently, four winter measurement campaigns were carried out at Ny Ålesund, Spitsbergen, in 1986, 1987, 1988, and 1989. The chemical analysis of more than 40 trace metals in samples from the ground stations was used to assess the origin of the arctic aerosol, in addition to other chemical and meteorological interpretations of the results obtained from the aircraft measurements.

In order to investigate the applicability of multivariate techniques for resolving sources and/or source regions for aerosol in the Norwegian Arctic, Maenhaut *et al.*[22] applied the absolute principal component analysis (APCA) and the chemical mass balance (CMB) methods to data sets from Ny Ålesund and Vardø (the northernmost station in mainland Norway). The APCA indicated that there were three components: their elemental profiles in relative concentration units and associated standard deviations are given in Table 4.1. The first component is basically an anthropogenic component, but the relative proportions of the crustal elements indicate that there was some crustal-like material mixed into it. In the second component, the crustal elements Al, Si, K, Ca, Sc, Mn, Fe, La, Sm, and Th occur in proportions similar to those in average crustal rock.[23] Thus, this component was identified as soil dust. However, most of the other elements in the profile were highly enriched and this fact suggests that the second component also contains some material from anthropogenic sources. The composition of the third component in Table 4.1 is fairly similar to that of bulk seawater, indicating that this component is essentially sea-salt. The average apportionments for the elements at Ny Ålesund are given in Table 4.2.

The three-component solution for the Ny Ålesund winter aerosol by Maenhaut *et al.*[22] is similar to the solution obtained by Lowenthal & Rahn[12] for the Barrow aerosol. However, the separation of the crustal

Table 4.1 Source profiles (in relative concentration units[a] and associated standard deviations) for the Ny Ålesund winter samples, as obtained by APCA[22]

	Component 1 Pollution		*Component 2 Crust*		*Component 3 Sea-salt*	
Na					139	±15
Mg					57	± 3
Al	11·5	± 0·9	24	± 1	3·1	± 0·9
Si	36	± 9	107	± 9	42	± 9
S	400	±40	210	±40		
K	12·3	± 0·6	11·7	± 0·6	15·6	± 0·6
Ca	4·5	± 1·2	17·7	± 1·2	23	± 1
Sc	0·0026	± 0·0002	0·0035	± 0·0002		
V	1·11	± 0·03	0·23	± 0·03		
Mn	1·01	± 0·03	0·49	± 0·03		
Fe	21	± 1	12·7	± 0·8		
Zn	5·0	± 0·2	1·98	± 0·16		
As	0·96	± 0·04	0·32	± 0·04		
Se	0·078	± 0·006	0·0178	± 0·0055		
Br					6·3	± 0·9
In	0·0031	± 0·0001	0·00079	± 0·00013		
Sb	0·102	± 0·004	0·045	± 0·004		
I	0·33	± 0·03	0·146	± 0·034		
Cs	0·0059	± 0·0003	0·0042	± 0·0003		
La	0·0089	± 0·0007	0·0103	± 0·0007		
Sm	0·00093	± 0·00016	0·00165	± 0·00016		
Th	0·00173	± 0·00016	0·0030	± 0·0002		

[a]Elemental concentrations of less than three standard deviations are not included in the table

and anthropogenic elements in the soil and pollution factors was much better in the analysis from Barrow. Maenhaut *et al.*[22] gave the alternative explanation that the differences in factors may result from differences in sample collection.

It was also concluded by Maenhaut *et al.*[22] that APCA does not allow one to obtain a fine resolution of the contributions from various distant anthropogenic source regions to the chemical composition of the arctic aerosol. To attempt this resolution, CBM source apportionment was performed using a set of elemental signatures obtained by Lowenthal & Rahn.[12] It was first assessed that the North American sources contribute at most 10% of the pollution at Ny Ålesund, and they were excluded from further analysis. The apportionment contained information on the

Table 4.2 Average contributions from various sources[a] to the atmospheric concentrations of elements in the Ny Ålesund winter samples[22]

	Component 1 Pollution	*Component 2 Crust*	*Component 3 Sea-salt*	*Sum of listed contributions*
Na			70	70
Mg			129	129
Al	27	49	10	86
Si	19	49	32	100
S	45	20		65
K	34	28	62	124
Ca	11	38	81	130
Sc	53	60		113
V	121	22		143
Mn	95	40		135
Fe	85	45		130
Zn	96	33		129
As	115	33		148
Se	46	9		55
Br			78	78
In	109	24		133
Sb	87	33		120
I	35	14		49
Cs	60	40		100
La	48	48		96
Sm	32	49		81
Th	38	56		94

[a]The source contributions are expressed as percentages of the average observed atmospheric concentrations. Contributions are not listed when the elemental concentration was excluded from the profile in Table 4.1.

pollutant contributions from four source regions: the UK (UKS), Western Europe (WEURS), Eastern Europe (EEURS), and central USSR (CUSSR). Average regional apportionments of elements for the above four regions in the Ny Ålesund winter and summer aerosol are presented in Tables 4.3 and 4.4, respectively. The use of three European signatures was supposed to provide a fine resolution of the contributions from the various parts of Europe. This attempt failed due to high collinearity of the European signatures. Thus, the final apportionment consisted of two source regions: Europe and central USSR. The results are presented in Table 4.5. It was concluded that during winter 60% of the pollution at Ny Ålesund is transported from sources in the USSR. The rest is transported from Europe. In summer, the contributions from the USSR

Table 4.3 Average regional apportionments of elements in the Ny Ålesund winter aerosol[22]

	Apportionment,[a] percent of mean predicted value			
	UKS	*WEURS*	*EEURS*	*CUSSR*
As	7	3	2	88
Sb	28	10	4	58
Se	31	10	2	57
V(X)	29	9	2	60
Zn	28	9	3	60
Mn(X)	39	10	4	46
In	17	4	1	78
Pb	27	9	4	59

V(X) and Mn(X) indicate non-crustal V and non-crustal Mn, respectively.
[a]UKS—United Kingdom WEURS—Western Europe EEURS—Eastern Europe CUSSR— Central USSR

Table 4.4 Average regional apportionments of elements in the Ny Ålesund summer aerosol[22]

	Apportionment, percent of mean predicted value			
	UKS	*WEURS*	*EEURS*	*CUSSR*
As	4	35	12	50
Sb	7	64	10	18
Se	9	67	6	19
V(X)	9	66	5	21
Zn	0	43	10	47
Mn(X)	9	67	13	12

V(X) and Mn(X) indicate non-crustal V and non-crustal Mn, respectively.

and Europe are 25% and 75%, respectively. These results are very similar to the results obtained by Lowenthal & Rahn[12] for Barrow.

Further improvement of the receptor modelling method to assess the origin of the arctic aerosol was obtained by applying APCA to aerosol elemental concentration measurements in separate particle size fractions. Li & Winchester[24] studied the contributions of elements from four source groups to the chemical composition of aerosols at Barrow during the

Table 4.5 Average regional apportionments of elements in the Ny Ålesund winter and summer aerosol[22]

	Winter		*Summer*	
	Apport.,[a] percent of mean pred. value		*Apport., percent of mean pred. value*	
	EURAVE	*CUSSR*	*EURAVE*	*CUSSR*
As	29	71	71	29
Sb	46	54	84	16
Se	44	56	83	17
V(X)	46	54	84	16
Zn	48	52	57	43
Mn(X)	59	41	90	10
In	24	76		
Pb	55	45		

V(X) and Mn(X) indicate non-crustal V and non-crustal Mn, respectively. Se was not included in the CMB fit for the 1984 summer samples.

[a]EURAVE—European average, constructed from the signatures for Western and Eastern Europe by Rahn & Lowenthal[62]
CUSSR—Central USSR

so-called 'megahaze' event during the period from March through May 1986. These groups represented sea-salt aerosol, crustal particles, carbonaceous fuel combustion pollutants (e.g. from coal), and a group for marine biogenic gas emissions that contains most of the bromine and significant amounts of sulphur and chlorine. Using Na as a reference element, Li & Winchester[24] concluded that sulphur was enriched in fine particles of 0·25–1·0 μm EAD (equivalent aerodynamic diameter) particles but not in coarse particles; chlorine was depleted in 0·5–1·0 μm EAD particles, but was about 50% enriched in other fine and coarse fractions, suggesting gas-to-particle interactions during aerosol ageing. Crustal particle components were found in all size fractions, with a maximum concentration in 1–2 μm EAD particles. Carbonaceous fuel combustion pollutants were indicated by the presence of Si and Cl and the absence of Al in all size fractions, and usually rich in sulphur, with the highest concentrations in the finest particles $<0{\cdot}25\ \mu$m EAD. Of particular interest was Si, often described as a crustal element. However, combustion of coal with high ash content may release volatile SiO from reduction of SiO_2 by carbon, which then forms a fine aerosol after SiO oxidation back to SiO_2. The Al is non-volatile during coal combustion.

Its absence when Si and other metals are present may therefore be an indicator of aerosol from burning high ash content coal.[25]

The last bromine–sulphur component in the Li & Winchester study was found in every fraction and contained most of the measured bromine. The greatest mass was measured in the $<0{\cdot}25\,\mu m$ EAD fraction of particles, suggesting a gas phase origin of this particle type.

The application of APCA presented by Li & Winchester[24] gives the basis for interpretation of coupled chemical reactions and physical processes in the Arctic as well as giving information concerning atmospheric ageing processes and, therefore, the history of the components.

Factor analysis techniques have been applied to several years of data collected at various stations in the Canadian Arctic.[10] Principal component analysis was employed to data sets from Igloolik, Mould Bay and Alert and the results indicated three groups of aerosols: sea-salt, crustal, and anthropogenic.

Principal component analysis (PCA) has also been applied to assess the origin of arctic aerosol on the basis of results from automated microanalysis of individual particles. Anderson *et al.*[26] employed a modified version of the scanning electron microscope (SEM) analysis of particles to study the origin of particles collected at Ny Ålesund in the period May through June, 1987. They concluded that the aerosol contained a mixture of altered and unaltered particle types from a variety of sources. According to the authors silicon-type particles appeared as relatively unaltered particles, as particles with sulphate coatings, and as particles that had reacted with Br. Most silicate particles were ascribed to originate from crustal material, while some may be emitted during coal combustion. Metal-rich particles of anthropogenic origin were varied in composition. Only a few of them were associated with an episode of polluted air from northern Eurasia. Chlorine-rich particles were of marine origin, but extensive fractionation and reaction of the marine aerosol components was suggested. Abundant S-rich particles were of undetermined origin, probably of multiple origins, but in the early period of highest concentration were associated with the pollution episode. After the episode, a substantial fraction of the S-fine-fraction may have been biogenic in origin. The above results are consistent with data obtained earlier by Saucy *et al.*[27]

The value of Individual Particle Analysis (IPA) in arctic aerosol research has been discussed by Sheridan.[28] He observed that particles emitted by anthropogenic sources, such as carbon soot and coal combustion spheres, occurred along with the highest concentrations of H_2SO_4 droplets. It was verified that graphite is the mineralogical phase of carbon present in arctic soot particles. A small fraction of the H_2SO_4 droplets were observed to contain detectable nitrogen, which suggests that partial neutralization of acid in the particles had taken place. Liquid H_2SO_4 was found to be the major particulate component during the AGASP-II measurements in the spring of 1986. During the arctic summer this may not, however, be the case, as shown by Anderson *et al.*[26]

The origins of the arctic aerosol have been studied with the help of not only trace metals but also of isotopes, halogens, graphitic carbon, and organic compounds. Stable lead isotope ratios were used by Sturges & Barrie[29] to provide evidence for the origin of atmospheric aerosols collected in the Canadian Arctic. The $^{206}Pb/^{207}Pb$ ratios measured at three stations were very uniform over time. The ratios between these isotopes in aerosols in the USA and Western Canada were found to be much higher than those in the arctic aerosol, while for Europe they were lower. The ratios in Eastern Canada were similar to the arctic data, but the meteorology argues against this region being a major contributor to the arctic air pollution. The $^{206}Pb/^{207}Pb$ ratios measured at Spitsbergen during the transport of polluted air from the USSR were very similar to the ratios measured in the Canadian Arctic. Thus, Sturges & Barrie[29] concluded that the USSR may be an important source of lead-bearing aerosols in the whole Arctic.

Sturges & Barrie[30] also studied the behaviour of chlorine, bromine, and iodine in aerosols at three stations in the Canadian Arctic. Bromine was found to peak every year just after the arctic dawn. The Br concentrations were so high that they could not be explained by emissions from known marine, automotive and crustal sources. According to the authors, this spring time excess of Br concentrations might originate from low-level biological production and show conversion of gaseous Br to particulate material, or an unidentified anthropogenic source, such as coal combustion. Chlorine, as an indicator of marine aerosols was correlated with Na but was depleted during periods of high sulphate concentrations. As concluded already by Pacyna & Ottar,[31] the Cl decrease can be explained as follows: during episodes of long-range transport of sulphur compounds, hydrochloric acid may be driven off from sea-salt particles by reaction with pollutant nitric and sulphuric acids. Iodine was found to be enormously enriched relative to seawater. The authors speculated that either some oceanic biological cycle might be involved or some unidentified pollution source.

Large concentrations of graphitic carbon particles in arctic air were already reported at the beginning of the 1980s,[32,33] mostly in connection with their role in the absorption of solar radiation in arctic haze. Direct identification of these particles was used to conclude that arctic haze has a major combustion derived component. Further work has been focused on assessing the vertical distribution of graphitic carbon particles and their associated absorption coefficients.[34] Although this research is very important for modelling the effects of arctic haze on the solar radiation balance, it is not very useful in assessing the origin of the haze.

Several episodes of elevated concentrations of organic pollutants were observed in the Norwegian Arctic[35] as a result of long-range transport of air masses from regions with intense use of pesticides, such as Asia and Europe. The following compounds were determined in samples collected at four stations in the Norwegian Arctic during the period August 1982–July 1984: hexachlorobenzene, α- and γ-hexachlorocyclohexane (α-HCH and γ-HCH), penta-, hexa-, hepta-, and octachlorobiphenyls, α- and γ-chlordane, and hepta-chloroepoxide. Substances from the DDT-group could not be detected in any sample. In addition, fluoranthene and the sum of phenanthro- and anthraquinones were measured. The fluoranthene and quinones are relatively volatile polycyclic aromatics, emitted during combustion processes, as well as being released from soil due to decomposition of various biological material. In order to discuss the origin of organic compounds in arctic air, Pacyna & Oehme[35] examined the concentration ratios of α-HCH to γ-HCH. The authors concluded that emissions from the USSR are the major sources for episodic increases in concentrations of organic air pollutants in the Norwegian Arctic during both the wintertime and summertime. In addition, emission from North American sources may also sporadically contribute. The HCH concentration levels found in summertime were higher than in wintertime, mainly due to the use of pesticides in the spring season and re-evaporation of material deposited in the Arctic. The occasionally observed high concentrations of fluoranthene and quinones during wintertime were explained as originating from combustion processes both in the Svalbard Archipelago and far away regions, mainly in the USSR.

The concentrations from the Pacyna & Oehme studies are in good agreement with recently presented data on airborne organochlorines in the Canadian Arctic.[36] It was concluded that the relative abundance of organochlorines in arctic air is as follows: α-HCH > hexachlorobenzene > γ-HCH $\sim$polychlorinated camphenes (e.g. toxaphene) > chlordanes $\sim$polychlorinated biphenyls > DDTs. Although Patton *et al.*[36] have not made any attempt to assess the origin of organochlorines in the Canadian Arctic, they presented the information on the use of pesticides in North America and Europe. The α-HCH to γ-HCH ratios in both the above studies have shown similar patterns with the winter–spring maximum and higher ratios in older air masses due to the photochemical transformation of γ-HCH to α-HCH. The occurrence of organic compounds in the Arctic is discussed further in Chapter 8.

Natural origins of the arctic aerosol have also been studied. Crustal elements have been measured at various altitudes in the Arctic during both winter and summer. Pacyna & Ottar[37] concluded that long-range transport of airborne loess from deposits in Asia and Africa can reach the Norwegian Arctic and contribute to the formation of haze layers, particularly at altitudes higher than 2 km. The authors analysed the chemical composition of particles with diameters larger than 2 μm collected during the March/April 1983 measurement campaign at Ny Ålesund. The area was covered by snow and ice, so local formation of mineral dust was excluded.[38] Pacyna & Ottar[37] used the concentration ratios of Si/Al, Si/Ca, Si/Fe, K/Ca, and K/Fe to show the relationship between the composition of the coarse particles measured at Ny Ålesund and dust deposits from the deserts of Asia. In Table 4.6 the chemical composition of aerosols at Ny Ålesund in the form of the above mentioned ratios is compared with the relevant data for aerosols measured in rural areas in North China at the beginning of April 1980,[39] and with possible source compositions: geochemical average shales and sandstones,[40] geochemical average Earth crust,[41] the stratospheric samples of El Chichon volcanic ash,[42] and geochemical average north-west Spitsbergen rocks.[43] Relative to the Chinese aerosol, the arctic aerosol is depleted in Fe. The Si/Al ratio is known to be non-conservative during long-range transport due to gravitational settling of Si rich coarse particles. Rahn[44] suggested that the Si/Al ratio in remote aerosols can be two or more times lower than in average crust and clay minerals. This fits well with the data in Table 4.6, which indicate that loess deposits in Asia can contribute to the Si and Al content of the aerosol at Ny Ålesund. Based on the large-scale wind field climatology at 700 mbar (about 3 km), there is very strong westerly flow north of about 20°N.[45] The 700 mbar trajectories to the study area proved that this air could have reached Ny Ålesund at the time of the measurements. In addition, the aircraft measurements in the Norwegian Arctic, carried out at the same time, have shown transport of coarse particles into the Arctic at up to 4 km altitude.[46] Thus, Pacyna & Ottar[37] concluded that dust raised over China could be transported to the Arctic. This conclusion has also been extended to the transport of loess deposits from the Sahara to the Arctic.

Table 4.6 Ratios of earth crust elements in coarse particle fractions

Sample	*Si/Al*	*Si/Ca*	*Si/Fe*	*K/Ca*	*K/Fe*
March–April 1983 Ny Ålesund[a]	1·3	2·5	13·5	0·22	1·15
Coarse mode aerosol from China[b]	2·9	2·0	4·0	0·22	0·43
Shales[c]	0·91	3·3	1·5	1·20	0·56
Earth crust[d]	3·41	7·6	5·5	0·71	0·52
Volcanic ash[e]	—	—	—	1·46	2·55
Sandstones[c]	14·7	9·4	37·5	0·27	1·09
North west Spitsbergen rocks[f]	4·6	15·8	7·1	1·42	0·64

[a] Average values for samples during non-episodic period.
[b] Averages of Winchester *et al.*[39]
[c] Averages of Turekian & Wedephol.[40]
[d] Averages of Mason & Moore.[41]
[e] Stratospheric ash from El Chichon eruptions May 1982.[42]
[f] Averages of Hjelle.[43]

The coarse dust particles, measured at Barrow by Li & Winchester,[24] had a mass size distribution similar to that expected from wind erosion of eolian dust, again suggesting long-range transport of dust particles to the Arctic.

The coarse particles may be transported within air masses for a few thousand kilometres partly because they are very irregular in shape. Bailey *et al.*[47] have made measurements on several flights over Barrow in April 1983 suggesting that some of the $>5\ \mu m$ particles were angular and undoubtedly crystalline, others plate-like and still others semi-fibrous. The 2–5 μm particles were more uniform in shape, many of them also crystalline. Irregularities in shape decrease the fall speed of a particle compared to that of a spherical particle with the same maximum dimension. Bailey *et al.*[47] concluded that the coarse particles were distributed approximately uniformly with height up to at least 5 km regardless of the presence or absence of haze layers.

Natural origins of the arctic aerosol have also been studied through the examination of organic anions, such as formate, acetate, propionate, pyruvate and methanesulphonate in coarse and fine fractions at Barrow in the late winter of 1986. Li & Winchester[48] concluded that the mass contributions of these anions to winter arctic haze are comparable to inorganic ions, such as chloride, bromide, nitrate, and sulphate. According to the authors, the carboxylic acid anions at Barrow originated mostly from natural sources at lower latitudes and were transported to the Arctic. Relative abundances of formate, acetate and propionate suggested their origin to be similar. Based on the good correlations among the above ions, similar to those reported from lower latitudes in gas phase and precipitation, Li & Winchester[48] suggested a common origin for all three ions, e.g. direct emissions from lower latitude vegetation.[49] About 20% of total measured non-sea-salt sulphate was found to be associated with methanesulphonate and attributed to natural marine biogenic sulphur precursors. A similar conclusion was made by Maenhaut *et al.*[22] on the basis of measurements of air pollutants in the Norwegian Arctic.

4.3 ORIGIN OF GASEOUS POLLUTANTS IN THE ARCTIC

The behavior of SO_2 and NO_2 in arctic air is not presented here, but is discussed in Chapter 6. These two gases are usually included in a group of gases that may be precursors of some of the arctic haze aerosol.

In addition to these two haze-causing gases, the other two groups of trace gases include: (1) trace gases that can be used as indicators of the origin of arctic haze, although they do not cause the haze; and (2) trace gases which are important for the radiation balance. The former group of gases has been studied more extensively and will be the focus of this section.

Major studies on the behaviour of trace gases in the Arctic and their use to study the origin of arctic haze were launched in the beginning of the 1980s. Khalil & Rasmussen[50,51] showed that concentration patterns of several chlorofluorocarbons, chlorocarbons and carbon monoxide at Barrow were similar to the concentration patterns of particulate anthropogenic pollutants such as non-crustal vanadium. The authors demonstrated that trace gases, such as CCl_3F (F-11), CCl_2F_2 (F-12), $CHClF_2$ (F-22), $C_2Cl_3F_3$ (F-113), CH_3CCl_3, C_2Cl_4 (PCE), C_2HCl_3 (TCE), CH_3Cl, $C_2H_4Br_2$, $CBrClF_2$, CH_2BrCl and CH_2Br_2 can be subject to episodic transport with air masses from source regions to the Arctic over several thousands of kilometres. The residence time of these gases in the air is long enough to ensure their persistence. Khalil & Rasmussen[51] suggested that F-113 and the fire-extinguishing compound CF_3Br, can be tracers of polluted air arriving in the Arctic from the USA, as the high technology industries of the USA are the major users of these two chemicals. Then CF_2BrCl, which is primarily used in Europe, was found to be a good tracer of anthropogenic pollutants originating in this part of the world. None of these three gases are believed to be used in the USSR on a wide scale. In order to trace the contribution to arctic air pollution of the anthropogenic emissions from the USSR, Hov *et al.*[52] used as signatures the ratio of light hydrocarbons ethane and propane to chlorinated ethenes C_2Cl_4 and C_2HCl_3. Ethane and propane originate mainly from natural gas exploitation, while C_2Cl_4 (tetrachloroethylene) and C_2HCl_3 (trichloroethylene) are industrial solvents. On the basis of measurements at Ny Ålesund , Spitsbergen, Hov *et al.*[52] concluded that air reaching the study area from an easterly direction (e.g. central USSR) is much richer in alkanes and much less abundant in chlorinated ethenes than the air sampled at Barrow. It was also indicated that the use of chlorinated ethenes is primarily confined to western industrialized countries.

4.4 A DISPERSION MODELLING TECHNIQUE APPLIED TO STUDY THE ORIGIN OF ARCTIC AIR POLLUTION

A receptor-oriented Lagrangian transport model has been developed to provide qualitative information on the contributions of emissions from various source regions in the USSR to the contamination of arctic air. In this connection the main sources of air pollution and source regions in the USSR were presented by Pacyna *et al.*[53] The major source regions are shown in Fig. 4.1, while the anthropogenic emissions of As, Cd, Cr, Mn, Ni, Pb, Sb, Se, V, and Zn from various source categories in the northern USSR are given in Table 4.7. These emission data have been used to estimate concentrations of the above trace elements with the help of a trajectory model similar to the model used in the Organization for Economic Co-operation and Development (OECD) and Economic Commission for Europe (ECE) programmes on long-range transport of air pollution in Europe.[54] Generally, the model describes the horizontal dispersion of trace elements within the atmospheric boundary layer. Calculations were based on two sets of trajectories: a set of ten-day back trajectories (850 mbar, isobaric) by Harris[55] and a set of four-day back trajectories (850 mbar, both isobaric and isentropic) by Iversen at the Norwegian Institute for Air Research (NILU). The major part of the work has been carried out with the latter set of trajectories as the usefulness of trajectory models applied to the study of arctic haze is

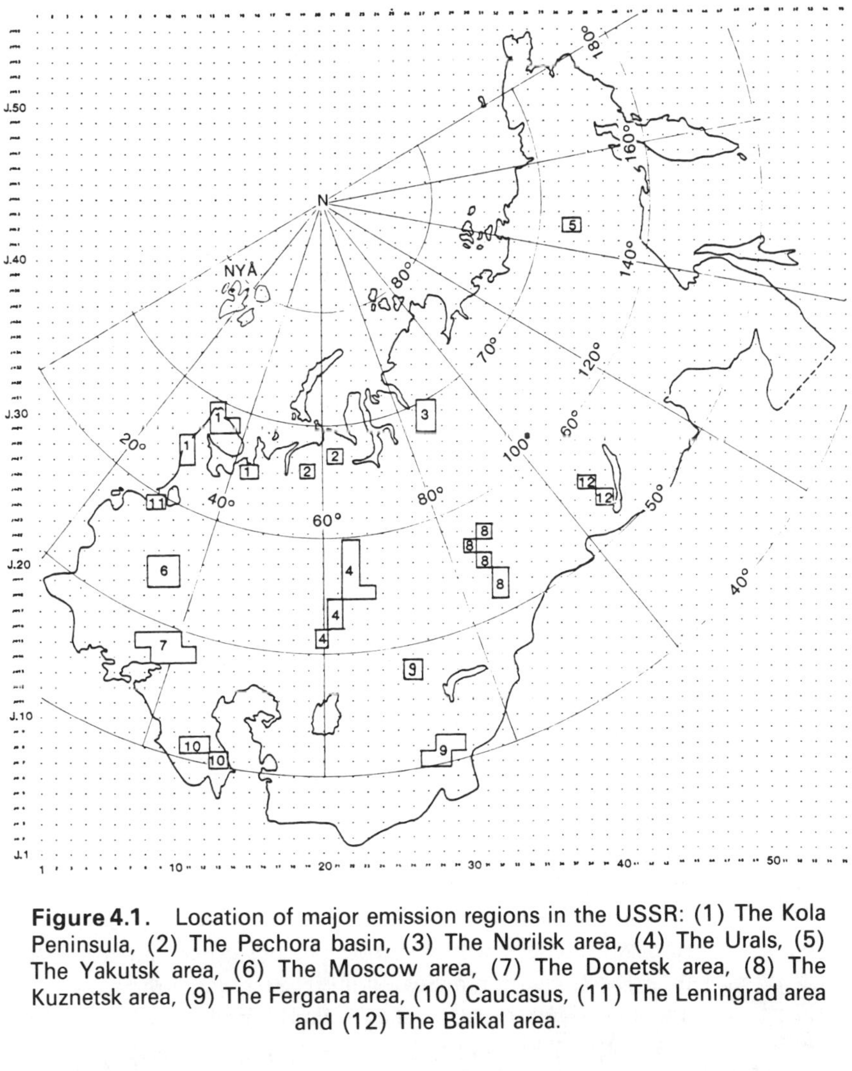

Figure 4.1. Location of major emission regions in the USSR: (1) The Kola Peninsula, (2) The Pechora basin, (3) The Norilsk area, (4) The Urals, (5) The Yakutsk area, (6) The Moscow area, (7) The Donetsk area, (8) The Kuznetsk area, (9) The Fergana area, (10) Caucasus, (11) The Leningrad area and (12) The Baikal area.

limited after five days.[56] They compared three long-range trajectory models and concluded that median horizontal displacements between five-day trajectory end-points were mostly in the 800–1000 km range.

The model used by Pacyna *et al.*[53] was based on the mass-balance equation

$$dq/dt = (1 - \alpha)Q/h - kq \tag{4.1}$$

Table 4.7 Emissions of air pollutants in the northern USSR (in tonnes y^{-1})

Region and source	*As*	*Cd*	*Cr*	*Mn*	*Ni*	*Pb*	*Sb*	*Se*	*V*	*Zn*
The Kola Peninsula										
Copper–nickel production	154	15	2	2	535	412	14·0	16·0		61
Fossil fuel combustion	2	1	34	30	40	54	2·0	4·0		10
Steel and iron	3	1	47	37	44	13	2·3	1·5		19
Phosphate fertilizer	—	10	—	—	21	3	—	2·4		80
Wood processing	6	2	39	37	5	26	4·8	1·7		10
Gasoline combustion	—	—	—	—	—	237	—	—		—
Total	165	29	122	106	645	745	23·1	25·6	122	180
The Pechora basin										
Coal mining and combustion	8·5	2·1	60	55	62	27	7	17·5	49	40
Wood processing	3	1·4	21	19	11	13	4·8	4·7	17	16
Gasoline combustion						158				
Total	11·5	3·5	81	74	73	198	11·8	22·2	66	56
The Norilsk area										
Copper–nickel production	242	24	3	2·5	900	650	22	24·9		235
Fossil fuel combustion	1	0·5	8	7·0	11	27	1	3	113	8
Wood processing	3	1·0	20	18·5	24	13	2·4	4·7	17	19
Gasoline combustion						53				
Total	246	25·5	31	28·0	935	743	25·4	32·6	130	262
The Urals										
Copper–nickel production	462	70	5	5	585	1 220	41·5	47		444
Fossil fuel combustion	42	33	150	130	790	173	13·6	57	2 730	180
Steel and iron	47	42	1 230	998	240	500	38·5	76	267	3 300
Gasoline combustion						7 630				
Total	551	145	1 385	1 133	1 615	9 523	93·6	180	2 997	3 924

where

q = trace element concentration in the air ($ng\ m^{-3}$)
t = time (s)
Q = trace element emission per unit area and time ($ng\ m^{-2}\ s^{-1}$) referred to the grid system of 150 km × 150 km used in the European Monitoring and Evaluation Programme (EMEP)
h = height of mixing layer (m)
k = decay rate for the trace element considered (wet and dry deposition) (s^{-1})
α = part of trace element emission deposited in the same grid element from which it is emitted. This local deposition supplements the deposition included in the decay rate, k.

The EMEP grid system of 150 km × 150 km was used to present the spatial distribution of emissions in the USSR. A constant mixing height (1000 m) was assumed. Only periods with no precipitation were considered. The dry deposition velocities for aerosols containing the respective trace elements varied from 0·1 $cm\ s^{-1}$ for Cd and Se to 0·5 $cm\ s^{-1}$ for Mn and Ni. Local deposition accounted for 11% of the trace element emission from a given grid element.

Two important assumptions were made when calculating the trace element concentrations: (1) no exchange of pollution between the boundary layer and the free troposphere; and (2) uniform concentrations within the mixing layer. The latter assumption was particularly important. Vertical profiles of particle light-scattering coefficients δ_{sp}, number concentrations of particles and trace element concentrations have proven their uniform distribution within the mixing layer.[57]

The mass balance equation (4.1) was integrated along the trajectories to calculate the concentrations of trace elements. The calculated concentrations were then compared with measurements. The comparison for the March 1983 data at Ny Ålesund is shown in Fig. 4.2.

The most interesting feature of the comparison presented in Fig. 4.2 is that the estimates follow the measurements for all metals and samples. The proportions between the measured concentrations of the metals are very similar to those between the calculated concentrations. Generally, the model has proven useful in calculating trace element concentrations at Ny Ålesund during periods with only slowly changing air trajectories. The best quantitative agreement between measurements and estimates was obtained for Sb and Cd. Anthropogenic emissions were overestimated for Cr, Ni, Pb, Se, and V. Zn and As emissions, on the other

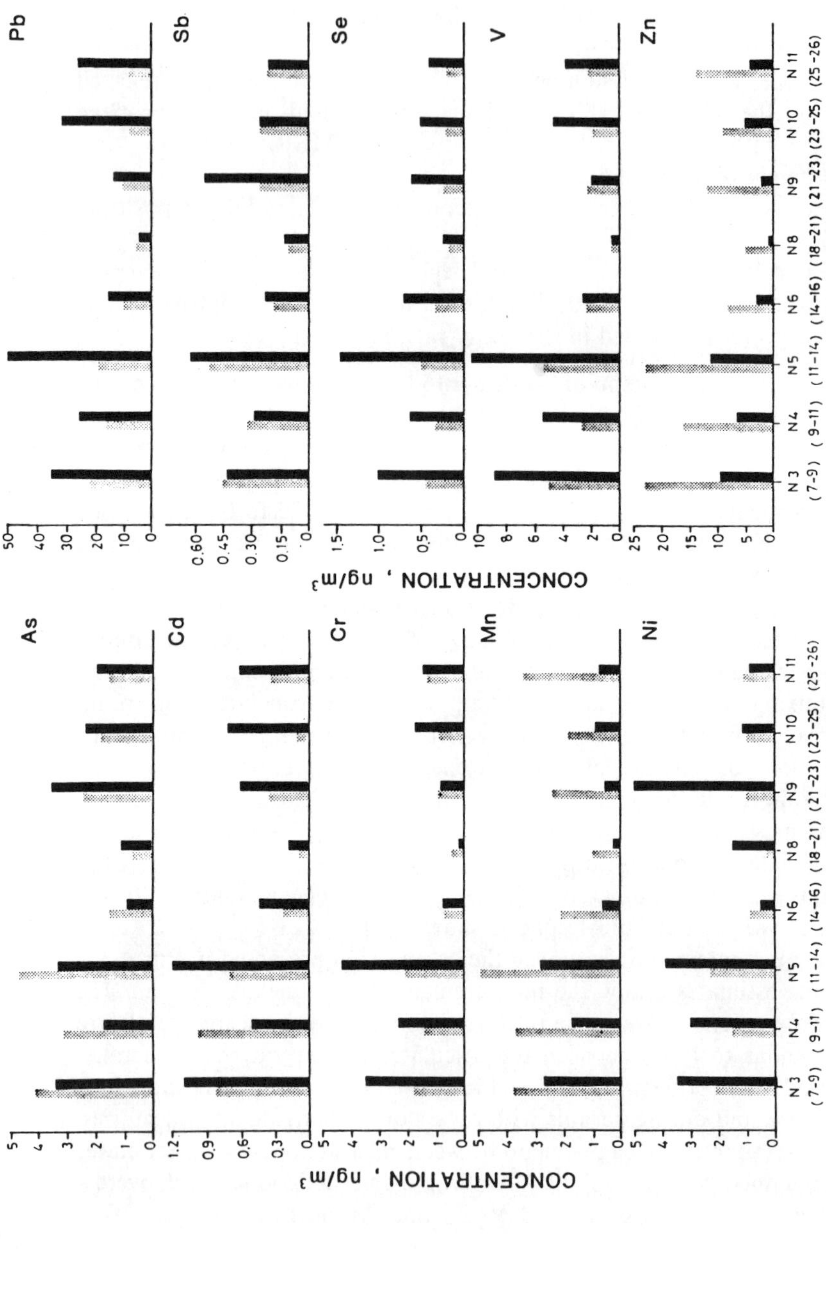

Figure 4.2. Measured (dotted bars) and calculated concentrations (full bars) of trace elements at Ny Ålesund in March 1983.

hand, seem to have been underestimated, most probably because of incomplete information on the emissions during copper–nickel production. The major reasons for the differences between measurements and estimates are: incomplete emission inventories, improperly assumed dry deposition velocities, disregard of wet deposition processes, uncertainties in mixing depth, and inaccuracies in the trajectory calculations.

Further development of the model will include the elaboration of source–receptor matrices for major trace elements emitted in various source regions and measured in the Arctic. This climatological model will serve to compute deposition of trace elements in the Arctic as tracers of deposition of other air pollutants in this area.

Recently Akeredolu *et al.*[58] have modelled the flux of anthropogenic trace metals into the Arctic from the mid-latitudes with the use of a similar approach to that described above. Comparison of the model-predicted concentrations with limited observations showed that they were within a factor of two or three. The total annual flux of Sb, As, Cd, Pb, Zn and V into the Arctic from Eurasia represented 3·4%, 6·0%, 4·2%, 3·0%, 3·1% and 1·7% of the source emissions, respectively. Concerning the source regions, for lead there was a nearly equal arctic input from Western Europe, Eastern Europe and the USSR, but for the rest of the above mentioned trace elements, the highest arctic input originated from Eastern Europe. These findings contradicted those reported by Lowenthal & Rahn[12] who for the same time period attributed 75–90% of elements in aerosols at Barrow to sources in the USSR.

4.5 CONCLUDING REMARKS

During the last decade several attempts have been made to explain the origin of arctic haze, with the main efforts being the US National Oceanic and Atmospheric Administration (NOAA) Arctic Gas and Aerosol Sampling Project (AGASP) measurement campaigns and the British Petroleum (BP) programme at the Norwegian Institute for Air Research (NILU). The results of these and several other measurement campaigns were presented at four international symposia on Arctic Air Chemistry, and recently in the excellent reviews by Barrie[59] and Shaw & Khalil.[20] These results conclude that the emissions from anthropogenic sources are the main contributors to the pollution phenomenon called arctic haze. The emissions from sources in Eurasia dominate the pollution measured in the Arctic during winter, as can be expected from meteorological studies of transport paths of air masses to the region.[16] The strong transport into the Arctic from Eurasia during winter is caused by the presence of the climatologically persistent Siberian high pressure region. During summer, transport of pollutants to the Arctic from sources in Europe seems to be more important than that from the rest of Eurasia. A North American source contribution is also more evident at this time.

The chemical composition of the air in the Arctic can be used to trace the pollution sources and to assess to what extent anthropogenic emissions have altered the geochemical cycles of various compounds. The latter issue can be studied with the help of the enrichment factor method as suggested by Pacyna & Winchester.[60] They concluded that the emissions from anthropogenic sources have altered the natural cycles for Ag, As, Cd, Pb, Sb, Se, and Zn as well as possibly for Cr, Cu, In, Mo, Ni, V, and W.

Although there have been a number of successful studies carried out to explain the origin of arctic haze, a quantitative assessment of contributions of pollution from various source regions is needed. In this connection, further development of source–receptor models is highly recommended and should be considered together with the question of the accuracy required in source–receptor relationships. The major goal of this research should be focused on the role of the models in developing control strategies to reduce the pollution of the arctic environment. For example, we should carefully assess the origin of lead entering the Arctic within air masses and recommend policies on lead emissions. Today, the lead concentrations in blood of those living in Greenland are comparable with mean normal values for non-occupationally exposed persons in industrialized areas in Europe and North America.[61] No evidence has been found to show that the present lead exposure level in Greenlanders could be of natural origin.

In future the chemical tracers of arctic air pollution could become the chemical tracers of the global contamination of our environment.

ACKNOWLEDGEMENT

I dedicate this work to the memory of my friend, the late Dr. Brynjulf Ottar.

REFERENCES

1. Mitchell, M., Visual range in the polar regions with particular reference to the Alaskan Arctic. *J. Atmos. Terr. Phys.*, (Special supplement) (1956) 195–211.

2. Shaw, G. E. & Wendler, G., Atmospheric turbidity measurements at McCall Glacier in northeast Alaska. Paper presented at the Conference on Atmospheric Radiation, Fort Collins, Colorado. *Proc. Amer. Met. Soc.*, Boston, (1972) 181–7.
3. Holmgren, B., Shaw, G.E. & Weller, G., Turbidity in the Arctic atmosphere. *AIDJEX Bull.*, **27** (1974) 135–48.
4. Rahn, K. A., Borys, R. & Shaw, G. E., The Asian source of Arctic haze bands. *Nature*, **268** (1977) 713–5.
5. Larssen, S. & Hanssen, J. E., Annual variation and origin of aerosol components in the Norwegian Arctic – Subarctic region. Paper presented at the WMO Technical Conference on Regional and Global Observation of Atmospheric Pollution Relative to Climate, Boulder, Colorado, 20–24 August 1979.
6. Rahn, K. A., The Arctic air-sampling network in 1980. *Atmos. Environ.*, **15** (1981) 1349–52.
7. Heidam, N. Z., On the origin of the Arctic aerosol: A statistical approach. *Atmos. Environ.*, **15** (1981) 1421–7.
8. Ottar, B., Gotaas, Y., Hov, O., Iversen, T., Joranger, E., Oehme, M., Pacyna, J. M., Semb, A., Thomas, W. & Vitols, V., Air pollutants in the Arctic. NILU OR Rept. No. 30/86. The Norwegian Institute for Air Research. Lillestrøm, Norway, 1986.
9. Heintzenberg, J., Hansson, H. -C., Covert, D. S., Blanchet, J. -P. & Ogren, J. A., Physical and chemical properties of arctic aerosols and clouds. In *Arctic Air Pollution*, ed. B. Stonehouse. Cambridge University Press, Cambridge, 1986, pp. 25–35.
10. Barrie, L. A. & Hoff, R. M., Five years of air chemistry observations in the Canadian Arctic. *Atmos. Environ.*, **19** (1985) 1995–2010.
11. Shaw, G. E., Evidence for a central Eurasian source area of Arctic haze in Alaska. *Nature*, **299** (1982) 815–18.
12. Lowenthal, D. H. & Rahn, K. A., Regional sources of pollution aerosol at Barrow, Alaska during winter 1979–80 as deduced from elemental tracers. *Atmos. Environ.*, **19** (1985) 2011–24.
13. Schnell, R. C., Arctic haze and the Arctic Gas and Aerosol Sampling Program (AGASP). *Geophys. Res. Lett.*, **11** (1984) 361–4.
14. Herbert, G. A., Harris, J. M. & Bodhaine, B. A., Atmospheric transport during AGASP-II: The Alaskan flights (2–10 April 1986). *Atmos. Environ.*, **23** (1989) 2521–35.
15. Rahn, K. A., The Mn/V ratio as a tracer of large-scale sources of pollution aerosol for the Arctic. *Atmos. Environ.*, **15** (1981) 1457–64.
16. Raatz, W. E. & Shaw, G. E., Long-range transport of pollution aerosols into the Alaskan Arctic. *J. Clim. App. Met.*, **23** (1984) 1052–64.
17. Sheridan, P. J. & Zoller, W. H., Elemental composition of particulate matter sampled from the Arctic haze aerosol. *J. Atmos. Chem.*, **9** (1989) 363–82.
18. Heidam, N. Z., The components of the Arctic aerosol. *Atmos. Environ.*, **18** (1984) 329–43.
19. Benson, C. S., Problems of air quality in local arctic and sub-arctic areas, and regional problems of arctic haze. In *Arctic Air Pollution*, ed. B. Stonehouse. Cambridge University Press, Cambridge, 1986, pp. 69–84.

20. Shaw, G. E., & Khalil, M. A. K., Arctic haze. In *The Handbook of Environmental Chemistry*, Volume 4/Part B, ed. O. Hutzinger. Springer, Berlin, 1989, pp. 69–111.
21. Heidam, N. Z., The Greenland aerosol: Elemental compositions, seasonal variations and likely sources. In *Arctic Air Pollution*, ed. B. Stonehouse. Cambridge University Press, Cambridge, 1986, pp. 37–52.
22. Maenhaut, W., Cornille, P., Pacyna, J. M. & Vitols, V., Trace element composition and origin of the atmospheric aerosol in the Norwegian Arctic. *Atmos. Environ.*, **23** (1989) 2551–70.
23. Mason, B. J., *Principles of Geochemistry* (3rd Edn). John Wiley, New York, 1966.
24. Li, S. -M. & Winchester, J. W., Haze and other aerosol components in later winter Arctic Alaska, 1986. *J. Geophys. Res.*, **95**, **D2**, (1990) 1797–810.
25. Li, S. -M., Winchester, J. W. & Pacyna, J. M., Arctic aerosol sulphur with silicon, but no aluminium: Evidence for long range transport of coal smoke. Paper presented at International Conference on Global and Regional Environmental Atmospheric Chemistry, Beijing, China, 3–10 May, 1989.
26. Anderson, J. R., Buseck, P. R., Saucy, D. A. & Pacyna, J. M., Compositions, size distributions, and principal component analysis of individual fine-fraction particles from the Arctic aerosol at Spitsbergen, May–June, 1987. *Atmos. Environ.*, (in press).
27. Saucy, D. A., Anderson, J. R. & Buseck, P. R., Cluster analysis applied to atmospheric aerosol samples from the Norwegian Arctic. *Atmos. Environ.*, **21** (1987) 1649–57.
28. Sheridan, P. J., Characterization of size segregated particles collected over Alaska and the Canadian high Arctic, AGASP-II flights 204–206. *Atmos. Environ.*, **23** (1989) 237–86.
29. Sturges, W. T. & Barrie, L. A., Stable lead isotopes in arctic aerosols: Evidence for the origin of arctic air pollution. *Atmos. Environ.*, **23** (1989) 2513–19.
30. Sturges, W. T. & Barrie, L. A., Chlorine, bromine and iodine in Arctic aerosol. *Atmos. Environ.*, **22** (1988) 1179–94.
31. Pacyna, J. M. & Ottar, B., Transport and chemical composition of the summer aerosol in the Norwegian Arctic. *Atmos. Environ.*, **19** (1985) 2109–20.
32. Rosen, H., Novakov, T. & Bodhaine, B. A., Soot in the Arctic. *Atmos. Environ.*, **15** (1981) 1371–4.
33. Heintzenberg, J., Size-segregated measurements of particulate elemental carbon and aerosol light absorption at remote Arctic locations. *Atmos. Environ.*, **16** (1982) 2461–8.
34. Rosen, H. & Hansen, A. D. A., Role of combustion-generated carbon particles in the absorption of solar radiation in the Arctic haze. *Geophys. Res. Lett.*, **11** (1984) 461–4.
35. Pacyna, J. M. & Oehme, M., Long-range transport of some organic compounds to the Norwegian Arctic. *Atmos. Environ.*, **22** (1988) 243–57.
36. Patton, G. W., Hinckley, D. A., Walla, M. D., Bidleman, T. F. & Hargrave, B. T., Airborne organochlorines in the Canadian High Arctic. *Tellus* (*Series B*), **41B** (1989) 243–55.
37. Pacyna, J. M. & Ottar, B., Origin of natural constituents in the Arctic aerosol. *Atmos. Environ.*, **23** (1989) 809–15.

38. Semb, A., Brækkan, R. & Joranger, E., Major ions in Spitsbergen snow samples. *Geophys. Res. Lett.*, **11** (1984) 445–8.
39. Winchester, J. W., Weixin, L., Lixin, R., Mingxing, W. & Maenhaut, W., Fine and coarse aerosol composition from rural area in North China. *Atmos. Environ.*, **15** (1981) 933–7.
40. Turekian, K.K. & Wedepohl, K. H., Distribution of the elements in some major units of the Earth's crust. *Bull. Geol. Soc. Amer.*, **72** (1961) 175–92.
41. Mason, B. & Moore, C. B., *Principles of Geochemistry* (4th Edn). John Wiley, New York, 1982, pp. 46–7.
42. Kotra, J. P., Finnegan, D. L., Zoller, W. H., Hart, M. A. & Moyers, J. L., El Chichon: Composition of plume gases and particles. *Science*, **222** (1983) 1018–21.
43. Hjelle, A., The geology of Danskøya and Amsterdamøya. *Norsk Polarinstitutts Skrifter*, **158** (1974) 7–38.
44. Rahn, K. A., Silicon and aluminium in atmospheric aerosols: Crust–air fractionation? *Atmos. Environ.*, **10** (1976) 597–601.
45. Duce, R. A., Unni, C. K., Ray, B. J., Prospero, J. M. & Merill, J. T., Long-range transport of soil dust from Asia to the tropical North Pacific: Temporal variability. *Science*, **209** (1980) 1522–3.
46. Ottar, B., Pacyna, J. M. & Berg, T. C., Aircraft measurements of air pollution in the Norwegian Arctic. *Atmos. Environ.*, **20** (1986) 87–100.
47. Bailey, I. H., Radke, L. F., Lyons, J. H. & Hobbs, P. V., Airborne observations of Arctic aerosols. 2, Giant particles. *Geophys. Res. Lett.*, **11** (1984) 397–400.
48. Li, S. -M. & Winchester, J. W., Geochemistry of organic and inorganic ions of late winter arctic aerosols. *Atmos. Environ.*, **23** (1989) 2401–15.
49. Keene, W. C. & Galloway, J. N., Considerations regarding sources of formic and acetic acids in the troposphere. *J. Geophys. Res.*, **91** (1987) 14466–74.
50. Khalil, M. A. K. & Rasmussen, R. A., Gaseous tracers of Arctic haze. *Environ. Sci. Technol.*, **17** (1983) 157–64.
51. Khalil, M. A. K. & Rasmussen, R. A., Statistical analysis of trace gases in Arctic haze. *Geophys. Res. Lett.*, **11** (1984) 437–40.
52. Hov, O., Penkett, S. A., Isaksen, I. S. A. & Semb, A., Organic gases in the Norwegian Arctic. *Geophys. Res. Lett.*, **11** (1984) 425–28.
53. Pacyna, J. M., Ottar, B., Tomza, U. & Maenhaut, W., Long-range transport of trace elements to Ny Ålesund, Spitsbergen. *Atmos. Environ.*, **19** (1985) 857–65.
54. Eliassen, A. & Saltbones, J., Modelling of long-range transport of sulphur over Europe: A two-year model run and some model experiments. EMEP/MSC-W Rept. 1/82. The Norwegian Meteorological Institute, Oslo, 1982.
55. Harris, J. M., Trajectories during AGASP. *Geophys. Res. Lett.*, **11** (1984) 453–6.
56. Kahl, J. D., Harris, J. M., Herbert, G. A. & Olson, M. P., Intercomparison of three long-range trajectory models applied to Arctic haze. *Tellus* (*Series B*), **41B** (1989) 524–36.
57. Pacyna, J. M. & Ottar, B., Vertical distribution of aerosols in the Norwegian Arctic. *Atmos. Environ.*, **22** (1988) 2213–22.

58. Akeredolu, F. A., Barrie, L. A., Olson, M. P. & Pacyna, J. M., The flux of anthropogenic trace metals into the Arctic from the mid-latitudes in 1979/80. *Atmos. Environ.*, (in press).
59. Barrie, L. A., Arctic air pollution: An overview of current knowledge. *Atmos. Environ.*, **20** (1985) 843–83.
60. Pacyna, J. M. & Winchester, J. W., Contamination of the global environment as observed in the Arctic. *Palaeography, Palaeoclimatology, Palaeoecology (Global and Planetary Change section)*, **82** (1990) 149–57.
61. Hansen, J. C., Exposure to heavy metals in Greenland from natural and man-made sources. In *Arctic Air Pollution*, ed. B. Stonehouse. Cambridge University Press, Cambridge, 1986, 249–58.
62. Rahn, K. A. & Lowenthal, D. H., Elemental tracers of distant regional pollution aerosol. *Science*, **223** (1984) 132–9.

Chapter 5

Physical Properties and Physical Chemistry of Arctic Aerosols

GLENN E. SHAW

Geophysical Institute, University of Alaska Fairbanks, Fairbanks, Alaska 99775-0800, USA

5.1 INTRODUCTION

Just as many macroscopic properties of matter are fathomed by considering atomic and molecular structure, certain features of the phenomenon of arctic air pollution can be unfolded by considering its microphysics and chemistry. In this chapter we discuss the physical and physical–chemical properties of the particulate matter that builds up in winter all through the northern polar air mass, an air mass about the size of Africa. It is demonstrated that the particles are agglomerated together, probably coated with a liquid film, and congregated in three major size modes. The smallest particle mode materializes when sunlit air masses intrude into the Arctic, but this mode disappears within a few days in the polar darkness. The majority of the particulate mass resides in a mode about a half micron diameter, which interacts strongly with light to cause certain perturbations on climate. This mode may represent the removal-resistant debris from sources thousands of kilometers upwind. Finally, one sometimes finds a supermicron mode of dust particles which evidently originates from strong convective storm activity over the northern deserts of Asia.

5.1.1 Large scale synoptic properties of arctic air pollution

'Arctic haze' is the pseudonym for a complex mixture of particles and pollution gases affecting the central region of the northern polar airshed. The polar airshed, or polar air mass, is defined as being that within the boundaries of the arctic front[1,2] (see Fig. 5.1). In mid to late winter, the arctic air mass has not only expanded to its maximum dimension, about the size of Africa, but it is also at this time that the air mass is most severely polluted. Since one is speaking of radiatively significant pollution levels of large areal extent, the late winter arctic air pollution phenomenon may contribute to global-scale changes in climate.[3,4]

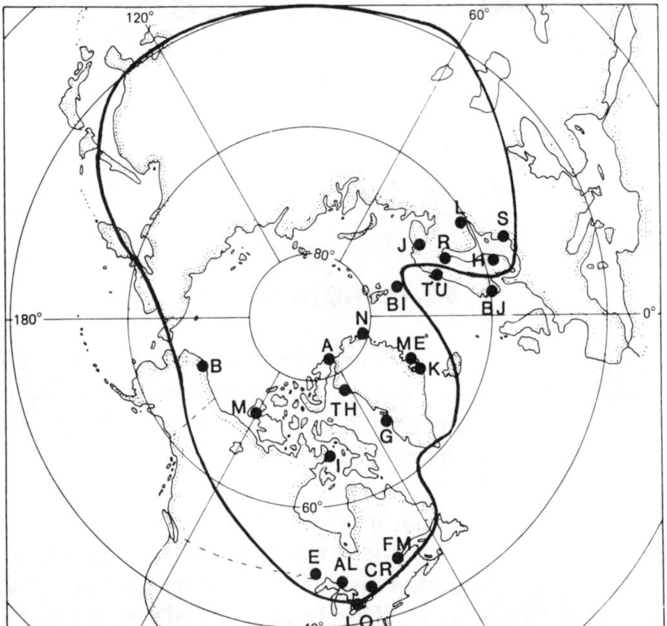

Figure 5.1. The mean position of the arctic front in January. The location of air monitoring stations: A—Alert; Al—Algoma; B—Barrow; BI—Bear Island; BJ—Birkenes; CR—Chalk River; E—ELA-Kenora; FM—Foret Montmorency; H—Hoburg; I—Igloolik; J—Jergul; K—Kap Tobin; L—Lesogorski; LO—Long Point; M—Mould Bay; ME—Mestersvig; N—Nord; R—Ricklea; S—Suwalik; TH—Thule; TU—Tustervaten (from Reference 5).

Two observations about the arctic air pollution seem especially compelling: (1) its fairly high mass intensity and optical depth, in spite of the general remoteness from pollution sources, and (2) its strong and asymmetric seasonal variation, offset in phase by about a month from polar sunrise.[5] Progress in understanding these two very basic features of the arctic pollution can be achieved by considering basic physics and physical–chemical behavior of the aerosol particles as they are found in what R. Schnell has called the 'arctic retort'.

5.2 SOURCE REGIONS OF THE ARCTIC AIR POLLUTION

By employing chemical 'fingerprinting'[6–13] and using meteorological hindcasting,[14,15] a consensus has slowly built that the largest fraction of so-called arctic haze is pollution-derived and that the major source regions are Central Eurasia, Europe and eastern North America. Since the industrial sources are primarily located in middle latitudes where westerly winds prevail, the most intensely polluted air on the planet is expected to lie eastward of the major sources; indeed, this is borne out by observation. Surprisingly, however, the mass concentrations of particles suspended in arctic air masses in the late winter months are nearly as large as in the air masses immediately downwind from industrial sources.

Given the remoteness of the Arctic from major industry, it is hard to imagine how its airshed can be so polluted. The difficulty is resolved, apparently, because there is a substantial lowering of the rate of gas and aerosol removal in the polar airshed. In this view, the Arctic is polluted not so much because of large pollution sources, but because of the nearly complete shut-down of atmospheric removal processes. The drop in particle removal has been attributed to darkness, mechanical stability, stagnation and stratification.

The build-up of arctic air pollution due to a shut-down in removal rate is analogous to an increase in a head of water in a tank when the output tap is closed down, as illustrated schematically in Fig. 5.2.

5.3 IMPLICATIONS OF THE LONG RESIDENCE TIME FOR PARTICLES

The inferred 'long residence time' for arctic air pollution has a number of interesting implications. For one thing, the usual notion of an aerosol's

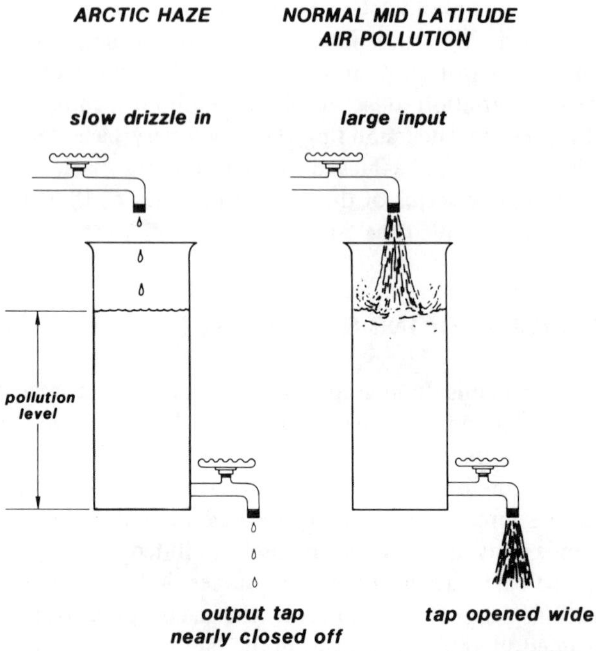

Figure 5.2. A hydrostatic analogy illustrating normal mid-latitude pollution and arctic air pollution. In the case of arctic air pollution the source strength is small, but there is heavy resistance to particle removal.

travel distance being a mere few hundred kilometers has to be replaced with one wherein the transport distance is imagined to reach from continent to continent. Another ramification is that since the material lingers in the air, environmental degradation due to surface deposition is not a problem; the acidity and conductivity of polar snows is low in comparison to that in the snows in the zones eastward of the major pollution sources.[16]

A further consequence of the long atmospheric lifetime of arctic air pollution is the absence of photolytically driven particle production during the polar night. We will demonstrate later that air pollution traveling for tens of days in the polar night gives rise to an unusual aerosol size distribution and, possibly, to an unusual cloud condensation nucleus spectrum.

5.4 MAJOR FEATURES OF ARCTIC AIR POLLUTION MICROPHYSICS

Particle size distributions are frequently approximated with a power law distribution[17] and are necessarily displayed on logarithmic graphs to encompass the huge dynamic range encountered. One loses a lot of intuitive grasp through such log scale representations. When replotted on linear scales, the particle size distribution looks like a delta function, for example.

A more convenient and insightful way to depict the atmospheric aerosol's probability distribution is to plot the mass or volume distribution function, $dn/d(\log m)$ instead of the number distribution $dn/d(\log r)$ (where m is particle mass and r is particle radius); this undergoes much less variability and indeed its salient features are amply and simply demonstrated on a plot with a linear ordinate scale, like that shown in Fig. 5.3. In Fig. 5.3, three modes of mass distribution are illustrated, using

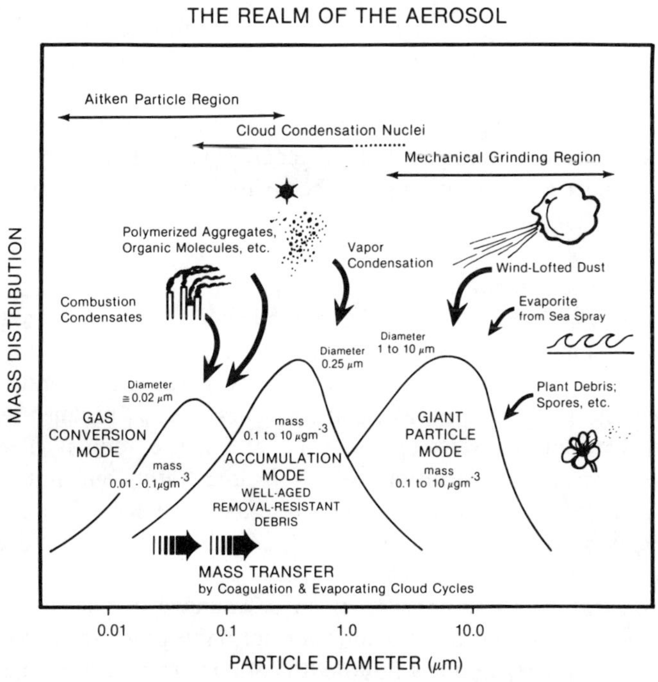

Figure 5.3. Schematic illustration of the three aerosol mass modes: the nucleation (gas-conversion) mode, accumulation mode and giant particle mode.

the terminology introduced some years ago[18] for air pollution sulfur aerosol. We have found that the three mode, log normal, volume distribution quite adequately describes the aerosol found at locations in Alaska and Antarctica, at the Mauna Loa observatory and at a mountain location in southern Arizona. The trimodal log normal distributions are much easier to work with and serve as a convenient tool to model the aerosol or to put hypotheses to a test. Volume modes have also been used to good advantage by Clarke and colleagues[19] in their research on the background oceanic aerosol.

The three modes, shown somewhat schematically in Fig. 5.3, are usually interpreted as arising from gas nucleation, accumulation of removal-resistant material, and entry onto the scene of supermicron, or 'giant', chunks of crustal material; hence the names 'nucleation mode', 'accumulation mode' and 'giant particle mode'. This interpretation should be considered as provisional; there has not been as much experimental confirmation of these suggestions as is popularly believed! Incidentally, a point of interest is that some of the best detective work in deducing that the smallest mode is probably due to sunlit-driven nucleation of vapors was done a hundred years ago by John Aitken in his studies of particles blowing in off the sea.[20]

Since we have found the trimodal mass distributions to fit the measured distributions at a large variety of geographic locations, including the Arctic, we next go on to discuss the behavior of the three modes one by one.

5.4.1 Nucleation mode

The nucleation mode of aerosols in arctic air pollution is frequently barely visible in comparison with the other two modes. Even if it is small, it cannot necessarily be ignored as it can contribute significantly to the total particle number concentration (of course this is because mass or volume distributions vary as the cube of particle radius). The total particle number concentration, which is frequently, but not always, dominated by the nucleation mode, is often an issue of interest; for example, it is needed in the theory of cloud activation or for calculating particle coagulation rates.

The nucleation mode consists of an assemblage of particles smaller than a tenth of a micron diameter, and normally possesses a geometric mean mass diameter of a few hundredths of a micron, or even smaller in the case of fresh air pollution. The nucleation mode undergoes considerable variation, changing from air mass to air mass, often in ways that are not very well comprehended.

An experiment was conducted (Fig. 5.4) at our clean air laboratory in central Alaska to assess how much of the variability in the nucleation mode is due to variation in particle size, in comparison with variation in particle concentration. The purpose was to determine to what extent the variability in particle number concentration (condensation nuclei, or CN) might be dependent on the possible appearance or disappearance of new particles of very small size, as opposed to uniform increases or decreases of particle concentration at all particle sizes. The diffusion screen in Fig. 5.4 is simply a parallel series of microscopic cylindrical pores through which the particle-laden air is forced. Particles possessing a Brownian motion sufficiently vigorous to random walk (diffuse) to the wall of the pore during their passage will attach and stick to the wall by polarization forces; thus the smallest class of particles is preferentially removed by diffusion, allowing the larger particles to pass through and be counted.

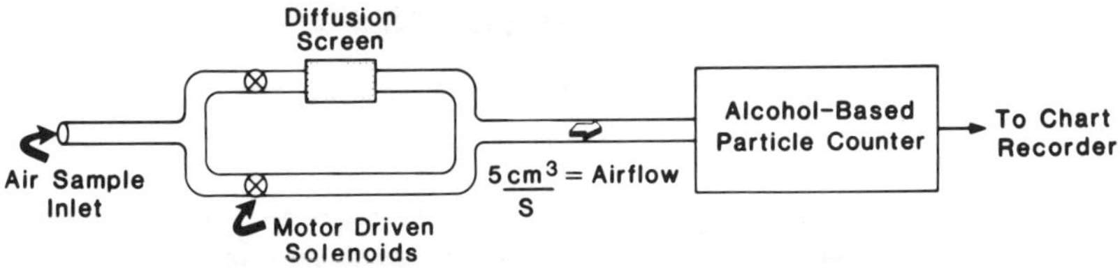

Figure 5.4. Experiment designed to provide a measure of the characteristic size of the condensation nuclei. Air alternately passes through the two parallel paths, one containing a diffusion screen that removes particles smaller than about 0·03 μm diameter.

We illustrate two data recordings from the experiment. Figure 5.5 shows pen recordings of the particle concentration during alternating passage of air through the diffusion screen (filtered) and through an unimpeded tube (unfiltered; see the drawing in Fig. 5.4). The two time periods shown were during stable conditions when no nearby sources of contamination could be implicated. Both records show similar concentrations of larger particles, namely those that survived passage through the diffusion battery, but the total particle concentration (all sizes) is greater in the top trace, which was accumulated during a time when there was strong air flow into the observing area from the Pacific Ocean to the south of Alaska; the other graph was taken during air flow from the arctic region to the north of Alaska.

The experiment provided repeated illustrations that arctic air is depleted in nucleation mode particles and that air masses entering Alaska

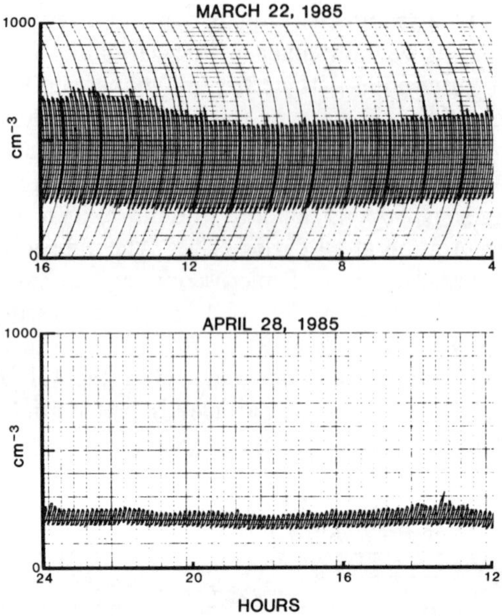

Figure 5.5. Chart recording of the experiment shown in Fig. 5.4 for two different meteorological situations: the upper trace was recorded during intrusion of a marine air mass from the northern Pacific region; the lower trace was recorded during a time when the arctic front was well equatorward of the sampling station and the station was immersed in an arctic air mass. The latter was depleted of small particles.

from the sunlit Pacific region to the south are strongly enriched in particles of very small size, smaller than approximately 0·03 microns. Evidently the presence of small particles (nucleation mode) requires a fairly recent (a few days) exposure to sunlit conditions; air from the 'arctic retort' tends to lose its nucleation mode.

Further evidence for the requirement of recent exposure to light as a prerequisite for a nucleation mode can be found in the seasonal behavior of the nucleation mode. The monthly mean condensation nuclei (CN) concentration (all particles larger than 0·01 micron diameter) at two mountain top locations in central Alaska is shown in Fig. 5.6; note that the concentration increased strongly at both stations around the time during and just following vernal equinox when the subarctic comes out of its polar dark period. The seasonal variation of solar radiation at the

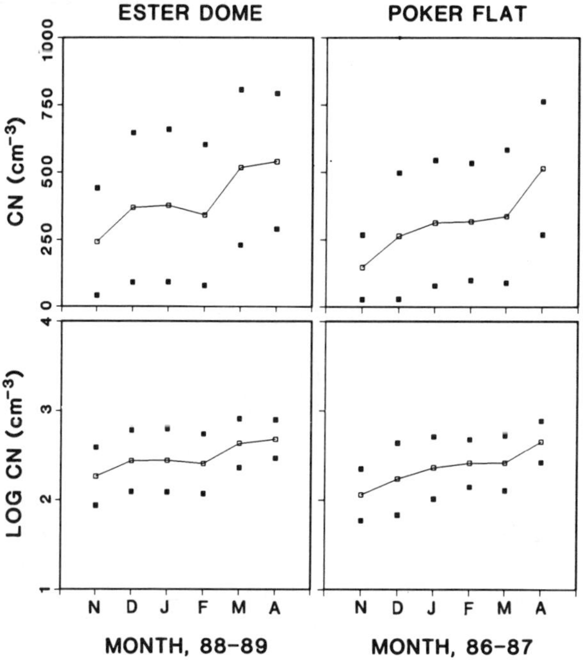

Figure 5.6. Monthly means (open squares) and standard deviations (closed squares) of condensation nuclei recorded at Ester Dome and at Poker Flat stations, both located in central Alaska.

central Alaska location increases very rapidly in late March and April (see Fig. 5.7) and might be responsible for producing new particles of small sizes by photolytic processes, such as through the photo-oxidation of SO_2. Figure 5.8 compares the frequency distribution of CN concentration for the months February (a dark month) and April (sun has risen); note the preponderance of higher CN concentration in the sunlit month.

The nucleation mode at our observatory in central Alaska is observed to increase in strength when air is advected rapidly from the sunlit regions of the northern Pacific Ocean. A measure of the strength of air importation from the Pacific region is simply the positive day-averaged temperature deviation from the multiseason mean for that date, provided that only the cold months are used so that complicating effects from localized solar radiation heating is nil. Surface air temperatures run 10–30°C higher than normal in central Alaska during strong intrusions

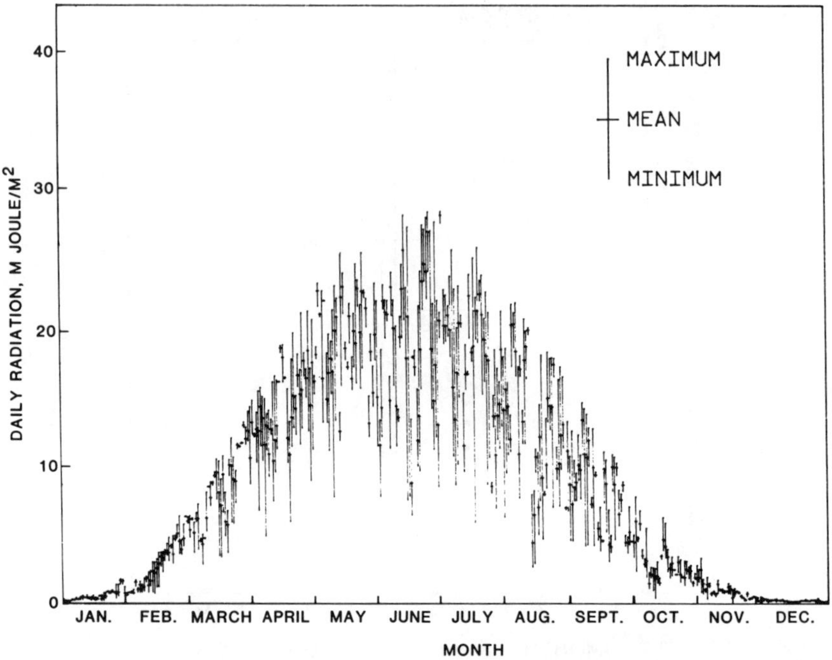

Figure 5.7. Solar radiation (global, downwelling) at Fairbanks (Courtesy of G. Wendler, Geophysical Institute, University of Alaska, Fairbanks).

of northern Pacific air, as can happen, for example, when a high-pressure ridge builds in southeast Alaska, along with placement of the Aleutian Low in the western Gulf of Alaska. A scatter plot of CN concentration versus air temperature deviation (Fig. 5.9) suggests, again, that the sunlit, warm, area of the northern Pacific is a source for new, very small, particles. We are quite confident that these particles are not connected with anthropogenic activities because the aerosolized black carbon (detected with an athelometer) was extremely small.

Similar trends existed between deviations of air temperature and particle *size*, similar to the result described for the episodes in Fig. 5.5. In other words, air masses originating in the clean Pacific region equatorward of the central Alaska sampling location are charged with high concentrations of very small particles. Arctic-derived air masses, however, have a strongly depleted nucleation mode. Colder than normal (e.g. arctic) air often contains no measurable nucleation mode.

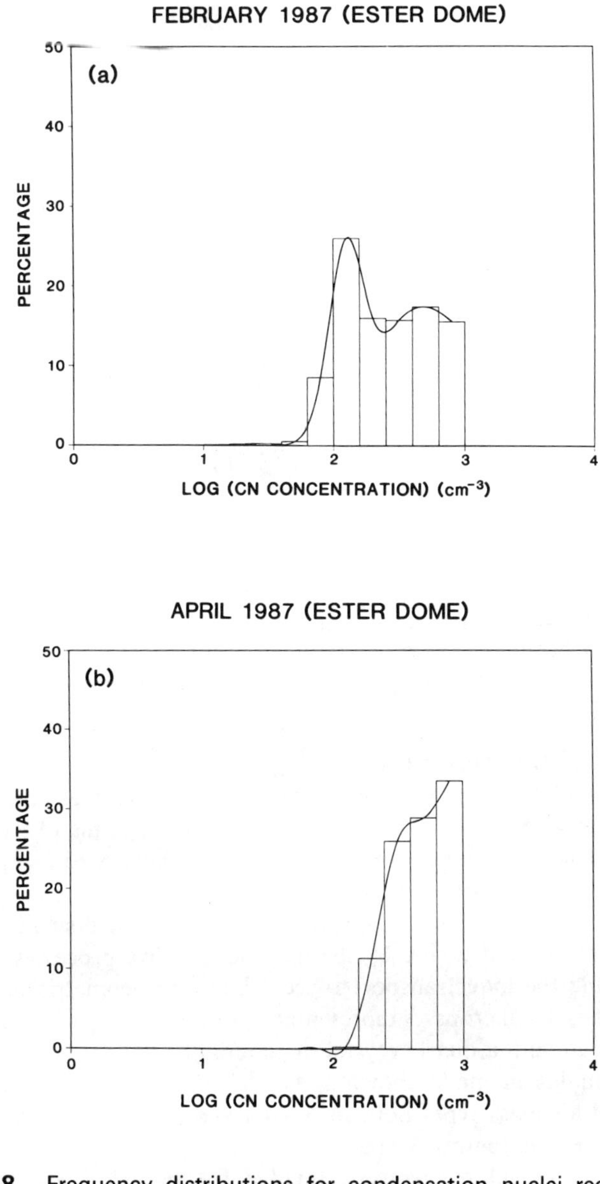

Figure 5.8. Frequency distributions for condensation nuclei recorded at Ester Dome in central Alaska (65°N) for the month of (a) February, a month with negligible amounts of solar radiation, and (b) April, a month with plentiful solar radiation.

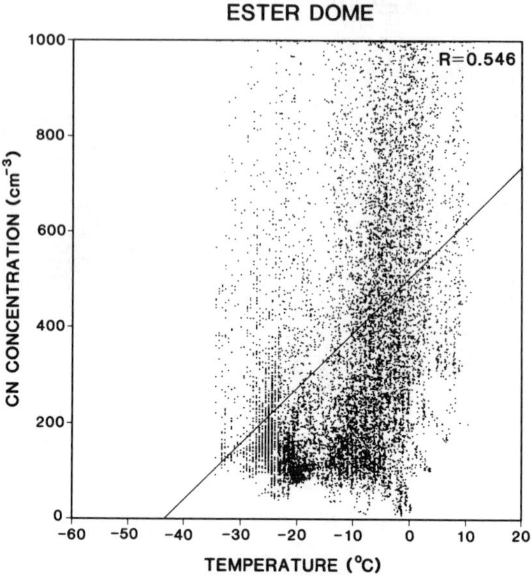

Figure 5.9. A scatter diagram showing the tendency for condensation nucleus concentration to be higher when warm, sunlit air from outside the Arctic Basin flows into interior Alaska.

5.4.2 Accumulation mode

The accumulation mode in the arctic air mass undergoes less variation, both in terms of its modal mean diameter and in terms of its strength, compared with the nucleation mode and the giant particle mode. Particles in this mode are probably the removal-resistant residue of particles injected into the atmosphere upwind; because larger and smaller particles have been removed earlier by inertial and diffusive processes that took place along the long transport route. The mean geometric diameter of this mode is 0·4 microns. A representative example of a mass distribution, as measured with a quartz crystal mass monitor,[21] is drawn in Fig. 5.10; the accumulation mode shown is typical of what is found in a wide variety of air mass types both in Alaska and elsewhere at clean locales such as over the central oceans.

Inertial removal mechanisms involve the particles skidding across aerodynamic flow lines and impacting and sticking on objects: examples are sedimentation through a viscous atmosphere, impaction of particles on snowflakes, ice crystals or water droplets, or impaction on surface

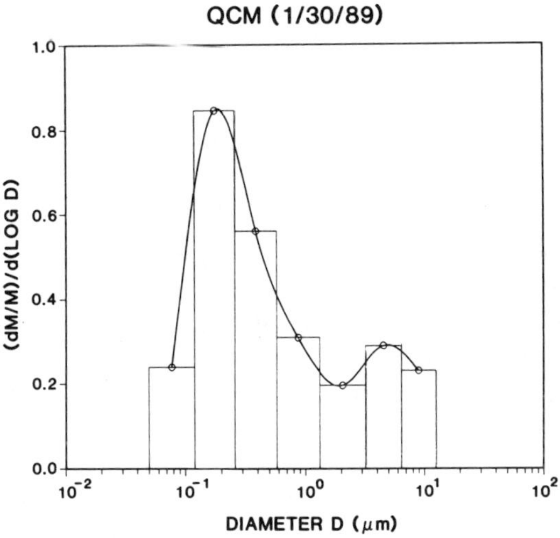

Figure 5.10. Mass size distribution for particles in an arctic air mass showing a well-developed accumulation mode and a small 'giant particle' mode. The distribution was made with a Quartz Crystal Mass detector. The ordinate is differential aerosol mass dm m^{-1} per unit of log *D*, where *D* is particle diameter.

obstacles as eddies bring particles into their vicinity. These processes of removal all depend on the dimensionless Stoke's number;[22] where the rate of removal is proportional to r^2, where r is particle radius.

Diffusive mechanisms of aerosol removal dominate at the smaller particle sizes. One may assign a diffusion coefficient, D, to a particle through the Einstein relationship, $D=kTB$, where k is Boltzmann's constant, B is the mechanical mobility (velocity per unit of applied force) of the particle moving through the viscous carrier gas, and T is temperature. Kinetic theory allows us to evaluate the mobility; for example in the Stoke's regime it is $(6\pi r\eta v)^{-1}$, where η is the gas viscosity and v is velocity. The particle removal rate under these diffusively controlled conditions then varies as r^{-a} where $1<a<2$.

When the two mechanisms, inertial and diffusive, are combined, the rate of removal of particles from the atmosphere can be written in the dimensionless form

$$-\mathrm{d}n/\mathrm{d}t=(R_{\mathrm{m}}/2)(r^2/r_{\mathrm{c}}^2+r_{\mathrm{c}}^2/r^2)n \tag{5.1}$$

where R_m represents the minimum rate of removal occurring at a critical particle radius r_c ($dn/dt = R_m$ when $r = r_c$). The concentration of particles (n) of radius r falls off exponentially, according to the first order removal mechanism expressed by eqn (5.1), which has the solution $n(t) = n(0) \cdot \exp(-t/T)$ where the folding time (T) would be

$$T = 1/[(R_m/2) \cdot (r^2/r_c^2 + r_c^2/r^2)] \tag{5.2}$$

The values of R_m and r_c for particles in the free atmosphere depend in complicated ways on the turbulence spectrum of the air, on topography, on the distribution and type of vegetation, on the frequency, extent and type of cloud formation, on precipitation events, etc. One possible interpretation to explain the ubiquitous accumulation mode is that it represents a kind of end point for an aerosol that originated from a wide variety of sources and that has eroded away in the atmosphere according to a law similar to eqn (5.1). In this interpretation, the value of the mean modal radius of the aerosol, r, provides a measure of r_c.

If one adopts the above interpretation, namely the assertion that the accumulation mode is the longest-lived aerosol mode in the free troposphere, then it follows that this mode would undergo less variability in time than particles in either larger or smaller modes, which are removed more rapidly from the atmosphere and hence would undergo greater fluctuation. This intuitive idea was tested by taking lagged autocorrelation of a several-month record of the three modes; the accumulation mode showed greater coherence over longer times than the other modes, suggesting that it is more robust and permanent.

In the Alaskan sector, the meteorological arctic front, which can usually be identified by crowding of isobars at the 850 mb level, has an average winter 850 mb temperature of approximately −10°C. Temperatures below −10°C are crosshatched in the time series of 850 mb temperatures shown in Fig. 5.11(a); these are associated with intrusions of arctic air masses into interior Alaska. Notice that strong incursions of arctic air affected central Alaska during March and April 1986. In Fig. 5.11(b), the particle concentration measured with an optical particle counter is illustrated; we have plotted a channel centered on the accumulation mode. There is an obvious synchronism, periods of cold arctic air being associated with high accumulation mode concentrations. We see that the polar air mass lying within the boundary of the arctic front has an amplified, robust, accumulation mode aerosol.

The accumulation mode tends to be much stronger in arctic-derived, polluted, air masses than in air masses flowing into Alaska from the

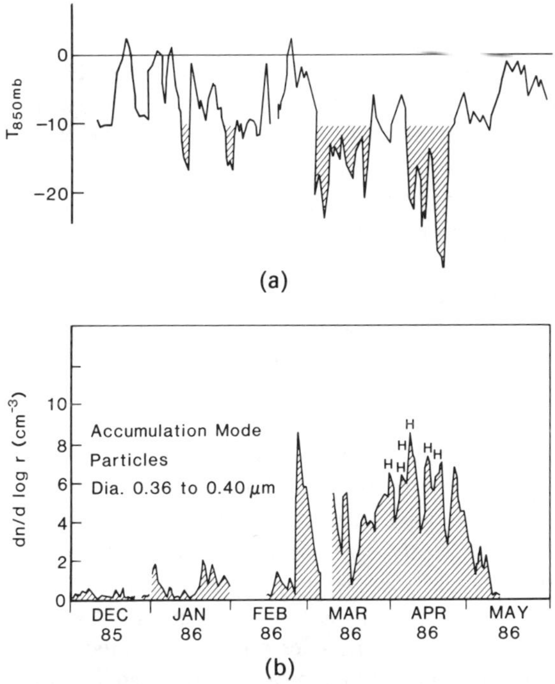

Figure 5.11. (a) The top curve is the day-averaged measure of the 850 mb temperature at Fairbanks; generally the arctic front had expanded equatorward of the station when the 850 mb level dropped below about −10°C (crosshatched); (b) The bottom curve represents a time series of accumulation mode particles as measured at Ester Dome, Alaska. H marks days when visual arctic haze was apparent. The ordinate is the number concentration (cm^{-3}) of particles in radius interval d*r*.

Pacific regions to the south. This is opposite to the behavior we discussed earlier for the nucleation mode particles: in that case, the highest concentrations were found in warm, oceanic air masses and the colder arctic air masses were frequently depleted in nucleation mode particles.

5.4.3 Optical effects attributable to the accumulation mode

Evidently the arctic air mass is charged with accumulation mode particles in late winter and spring months. The accumulation mode particles are so concentrated that it is perfectly feasible to tell when air flows down from the Arctic simply on the basis of the washed-out coloration of the sky, sometimes coupled with a distinct lowering of visibility of the mountains lying south of our station. The letter H in Fig. 5.11(b) designates days when haze was apparent to the naked eye; it is interesting to note that these correspond rather well to peaks in the accumulation mode aerosol as sensed with the optical particle counter (OPC). Incidentally, the OPC observations were analyzed several months after the haze records were recorded in a notebook, so there was no bias.

An interesting coincidence occurs between the optical interaction, on the one hand, and the residence time, on the other, of particles lying in the accumulation mode. It will be recalled that these particles are comparable in size to the wavelength of visible light; a resonance occurs. I sometimes call it the 'chaff effect'. Chaff, or cut-up pieces of aluminum foil, works by resonating with the radar's operating wavelength and providing a much larger than normal reflectance cross-section. It has obvious tactical usage in warfare. Similarly, the accumulation mode particles, being, as we have seen, around a third of a micron in diameter, resonate rather efficiently with optical band radiation in the central visible spectrum where the majority of the Sun's electromagnetic radiation lies. But, for entirely fortuitous reasons, this size range is nearly coincident with that where particles remain for longest times, travel for greatest distances, and build up in remote regions of the planet's atmosphere. This is illustrated in Fig. 5.12.

The chaff effect would seem to have profound implications which, surprisingly, have not been much discussed in the literature so far. The idea is that a minority of aerosols entering the atmosphere, namely those of around a third or a half micron in diameter, have extraordinary characteristics; they interact strongly with solar radiation, resulting in visible haze, and have possible climatic implications. The effect is quasi-global in the case of arctic air pollution, which has a large fraction of its mass in the accumulation mode. It is surprising that this interesting coincidence has not gone on to become a conceptual point of interest in the atmospheric sciences; it could well have deep implications in the history and evolution of climate.

The build-up of the accumulation mode aerosol across the entire Arctic Basin in late spring causes serious interactions to occur with incoming solar radiation just after polar sunrise. Possible climatic interactions are obvious, and could result in heating of the Earth–atmosphere column due to the fact that even slight optical absorption by aerosols over highly reflecting pack ice significantly lowers albedo. We do not have the

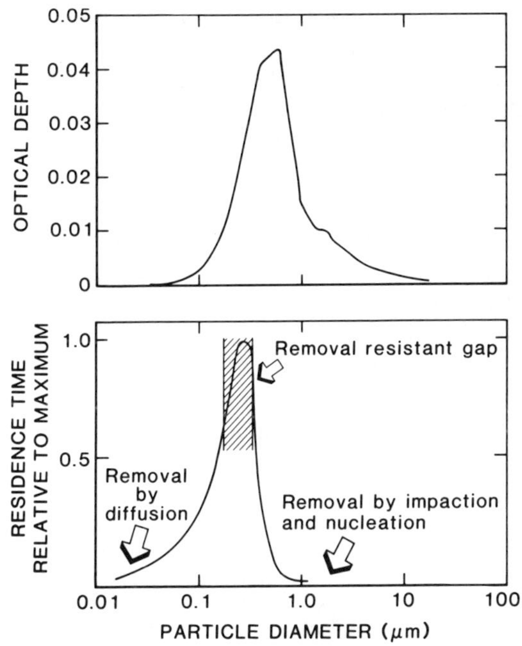

Figure 5.12. Top curve is the computed optical depth for sunlight (ordinate) for a hypothetical monodispersed aerosol of diameter *D* (abscissa). The aerosol column mass was held constant for the calculations. This illustrates that the largest optical effect occurs (for a given mass loading of aerosols) at a diameter of approximately one half of a micron. Optical depth is dimensionless. The bottom curve is a plot of undimensional residence time for particles being removed by first-order rate removal by diffusive and inertial mechanisms. We scaled this curve to peak at about one third of a micron diameter, to coincide with the accumulation particle mode observed for arctic air mass systems. Note that the longest-lived particles also interact strongly with solar radiation.

detailed data record on the microphysical parameters to justify elaborate modeling efforts on this climatic change subject. However, Shaw & Stamnes[4] estimated that a heating forcing rate of 20–30 W m^{-2}, mostly deposited in the lower regions of the troposphere, might be a representative climatic perturbation for the arctic air pollution spread out over the northern polar cap in late spring. Porch & MacCracken[23] and Cess[3] reached similar conclusions, while Blanchet & List[24] extended the model to include the infrared radiative effects, along with the radiative properties of the soot aerosol contained in the snow, as modeled by Wiscombe & Warren.[25] This modeling exercise indicated that the Earth–atmosphere system would heat up, but by about a factor of two lower than had been predicted in the earlier models.

The seasonal variation of optical depth of the arctic air pollution (see Fig. 5.13) has been deduced from climatological radiation measurements at Resolute (75°N, 95°W) for a ten-year data run[26], and is very similar to that deduced by the author for the Alaskan Arctic.[27] The optical depths are large enough to cause scattering effects easily detectable by eye, for instance bright aureoles of scattered light around the sun, haze bands, etc. Impressive reductions in direct solar radiation rays for the low sun can be achieved and the optical depths are easily large enough to provide a certain unnatural redness to the setting sun in the arctic regions. The writer has observed significant differences in cloud-free sunsets and sunrises in autumn and spring, for instance.

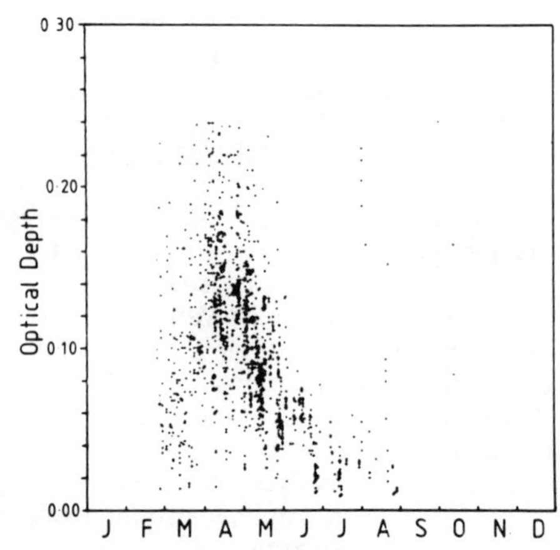

Figure 5.13. Scatter diagram of aerosol optical depths derived by delta-Eddington direct method (individual measurements, 1970–1979) (from Reference 26). Optical depth is dimensionless.

5.4.4 The giant particle mode

Particles larger than those in the accumulation mode are removed relatively rapidly by a plethora of atmospheric processes and hence, if detected in polar air masses, must have come in through very rapid transport or be due to local sources of particle generation. Wind erosion is one obvious source of the giant particles, and indeed our preliminary analysis of the giant particle mode measured in Alaska indicates that high correlations are found between the strength of the mode and the wind speed.

Giant particles, a micron or so in diameter, can, under certain conditions, travel across oceans;[28] such transoceanic transport episodes lasting for several weeks in late spring were reported as early as 1979.[29] Even though arctic haze is primarily pollution, dust from the great deserts in Asia does occasionally reach the Alaskan Arctic;[30,31] it does so, however, only rarely.

Before ending this section, it may be well to remark a little further on desert dust carried by winds out over the world's oceans on global scales. This may seem to be at variance with the arguments built up earlier, which suggested that only particles around the size of the accumulation mode have much probability of reaching global scales of distances. In the case of blowing desert dust, one has a special situation where deep convection and strong storminess can lead to massive injections of large particles to altitudes of 500 mb (roughly 5 km) and greater. At these altitudes, transport can be rapid and relatively unimpeded by cloud, one reason being the scarcity of water vapor over arid cold regions like the Gobi Desert.

In summary, we can say the following about the giant particle (supermicron) mode in arctic air masses:

1. Its presence in the Arctic requires rapid transport and vigorous meteorological conditions, such as deep convection, at or near the source.
2. The supermicron mode is usually transitory, at least in the higher latitudes, and requires rather rare meteorological synoptic patterns.
3. Though contributing minor quantities of particles by numbers, for example being probably trivial for cloud physics processes, they can sometimes be significant by mass, and can dominate the chemistry of refractory elements. It would in general be wise practice to exclude these particles or measure them separately when carrying out chemical tracing experiments.

5.4.5 The cross-mode idea

We have referred earlier to the way in which the strengths of the accumulation and nucleation modes varied with temperature or, more accurately, with the deviation in temperature of the air mass from the seasonal normal. The accumulation mode nearly disappears, but in warm, sunlit, maritime air masses, even though one finds rather large numbers (10^3–10^4 cm^{-3}) of small particles. Arctic air has a robust accumulation mode, but relatively few particles. The variation with temperature deviation (equivalent to air mass origin in this instance) is therefore opposite for the accumulation and nucleation modes.[32] The lack of awareness of this cross-mode effect has resulted in confusion: for instance, instruments sensing total aerosol number concentration, like CN counters, indicate that arctic air is very clean. However, instruments or methods that tend to be more directly responsive to aerosol mass, such as chemical mass determinations or instruments like nephelometers,[33] would detect an opposite kind of behavior; for instance according to these types of measurements arctic air would seem relatively dirty.

As a final example of what we call the cross-mode paradigm (the opposite temperature, or air mass, behavior of the accumulation and nucleation modes), we show in Fig. 5.14 the aerosol mass distribution in

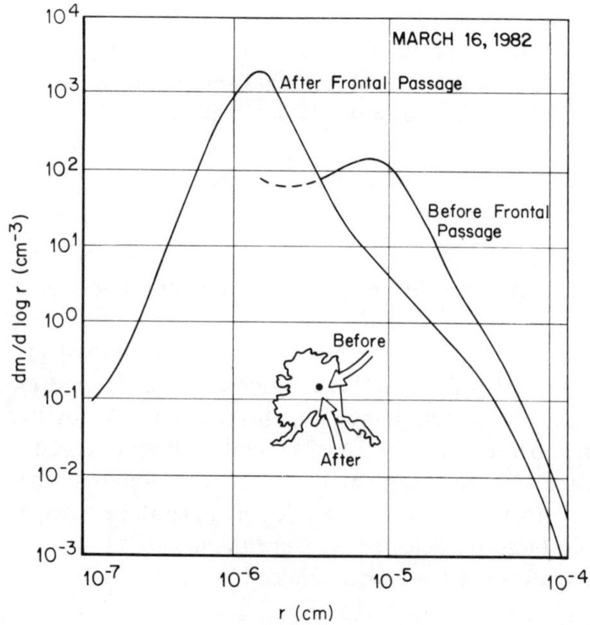

Figure 5.14. Graph of the size distribution of atmospheric aerosols before and after the passage of a meteorological front that separated a cold polar air mass from a warm northern Pacific air mass. This shows an example of what we find to be a general principle: arctic air masses are enriched in relatively large particles, while Pacific air masses contain an abundance of smaller particles, about a hundredth of a micron in diameter.

air before and following a frontal passage through the region of central Alaska, during which cold, arctic-derived air was replaced by warm Pacific marine air. Note that the accumulation mode at around 10^{-4} cm disappeared following the frontal passage, but that a strong increase in nucleation mode particles ensued; the latter are presumably naturally occurring particles nucleated from sunlit gases, probably organic sulphides, emitted from the northern Pacific Ocean. This figure is a rather comprehensive statement of the aerosol mass distribution in the two highly disparate Pacific marine and arctic air mass systems.

5.5 DISCUSSION OF THE AEROSOL RECORD AT BARROW, ALASKA

Figure 5.15 shows the seasonal variation of CN (frequently dominated by nucleation mode) and optical scattering coefficient, which is mainly sensitive to accumulation mode particles. These were measured at the NOAA clean air station at Barrow, on the northern tip of Alaska several hundred kilometers north of the Arctic Circle.[34] This station is immersed all winter in the arctic air mass. Our previous contention regarding the origin and behavior of the accumulation modes and nucleation modes, which were deduced from our studies in central Alaska where Pacific and

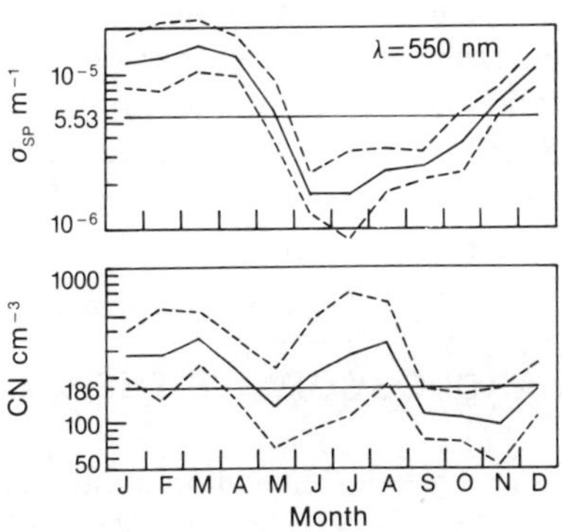

Figure 5.15. Geometric means by month of aerosol parameters measured at Barrow, Alaska, 72°N latitude. The top curve is nephelometer scattering coefficient (e.g. Reference 33), which is roughly proportional to the number concentration of accumulation mode particles. The lower curve is concentration of condensation nuclei (in cm^{-3}). Note log scale. Dashed lines represent plus and minus one standard deviation. (Courtesy of B. Bodhaine.)

arctic air masses are interchanging all winter, is borne out to a certain extent by these curves. Note the relatively high nephelometer readings (accumulation mode) in winter, and the deep minimum in summer; this is consistent with the idea that the polar air mass fills up with accumulation mode in late winter–early spring. The condensation nuclei at Barrow have a complex seasonal variation, undergoing weak maxima in late winter *and* in late summer! One, so far unproven, hypothesis, is that the late summer particles might be due to the presence of organic products[35,36] emitted by the late-blooming tundra, such as photo-oxidized terpene molecules. Another possibility is that the peak might consist of non-marine converted sulfates due to biogenic dimethylsulfide in the Arctic Ocean, which is at its warmest and most ice free in late summer.

The peaking of CN in late winter, coincident in time with the peaking of accumulation mode aerosol, is probably a manifestation of our previously alleged situation that the nucleation mode disappears during the polar night. In such a case, if extreme enough, the CN would become a measure of the accumulation mode number concentration and may peak when the accumulation mode reaches its maximum mass concentration.

To test the plausibility of the hypothesis stating that midwinter CN counts are in great part due to accumulation mode particles, we calculated the correlation coefficient between the time record of CN (the condensation nuclei, which represent total particle concentration integrated over all particle sizes) and the strength of the accumulation mode particles. The correlation coefficient of the variations was indeed highest in late spring ($r = 0{\cdot}8$) at the time of the maximum in accumulation mode aerosols, suggesting that the spring peak in CN is due simply to an amplified accumulation mode aerosol.

5.6 ELECTRON MICROSCOPY OF THE ARCTIC AEROSOL

One is struck by the preponderance of sulfur compounds when examining the microscopic view of arctic-derived aerosols. In Fig. 5.16 is shown a representative collection of accumulation mode particles sampled from an arctic air mass. The X-ray energy spectra, excited by bombarding electrons, for two particles taken at random are evidently quite different. This illustrates that one cannot determine very much about the composition of the accumulation mode aerosol by particle morphology. In

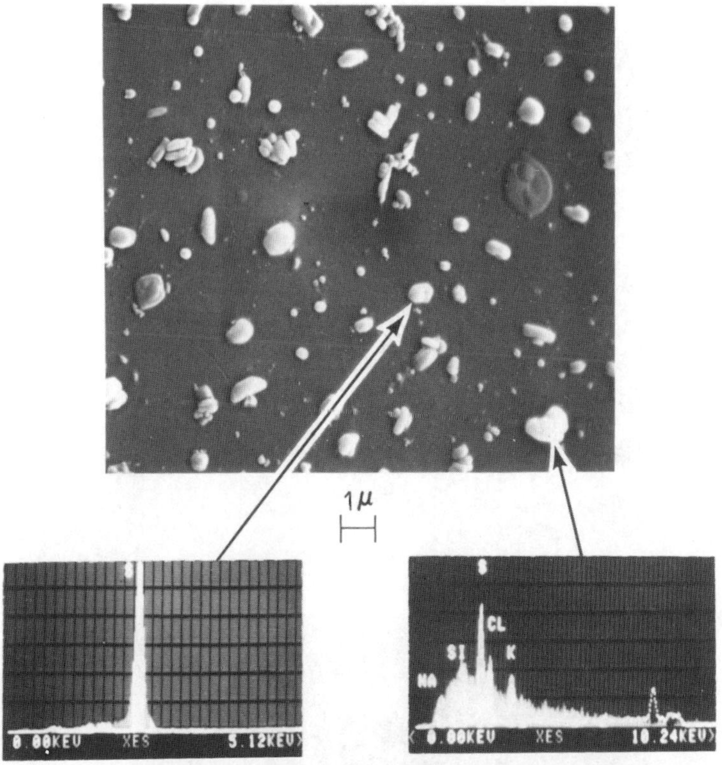

Figure 5.16. Electron microscope photograph showing particles from an arctic air mass. X-ray spectra are shown for two of the particles.

general the particles lack any very obvious crystalline habit and appear to be rather similar, though they can vary enormously in terms of composition. Figure 5.16 also illustrates that the particles are frequently agglomerates (internally mixed); careful probing with the electron beam shows that sometimes the agglomerates in contact are entities of differing chemical compositions. Notice that sulfur appears in both the spectra; this is very typical. Evidence of internal mixing among the submicrometer aerosol has also been found by Covert & Heintzenberg[37] and by Heintzenberg.[38]

Figure 5.17 is a map of particles impacted on collodion-coated, carbon-strengthened grids, showing accumulation mode particles, some of which demonstrate a liquid-like appearance. The lower map is the same field,

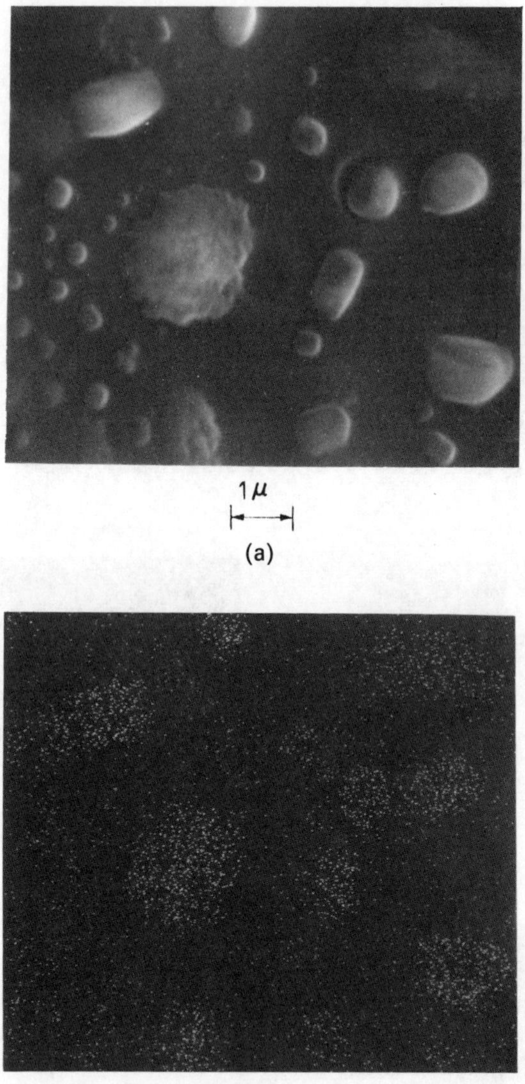

1 μ

(a)

(b)

Figure 5.17. The top photograph is an electron microscope photograph of particles sampled from an arctic air mass. The bottom photograph shows the same field seen in light of X-ray photons at an energy emitted by sulfur atoms. This indicates that all the particles are coated with a film of sulfur-containing material.

but illuminated by those photons possessing the energy emitted by sulfur atoms. It is immediately obvious that all the particles contain sulfur, with hardly any exceptions, even though individual probing with the scanning electron beam indicates that the particles have variable chemical compositions. We have found this to occur on virtually all the aerosol samples taken from arctic-derived air mass systems in Alaska, and conclude that the sulfur signature is most likely to arise from a thin and ubiquitous liquid coating on the particles.

Even though it is sometimes very hard to differentiate the particles by their morphology, as discussed above, and this is especially true when the particles are accumulation mode particles in polluted arctic air mass systems, there are occasions when the aerosol particles show quite striking microscopic morphology. The giant mode particles in Fig. 5.18(a) were collected during an intrusion of desert dust from the Gobi Desert,[30] while those in Fig. 5.18(b), (c) were collected during suspected periods of contamination from nearby pollution sources. Vapors have condensed on the particles during their transport to the Arctic.[39] Bigg[40] shows interesting comparisons of the particle morphology for samples taken from the four NOAA baseline stations, including one at Barrow and one at the South Pole.

5.7 AEROSOL–CLOUD INTERACTIONS

Mention was made of the fact that small aerosol particles are more mobile; that is, they have larger diffusion coefficients than larger particles. The high diffusivity leads to rapid removal of small, nucleation mode aerosols. If we consider particles with diffusion coefficient D, diffusing and sticking to a collection of large cloud water droplets of radius R, and concentration N_c (cm^{-3}), one can solve the diffusion equation to evaluate the rate of particle rcmoval (Maxwellian diffusion). It is

$$dn/dt = -4\pi R D n N_c \qquad (5.3)$$

where n is the time-varying concentration of nucleation mode particles (assumed for the moment to be all the same radius r, i.e. to all possess the same diffusivity). For the case of aerosol particles of radius r, D is proportional to r^{-2} (provided that $r \ll \lambda$, where λ is the mean free path length of the carrier gas molecules; $\lambda \simeq 0{\cdot}067$ μm for air at STP).

Because of the r^{-2} dependency in the diffusion coefficient, the smallest aerosols in a collection will be removed most rapidly: the particle size

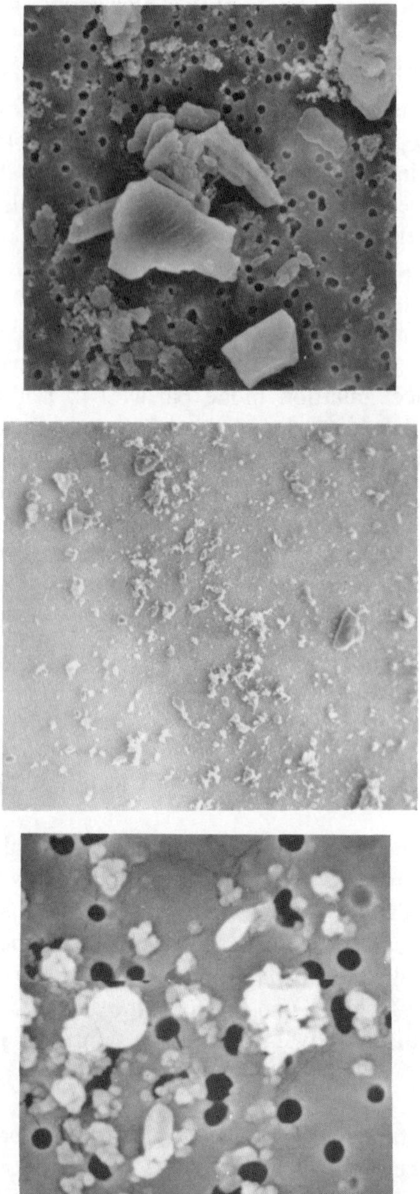
(a)
(b)
(c)

Figure 5.18. Electron microscope photographs showing: (a) supermicron particles of dust which were carried into Alaska from the Gobi Desert region; (b) particles of flyash from nearby contamination; (c) particles consisting of clumped spheres during an episode of arctic air pollution.

distribution will therefore suffer a constant erosion; the nucleation mode will eventually be eaten away. The author has constructed a model of this erosion process,[41] which is imagined to operate on the smallest size classification of aerosol particles when the air in which they are suspended reaches supersaturation and cloud forms. It is, furthermore, supposed that the air mass is traveling in darkness so that one can ignore the creation of new, small, nucleation mode particles by possible photo-oxidation processes; the model is therefore suitable for air masses traveling in the polar night region. Figure 5.19 shows the temporal behavior of the aerosol size distribution according to the modeled calculations: the distribution is drawn for time spent in the presence of cloudy air of 0·2, 0·5 and 4 days. In making the calculations for Fig. 5.19, we employed cloud microphysical parameters similar to those reported by Jayaweera & Ohtake.[42]

Pollution injected into the atmosphere at mid-latitudes, flowing northward into the arctic regions, will probably find itself in the presence of clouds simply due to the radiative cooling experienced by the air mass as it ages in the polar darkness. We see from Fig. 5.19 that coexistence of cloud and a polluted air mass for some tens of hours is sufficient to completely remove particles of a few hundredths of a micron in diameter, which is where the nucleation mode is found in non-arctic air mass systems. It seems likely that this diffusive-driven erosion of the nucleation

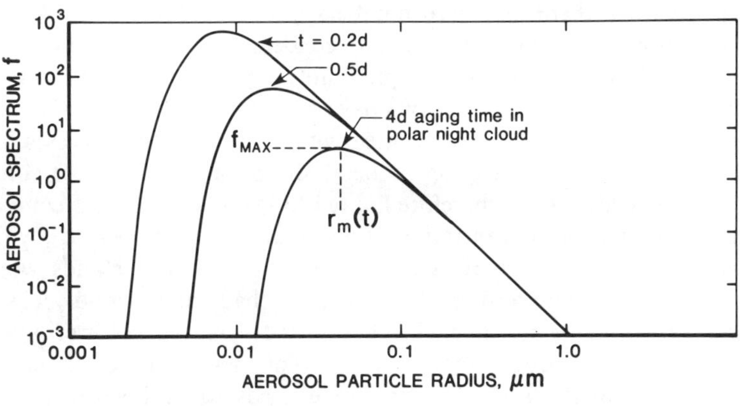

Figure 5.19. Plot of the idealized aerosol size distribution function, which would correspond to aging time, t_c, in arctic stratus cloud of approximately 14 h, 2·3 days and 15 days (from Reference 45).

mode particles by cloud is responsible for removing the nucleation mode from the arctic air mass systems.

One interesting ramification from the above argument pertains to the formation mechanism of clouds in arctic air masses. During the cloud activation process, the cloudy atmosphere passes through a maximum supersaturation.[43] For gentle updraft velocities of a few to a few tens of $cm\,s^{-1}$, typical for stratiform cloud found in the mid to high latitudes in winter, the maximum atmospheric supersaturation is somewhat less than a few tenths of one percent. Such low supersaturations are capable of activating only the largest and most soluble particles in the ensemble of atmospheric aerosols. For example the critical radius, r_c, for a dry soluble particle in a water vapor field of supersaturation S_c, is given approximately by the relationship

$$r_c = 1{\cdot}4 \times 10^{-6} S_c^{-2/3} \tag{5.4}$$

where the calculation was made for $(NH_4)_2SO_4$ (a common constituent of pollution aerosol), but does not change very significantly for other common salts.[22] For $S_c = 0{\cdot}5\%$, the critical radius is $0{\cdot}02\ \mu m$. We see, therefore, that the critical radius can quite easily get down to particle sizes small enough to have been eroded away by previous exposure to cloudy air; for instance, in the above example of one half percent maximum supersaturation, the critical radius of $0{\cdot}02\ \mu m$ corresponds to cloud erosion times of about two hours (see Fig. 5.19).

In a normal atmosphere, even in very clean locations such as above the central oceans, there are large numbers (i.e. hundreds cm^{-3}) of quite small particles for cloud droplets to condense on. New particles are evidently being replenished through a not very well understood gas to particle conversion process. As we have seen, transport times of tens of days in the polar night are associated with cloud erosion processes that effectively cause the nucleation mode to disappear. Under these conditions, cloud droplets are therefore limited to low concentration, comparable to the number concentration of accumulation mode aerosol, which sometimes is well below $100\ cm^{-3}$. Such coarse-grained clouds would precipitate easier by non-Bergeron processes than would clouds consisting of larger numbers of finer droplets,[44] and would have lower albedo than their higher-droplet counterparts. This sensitivity of the cloud microphysics to the accumulation mode needs to be investigated, especially since the accumulation mode in arctic air mass systems tends to be mainly pollution-derived.

The alteration of the microphysics of the polar clouds by arctic air pollution would perturb the infrared radiation budget of the winter polar region, since the clouds interact strongly with terrestrial-band radiation through Mie scattering.

SUMMARY

The arctic aerosol consists to a large extent of pollution debris, being especially rich in sulfate, which was injected as gases or particles from industrial sources several thousand kilometers upwind. The material lingers in the atmosphere probably for tens of days and loses nucleation mode particles through a combination of non-precipitating cloud diffusive erosion processes and lack of new particle production in the polar darkness.

On the other hand, the arctic air mass contains a strong so-called accumulation mode of particles, which is chemically complex, optically active, and which is interpreted to be the removal-resistant strain of widely heterodispersed aerosols, the larger and smaller ones having been removed by inertial or diffusive-controlled processes. In addition, the accumulation mode particles apparently serve as surfaces for condensation of vapors; there is evidence that the surfaces are covered with liquid films, perhaps with sulfuric acid, as all the accumulation mode particles exhibit a sulfur signature.

The noteworthy aspects of the aerosols in arctic air masses are due to long periods spent in the polar darkness. The persistent accumulation mode aerosol, around a half micron in diameter, interacts strongly with sunlight because of an optical Mie scattering resonance in the spring months during polar sunrise. This gives rise to the interesting consequence of a considerable influence on the radiation budget by a fairly small mass of aerosol material. In other words, the small mass of the arctic pollution aerosol belies its importance as a perturber of radiation. The climatic consequences, in springtime, consist of a radiative heating of the Earth–atmosphere column of approximately 10 W m^{-2}, or roughly comparable to a doubling of carbon dioxide. The area affected is roughly that of the continent of Africa and the radiative perturbation occurs only around the time of vernal equinox.

ACKNOWLEDGEMENTS

The author wishes to acknowledge insights that have come from hours of discussion with Kenneth Rahn at the University of Rhode Island. Ken has for years been the driving force behind the issue of arctic air pollution, even when it was an unpopular subject.

Some of the work reported here was supported by the National Science Foundation under grants DPP86-20395 and ATM86-11438, and by the Office of Naval Research under contract N00014-86-K-0055.

REFERENCES

1. Barrie, L. A., Arctic air chemistry: An overview. In *Arctic Air Pollution*, ed. B. Stonehouse. Cambridge University Press, Cambridge, UK, 1986, pp. 5–24.
2. Barry, R. G., Seasonal location of the arctic front over North America. *Geograph. Bull.*, **9** (1967) 79–95.
3. Cess, R. D., Arctic aerosols: Model estimates of interactive influences upon the surface–troposphere radiation budget. *Atmos. Environ.*, **17** (1983) 2555–64.
4. Shaw, G. E. & Stamnes, K., Arctic perturbation of the solar radiation budget. *Annals N.Y. Acad. Science*, **338** ed T. J. Kneip, 1980, pp. 533–9.
5. Barrie, L A. & Hoff, R. M., The oxidation rate and residence time of sulfur dioxide in the Arctic atmosphere. *Atmos. Environ.*, **18** (1984) 2711–22.
6. Barrie, L. A., Hoff, R. M. & Daggupaty, A. M., The influence of midlatitudinal pollution sources on haze in the Canadian Arctic. *Atmos. Environ.*, **15** (1981) 1407–19.
7. Heidam, N. Z., On the origin of the arctic aerosol: A statistical approach. *Atmos. Environ.*, **15** (1981) 1421–7.
8. Ottar, B., The transfer of airborne pollutants to the arctic region. *Atmos. Environ.*, **15** (1981) 1439–45.
9. Pacyna, J. M., Ottar, B., Tomza, U. & Maenhauat, W., Long range transport of trace elements to Ny Ålesund, Spitsbergen. *Atmos. Environ.*, **19** (1985) 857–65.
10. Rahn, K. A., Relative importance of North America and Eurasia as sources of arctic aerosol. *Atmos. Environ.*, **15** (1981) 1447–55.
11. Rahn, K. A., The Mn/V ratio as a tracer of large-scale sources of pollution aerosol for the Arctic. *Atmos. Environ.*, **15** (1981) 1457–64.
12. Rahn, K. A. & Lowenthal, D. H., Elemental tracers of distant regional pollution aerosols. *Science*, **223** (1984) 132–9.
13. Li, S. M. & Winchester, J. W., Size distribution and chemistry of late winter arctic aerosols. Submitted to *J. Geophys. Res.*
14. Raatz, W. & Shaw, G. E., Long-range tropospheric transport of pollution aerosols into the Alaskan Arctic. *J. Clim. Appl. Met.*, **23** (1984) 1052–64.
15. Pacyna, J. M., Semb, A. & Haussen, J. E., Emission and long-range transport of trace elements in Europe. *Tellus*, **36B** (1984) 163–78.
16. Barrie, L A. & Vet, R. J., The concentration and deposition of acidity, major ions and trace metals in the snowpack of eastern Canada during 1980–1981. *Atmos. Environ.*, **18** (1984) 1459–69.

17. Junge, C. E. In *Air Chemistry and Radioactivity*, Chapter 2. Academic Press, New York, 1963, pp. 382.
18. Whitby, K., The physical characteristics of sulfur aerosols. *Atmos. Environ.*, **12** (1978) 135–59.
19. Clarke, A. D., Ahlquist, N. C. & Covert, D. S., The Pacific Marine aerosol: Evidence for natural acid sulfates. *J. Geophys. Res.*, **92** (1987) 4179–90.
20. Aitken, J., On the number of dust particles in the atmosphere. (Originally published in *Tran. Roy. Soc. Edinburgh*, **35** (1888) 187–206). Reprinted in *Collected Scientific Papers of John Aitken*, ed. Cargill G. Knott. Cambridge University Press, Cambridge, UK, 1923.
21. Chuan, R. L., A new method for measuring particulate mass. *J. Aerosol Sci.*, **7** (1970) 111–14.
22. Twomey, S. In *Atmospheric Aerosols*, Chapter 3. Elsevier Scientific Publishing Company, Amsterdam, 1977, pp. 301.
23. Porch, W. M. & MacCracken, M. C., Parametric study on the effects of arctic soot on solar radiation. *Atmos. Environ.*, **16** (1982) 1365–71.
24. Blanchet, J. & List, R., On radiative effects of anthropogenic aerosol components in arctic haze and snow. *Tellus*, **39B** (1987) 293–317.
25. Wiscombe, W. J. & Warren, S. G., A model for the spectral albedo of snow. I: Pure snow. *J. Atmos. Sci.*, **37** (1980) 2712–33.
26. McGuffie, K., Cogley, J. G. & Henderson-Sellers, A., Climatological analysis of arctic aerosol; quantity and optical properties at Resolute, N.W.T. *Atmos. Environ.*, **19** (1985) 707–14.
27. Shaw, G. E., Atmospheric turbidity in the polar regions. *J. Appl. Meteor.*, **21** (1982) 1080–8.
28. Prospero, J. M., Eolian transport to the world ocean. In *The Sea*, Chapter 7, The Ocean Lithosphere, ed. C. Emiliani, Wiley Interscience, New York, 1981, pp. 801–74.
29. Shaw, G. E., Transport of Asian desert aerosol to the Hawaiian Islands. *J. Appl. Meteor.*, **19** (1980) 1254–9.
30. Rahn, K. A., Borys, R. D., Shaw, G. E. & Jaenicke, R., Long-range impact of desert aerosol on atmospheric chemistry: Two examples. In *Saharan Dust: Mobilization Transport, Deposition*, ed. C. Morales. John Wiley and Sons Ltd., Chichester, 1979, pp. 243–66.
31. Rahn, K. A., Borys, R. D. & Shaw, G. E., Asian dust over Alaska: Anatomy of an arctic haze episode. *Geol. Soc. of America Special Paper*, 186 Desert Dust: Origin, Characteristics and Effect on Man, ed. T. Péwé. The Geological Society of America, Boulder, CO, 1981, pp. 37–70.
32. Shaw, G., Aerosol size–temperature relationship. *Geophys. Res. Lett.*, **15** (1988) 133–5.
33. Charlson, R. J., Ahlquist, N. C., Selvidge, H. & MacCready, P. B., Monitoring of atmospheric aerosol parameters with the integrating nephelometer. *J. Air Pollut. Contr. Assoc.*, **19** (1969) 937–42.
34. Bodhaine, B., Aerosol measurements at four background sites. *J. Geophys. Res.*, **88** (1983) 10753–68.
35. Altshuller, A. P., Review: Natural volatile organic substances and their effect on air quality in the United States. *Atmos. Environ.*, **17** (1983) 2131–65.

36. Went, F. W., On the nature of Aitken condensation nuclei. *Tellus*, **18** (1966) 549–56.
37. Covert, D. S. & Heintzenberg, J., Measurement of the degree of internal/external mixing of hygroscopic compounds and soot in atmospheric aerosols. *The Science of the Total Environment*, **16** (1984) 347–52.
38. Heintzenberg, J., Chemically resolved submicrometric size distributions and external mixing of the arctic haze aerosols. *Tellus*, **39** (1987) 374–82.
39. Shaw, G. E., Production of condensation nuclei in clean air by nucleation of H_2SO_4. *Atmos. Environ.*, **23** (1989) 2841–6.
40. Bigg, E. K., Comparison of aerosol at four baseline atmospheric monitoring stations. *J. Appl. Meteor.*, **19** (1980) 521–33.
41. Shaw, G. E., On the physics of Aitken particles in polar air mass systems. *J. Aerosol Sci.*, **17** (1986) 73–8.
42. Jayaweera K. & Ohtake, T., Concentration of ice crystals in arctic stratus clouds. *J. Rech. Atmos.*, **7** (1973) 199–207.
43. Fletcher, N. H. In *The Physics of Rainclouds*, Chapter 4. Cambridge University Press, Cambridge, UK, 1966, pp. 389.
44. Twomey, S., Cloud nucleation in the atmosphere and the influence of nucleus concentration levels in atmospheric physics. *J. Phys. Chem.*, **84** (1980) 1459–63.
45. Ram, M., Gayley, R. I. & Shaw, G. E., Empirical functions for atmospheric size distributions. *J. Aerosol Sci.*, **20** (1989) 351–3.

Chapter 6

Sulphur and Nitrogen Pollution in the Arctic Atmosphere

LEONARD A. BARRIE and JAN W. BOTTENHEIM

Atmospheric Environment Service, 4905 Dufferin St., Downsview Ontario, Canada M3H 5T4

6.1 INTRODUCTION

Since the beginning of the 1970s, when there was a growing awareness that air pollution was not confined to small areas around urban or industrial sources but rather is transported long distances before being removed to the Earth's surface, studies have shown that human activities at mid-latitudes are changing the arctic atmosphere.[1-4] In particular, it was found that in the North American Arctic, reports of 'ice crystal haze' by Canadian flight lieutenant K. R. Greenaway observed in the late 1940s[5] and of 'arctic haze' by M. Mitchell in the early 1950's[6] were not just an indication of ice crystals or of wind blown dust but rather of air pollution originating from outside the northern polar region.

Sulphur and nitrogen compounds are major by-products of fossil fuel combustion and the smelting industry. The most common means of disposal is venting to the atmosphere. As acids they have been the cause of 'acid rain' and 'oxidant' environmental problems affecting lakes and forests in Europe, eastern North America and Asia on a regional scale of hundreds to thousands of kilometres. Now we know that the range of such pollution extends to tens of thousands of kilometres from sources well into the Arctic, where it is of concern not so much for its acidifying influence, which is considerably weaker than further south,[3] but for its impact on climate,[7] for the introduction of potentially toxic substances to the north and for its interference in atmospheric chemical reactions. In this chapter, the current state of knowledge of the nature, occurrence (past and present), origin and fate of sulphur and nitrogen pollution in the arctic atmosphere is reviewed.

6.2 ATMOSPHERIC SULPHUR AND NITROGEN COMPOUNDS IN THE ARCTIC

Sulphur and nitrogen compounds can exist in the atmosphere as solid or liquid particulate matter or as gases. Human activities contribute significantly to the atmospheric budget of S and N. Sulphur is mainly released in the form of gaseous sulphur dioxide (SO_2); a small fraction is released in the form of particulate sulphates (1–5%). Nitrogen is also released as a gaseous species. Depending on the type of source this takes the form of NO, NO_2, NH_3 and N_2O. In this review we will focus only on NO and NO_2, the sum of which is denoted as NO_x, and their products from atmospheric oxidation reactions. The sum of all oxides of nitrogen (NO_x + oxidation products, but excluding N_2O) is indicated by the term NO_y.

Once in the atmosphere, these sulphur and nitrogen compounds undergo chemical and physical transformation as well as removal to the Earth's surface by dry and wet deposition processes. Dry deposition is direct uptake by the Earth's surface. Wet deposition is removal to the Earth's surface either by incorporation into precipitation, which falls to the Earth as rain or snow, or by incorporation into fog drops which are impacted on the Earth's surface. The lifetime of a sulphur or nitrogen compound in the atmosphere is governed by these transformation and removal processes. In this section we will summarize the transformation chemistry as it pertains to the arctic atmosphere. For more detailed reviews, we refer the reader to the recent monographs by Finlayson-Pitts & Pitts[8] and Warneck.[9] Removal processes are discussed in detail in Chapter 3 of this volume.

6.2.1 Sulphur compounds

The transformation chemistry of sulphur oxides is considerably simpler than that of the nitrogen oxides. SO_2 is converted irreversibly in the atmosphere to SO_4^{2-}. Thus once SO_4^{2-} is formed, there is no reaction that leads to the reformation of SO_2 from SO_4^{2-}. SO_2 oxidation can occur either in the gas-phase, involving hydroxyl (OH)-radicals, or in cloud or rain droplets largely via reaction of dissolved SO_2 with H_2O_2 or O_3. Both SO_2 and SO_4^{2-} can be removed from the atmosphere via wet and dry deposition processes. Figure 6.1 summarizes the pollution sulphur pathways in the atmosphere and compares it to those of nitrogen oxides (described in the next section).

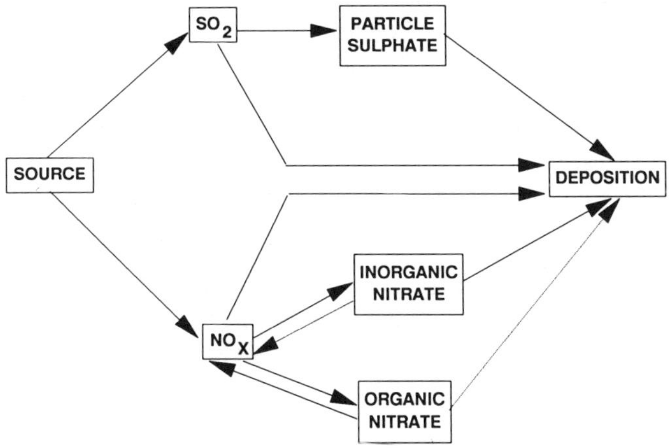

Figure 6.1. Simple schematic of the atmospheric chemical reservoirs of sulphur and nitrogen oxides between source and sink.

As far as the arctic atmosphere is concerned, the prevailing partitioning between SO_2 and SO_4^{2-} will depend largely on the efficiency of the transformation rate from SO_2 into SO_4^{2-}. During the winter this efficiency is low due to lesser amounts of sunlight and hence lower levels of the oxidants (OH, H_2O_2 and O_3). We therefore expect an annual cycle in the SO_2/SO_4^{2-} ratio, with a maximum in the winter and a minimum in the summer, as is actually observed.[25]

6.2.2 Nitrogen compounds

Budgets of the chemistry of nitrogen oxides in the troposphere have been reported in the literature,[10,11] but little attention has been paid to the Arctic. This is not surprising in view of the sparsity of observations in the north until about the mid-1980s. However since that time, considerable progress has been made in our knowledge of nitrogen compounds in the region.

Figure 6.2 summarizes the nitrogen oxide chemistry in the atmosphere. It is organized in three rows. In the centre are nitric oxide (NO), nitrogen dioxide (NO_2) and nitrogen trioxide (NO_3). $NO + NO_2$ are often designated NO_x and are the form in which nitrogen oxides enter the atmosphere, either from anthropogenic or natural processes. All three have a so-called free electron and are therefore paramagnetic. This property makes them quite reactive and therefore less stable than most other nitrogen oxides. The top row compounds are the inorganic nitrates, products of the reactions of the middle row oxides with O_3 and the highly reactive free radical species OH and HO_2 (known collectively as HO_x). On the bottom row are the organic nitrates. These compounds are produced in the atmosphere from reactions of the middle row nitrogen oxides with carbon containing free radicals (indicated in the figure by 'RO', 'RO_2' and 'RCO_3'). The latter are produced as a result of interactions between hydrocarbons in the air (either of anthropogenic or natural origin) and the previously discussed HO_x group. In the arctic troposphere, peroxyacetyl nitrate (PAN) and the alkyl nitrates are the major organic nitrates observed up to this date. Also indicated are the predominant reaction pathways between the different compounds.

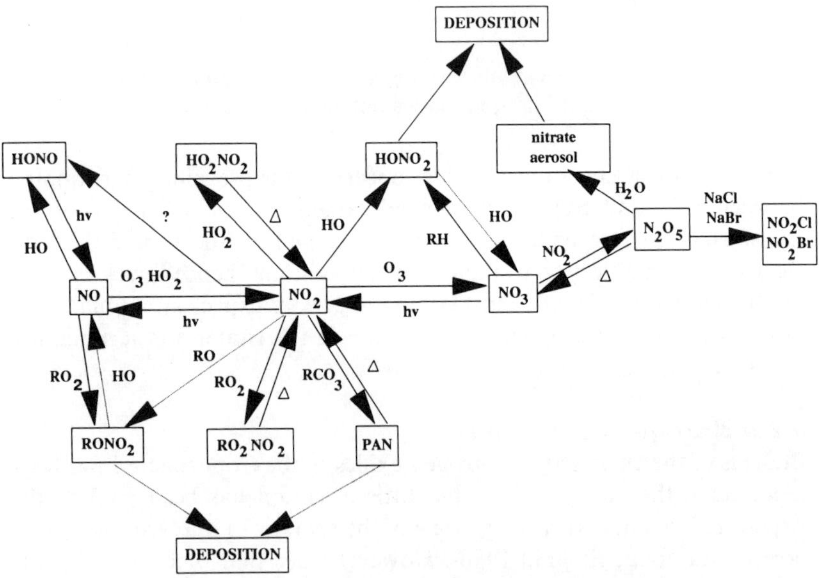

Figure 6.2. Schematic illustrating the complex chemistry of nitrogen oxides in the atmosphere.

A significant aspect of the chemistry of the nitrogen oxides, and one which contrasts with that of sulphur, is that it consists of a set of chemical equilibria. As a result, a reaction to convert a certain compound into another is usually counteracted by a reverse reaction that leads to the reformation of the original compound. The relative rates of reaction in each direction will determine which of the compounds is expected to be more prevailing in the atmosphere. The environmental conditions such as the temperature and the presence of sunlight have a great impact on these rates.

This picture of interactions has to be combined with a consideration of the source and sink terms. As stated earlier most nitrogen oxides enter the troposphere in the form of NO_x. Removal by deposition to the Earth's surface, either wet or dry, is predominantly in the form of the inorganic nitrates. This is because the inorganic nitrates are readily soluble in water while the organic nitrates are not. Once formed, the organic nitrates can function as a temporary reservoir of the nitrogen oxides. In this capacity, they extend the transport scale of NO_y. Figure 6.1 shows this atmospheric cycling of nitrogen oxides in a simplified form.

As long as an air mass does not reside in arctic regions, the daily presence of sunlight energy and moderate temperatures lead to reformation of NO_x from the organic reservoir compounds. This can be followed by inorganic nitrate formation and deposition removal. In contrast, in the winter Arctic where sunlight and temperatures are low, the organic nitrates are much more stable, and decomposition to NO_x is very slow. The nitrogen oxide equilibrium therefore shifts towards these compounds. Thus in the winter Arctic, remote from any major sources, we expect the air mass to contain mainly organic nitrates such as PAN and, to a lesser extent, NO_x(predominantly as NO_2) formed from slow decomposition of the organic nitrates. Such a qualitative picture is confirmed by modelling studies,[12] and by observations (Section 6.3.2).

It should be mentioned here that the arctic conditions described above also favour the production of N_2O_5 via the following reactions:

$$NO_2 + O_3 \rightarrow NO_3 \qquad (6.1)$$

$$NO_3 + NO_2 \leftrightarrow N_2O_5 \qquad (6.2)$$

NO_3 is highly unstable in sunlight, and decomposes back into NO_2. The equilibrium of reaction (6.2) is temperature dependent; low temperatures favour a shift to the right, i.e. the formation of N_2O_5. The importance of this will be discussed in the next section.

6.2.3 Aerosols

The study of arctic air pollution was driven originally by concerns over the 'arctic haze' phenomenon. The haze manifests itself in the form of visibility reduction due to small particles. It is therefore not surprising that attention has focused initially on the chemistry of the particulate (i.e. aerosol) component of arctic pollution. Both nitrogen and sulphur oxides are constituents of the aerosol, although particulate nitrate is now known to be a minor component, both in terms of the overall aerosol composition, and the partitioning between the many nitrogen oxides in the gas and aerosol phase.

A discussion of sulphur and nitrogen compounds in arctic haze is considerably simplified by understanding that chemical and physical processes of aerosols, their sources, interactions with gases, and sinks are highly dependent on particle size.

The aerosol mass is distributed in three modes[13] between 0·005 and 100 μm diameter called nucleation, accumulation and coarse modes (See Chapter 5). Each one is maintained by a dynamic equilibrium of production and destruction processes. The nucleation mode (from 0·005 to 0·05 μm diameter), at the lower end of the size range, originates from the condensation of unstable gases which are formed in the atmosphere (e.g. H_2SO_4 from oxidation of SO_2). Its sink is coagulation to form larger particles and removal by dry and wet deposition.

At mid-range is the accumulation mode (0·05–1 μm) consisting of particles (i) released during fuel combustion and metal processing activities, (ii) produced by coagulation of smaller nucleation mode particles, (iii) formed by liquid or solid phase gas to particle conversion and (iv) emitted from volcanoes. Its main sinks are wet and dry removal to the Earth's surface.

At the upper end is the coarse particle mode maintained by mechanical generation processes such as soil suspension in wind, sea-salt aerosol production by bursting air bubbles at the air–water interface, and by volcanic emissions. The main sinks are sedimentation at the upper end and wet removal at the lower end.

There is considerable overlap between the coarse and accumulation modes. Because large particles in the coarse mode sediment from the atmosphere faster than the small ones, the median of the coarse particle mass–size distribution shifts to smaller sizes as the distance from the source increases. Thus, wind blown dust transported into the Arctic from Asian deserts will have considerably lower mass median diameters than those in areas close to continents.

From the point of view of sulphur and nitrogen chemistry, the nucleation mode may be sustained by condensation products of gas phase oxidation of sulphur and nitrogen gases. The accumulation mode may be affected by gases reacting on existing particles to form particulate mass. Furthermore, nitrate in the form of HNO_3 and NH_3 gases may be driven off accumulation mode particles by acidic H_2SO_4 particles[14] to be recaptured by more alkaline larger particles at the lower end of the coarse particle mode. This accounts for observations that total NO_3^- mass is concentrated more in the coarse mode particles than total SO_4^{2-} mass.[15,16]

Heterogeneous chemical reactions between aerosol particles and gaseous compounds may have an important impact on the chemical composition of the arctic atmosphere. It has recently been suggested that such chemistry may be at least partly responsible for the depletion of tropospheric O_3 that is observed in the spring in the lower arctic boundary layer,[17-19] as discussed elsewhere in this volume (see Chapter 7). In this case, an apparently efficient reaction can occur between sea-salt aerosols or sea-salt crystals on the arctic ice-cap and N_2O_5 gas.[19] The abundance of N_2O_5 should be at a maximum during the winter and early spring in the Arctic (see Section 6.2.2). Its reaction at that time with sea-salt components such as NaCl and NaBr would lead to the production of nitryl-halide compounds (e.g. NO_2Cl and NO_2Br, respectively) that are expected to be highly unstable in the presence of sunlight. At the onset of arctic spring, this may be a source of Br atoms that are believed to play a central role in the O_3 depletion chemistry.

Other potentially important heterogeneous reactions of NO_2 or N_2O_5 with sea-salt aerosols have recently been suggested[18,19] as a source for arctic HONO. Such heterogeneous chemistry had been proposed earlier on the basis of laboratory experiments and might explain actual arctic observations.[37]

6.3 OCCURRENCE OF SULPHUR AND NITROGEN COMPOUNDS IN THE ARCTIC

6.3.1 Sulphur compounds

6.3.1.1 Spatial and temporal distributions

Sulphur compounds have been measured in arctic air at many locations in the northern hemisphere. These locations are shown in Fig. 6.3 in

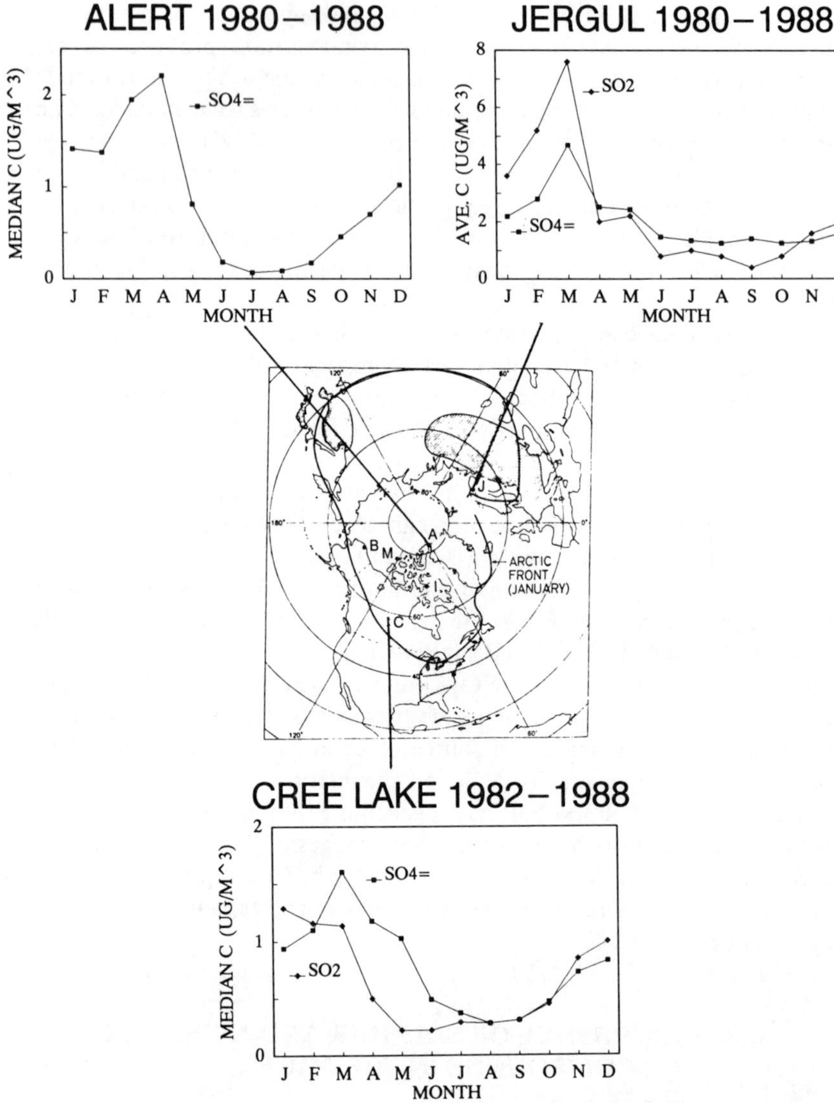

Figure 6.3. The annual variation of the concentration (C) of sulphur oxides in the arctic air mass. The dark line on the central diagram depicts the mean location of the arctic air mass in January. The shaded areas depict major pollution source regions in the northern hemisphere. I—Igloolik, M—Mould Bay and B—Barrow.

relation to the mean position of the arctic air mass in January (the boundary of which is known as the arctic front). This air mass is a unique meteorological feature of the troposphere characterized by (i) sub-zero temperatures much of the year, (ii) little precipitation, most of which falls during the warmer half of the year, (iii) stable stratification that prevents strong vertical mixing, and (iv) low levels of solar radiation especially during winter.[3] In winter, it expands to occupy a large fraction of the North American and Asian continents. In summer, it contracts to an area roughly inside the Arctic Circle.

Arctic haze, as indicated by atmospheric sulphates which form a major fraction of suspended particulate mass, undergoes a strong seasonal variation. This is illustrated by multi-year observations at Alert in the Canadian Arctic and Jergul in the Norwegian Arctic (Fig. 6.3). Haze peaks from January to April and is at a minimum from June to August. Long term observations of SO_2 in air are lacking in the high Canadian Arctic but are available from the Norwegian Arctic where it has a somewhat similar temporal variation to that of SO_4^{2-} (Fig. 6.3). Seasonal variations in arctic sulphur compounds are linked to those in south to north transport of sulphur between mid-latitudes (primarily Europe and Asia, see Section 6.4.1) and to those of precipitation scavenging as indicated by monthly precipitation amounts in the Canadian high Arctic (Fig. 6.4).

That arctic haze is widespread is illustrated by the spatial distribution of the mean January to April concentration of sulphate in air for 1980 (Fig. 6.5). Concentrations generally decrease from Eurasia into the high Arctic. Between the influence of Eurasian and eastern North American sources, there is a minimum (saddle) located in the North American–Greenland Arctic. Air coming out of the Arctic into mid-continental North America has an elevated 'background concentration' of sulphate and sulphur dioxide reflecting the influence of a polluted winter arctic air mass. This is illustrated by the mean seasonal variation of sulphate and sulphur dioxide at Cree Lake, Saskatchewan, Canada (Fig. 6.3). In winter, the arctic air mass is dominant at this location while in summer, northern Pacific air masses prevail.[20] The observations at Cree Lake show that:

(i) background concentrations of the winter arctic air mass are much higher than those of summer North Pacific air;
(ii) median sulphate concentrations at Cree Lake in winter arctic air (0·8–1·7 $\mu g\, m^{-3}$) are slightly lower than those observed in the high Arctic at Alert (1–2·5 $\mu g\, m^{-3}$); and

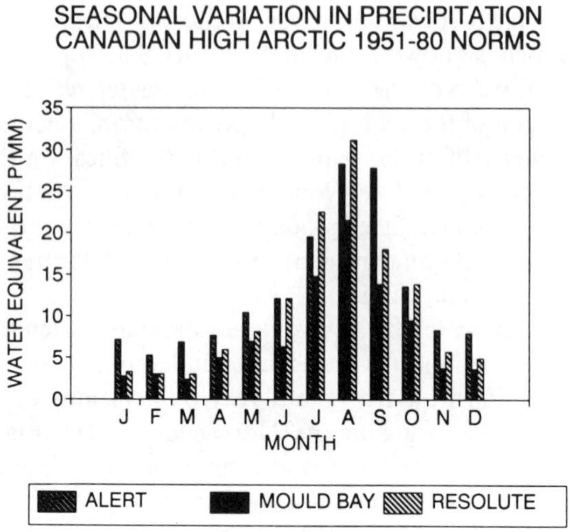

Figure 6.4. Mean annual variation of precipitation amount in the Canadian high Arctic (see Fig. 6.3 for locations). Results are based on climatic observations in Canada between 1951 and 1980.

(iii) sulphur dioxide concentrations in the winter arctic air mass of North America are 0·9–1·3 $\mu g\, m^{-3}$.

The last observation is consistent with those made during March and April of 1986 and 1988 at Alert.[21,22]

6.3.1.2 Sulphate in arctic haze particulate matter

The composition of the soluble constituents of arctic haze as a function of particle size, as measured at Igloolik, Canada, in February 1982[15] (Fig. 6.6), illustrates a relatively constant feature of arctic haze. In sub-micrometre accumulation mode particles, acidic ammonium salts dominate while in the super-micrometre coarse mode particles, salts of sodium, magnesium and chloride originating mostly from sea-salt dominate. Nitrates play a minor role in the total ion balance of soluble particulate matter.

Another study of the soluble composition of haze particles was at Barrow, Alaska, in winter–spring 1986.[23] The composition of soluble particle mass (in contrast to soluble ion equivalents in Fig. 6.6) for coarse super-micrometre particles and fine sub-micrometre particles is shown in Fig. 6.7. Note that at this coastal arctic location non-sea-salt sulphate

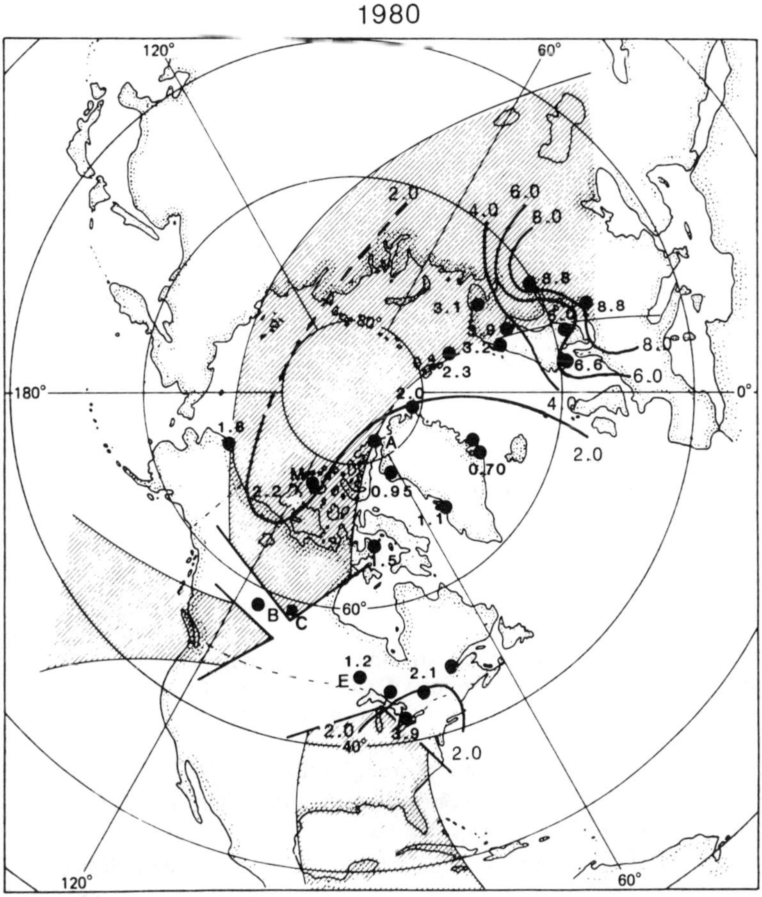

Figure 6.5. A schematic diagram illustrating the three main sources of air for the North American airshed (Arctic, Pacific, Caribbean) superimposed on the spatial distribution of the arithmetic mean SO_4^{2-} air concentrations ($\mu g\ m^{-3}$) for the period January–April when the arctic air mass is most polluted (from Barrie[3]).

and its associated ammonium ion dominates the mass of the fine particles while sea-salt (Na^+, Cl^-) and organic formate (Fo^-) and acetate (Ac^-) dominate the mass of the coarse particle fraction. The H^+ ion, although important in the ion balance, does not contribute significantly to the total mass. The observations at Barrow show that formate and acetate contribute significantly to the total anion equivalents. This implies that

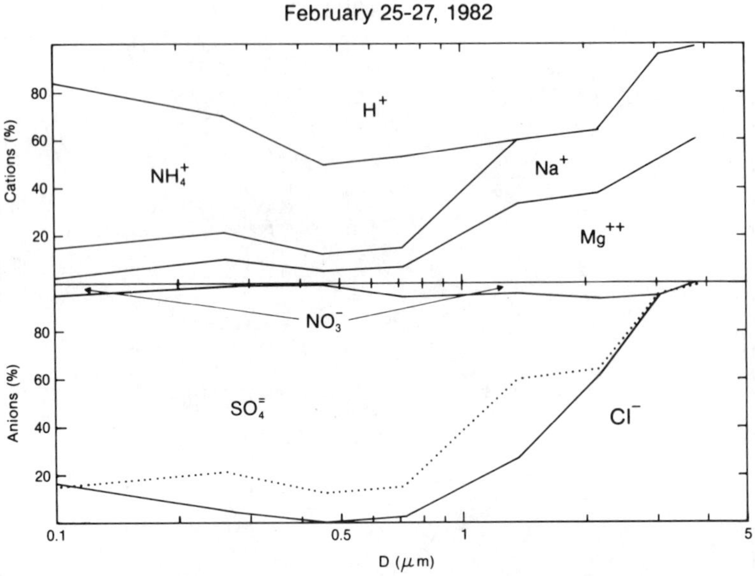

Figure 6.6. Cation and anion balance in suspended particles at Igloolik Canada (see Fig. 6.4) as a function of particle aerodynamic diameter (*D*).[15] Hydrogen ion is inferred from an ion balance that neglects calcium. The dotted line represents the fraction of anion equivalents required to balance sodium and magnesium ions.

the inferred H^+ concentration in the Igloolik results (Fig. 6.7) are actually underestimates of the true concentration since these organic ions were not measured in that analysis.

That sulphates contribute substantially to the accumulation mode particles which are involved in scattering of sunlight and hence in causing atmospheric haziness is illustrated by the correlation between the observed particle scattering parameter b_{scat} and sulphate concentrations at Igloolik and Mould Bay in the Canadian Arctic (Fig. 6.8).[24] From this relationship, it was inferred that the mean and standard deviation of the fraction of fine particle mass contributed by sulphate was $22 \pm 10\%$. Taking into account the greater contribution of sea-salt particles at the coastal location of Barrow, these results are consistent with the mass composition of fine particles in Fig. 6.7.

6.3.1.3 Conversion of sulphur dioxide to sulphates

From observations at Cree Lake and Jergul (Fig. 6.3), it is apparent that although similarities exist in the seasonal variations of SO_4^{2-} and SO_2,

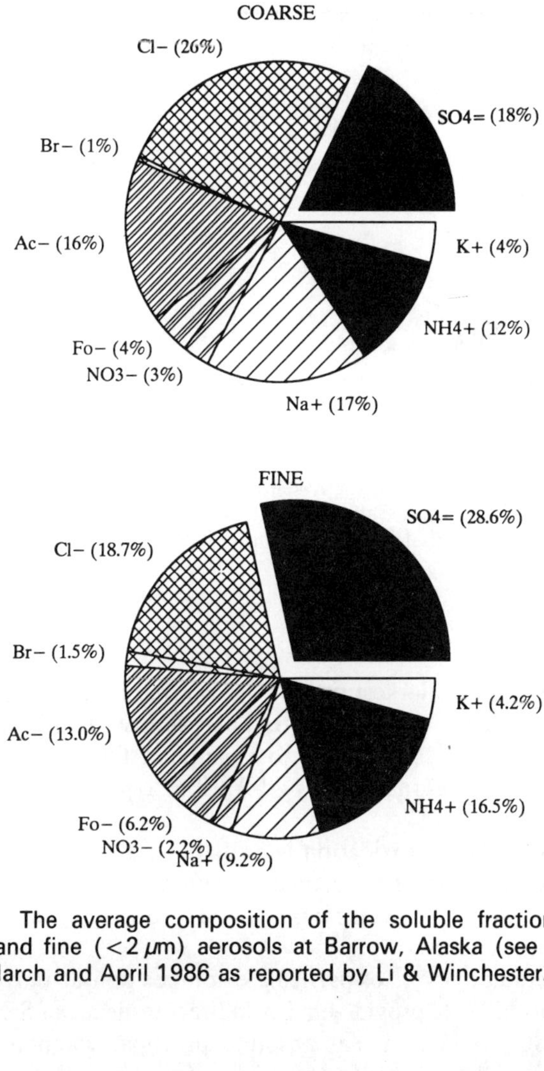

Figure 6.7. The average composition of the soluble fraction of coarse (2–10 μm) and fine (<2 μm) aerosols at Barrow, Alaska (see Fig. 6.3) in March and April 1986 as reported by Li & Winchester.[26]

there are marked differences which suggest that oxidation of SO_2 to SO_4^{2-}, particularly around polar sunrise between February and April, is playing a major role in the atmospheric sulphur chemistry. Using observations of the ratio of SO_4^{2-} to particulate vanadium (V) and of SO_4^{2-} to SO_2 in the Arctic and in mid-latitude source regions as input to a simple chemical transport model, it was deduced[25] that the mean oxidation rate of SO_2 between the sources and the Canadian Arctic was

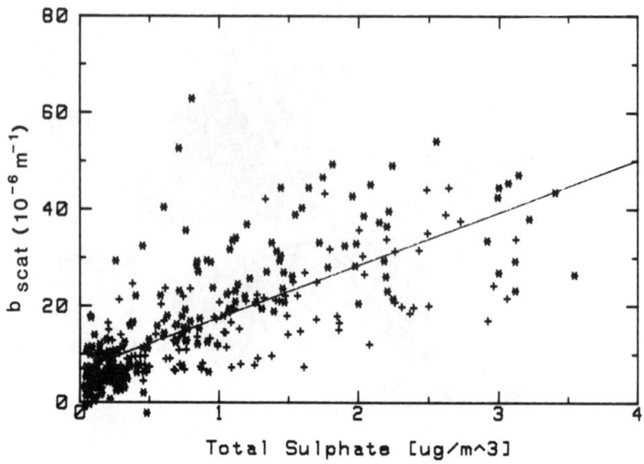

Figure 6.8. The relationship between weekly average aerosol scattering (b_{scat}) and SO_4^{2-} at Mould Bay and Igloolik (M and I in Fig. 6.3) from 1979 to 1983 (from Barrie & Hoff[24]).

0·04–0·1% h^{-1} in the darkest part of winter (December to February) and increased to 0·2% h^{-1} into April after polar sunrise. This is illustrated by the observed increase of airborne total sulphur, that is SO_4^{2-} in air, at Alert between late February and mid-April as typified by results in Fig. 6.9.

Enhanced springtime oxidation is probably a major factor influencing the difference between the seasonal variations observed at Cree Lake (Fig. 6.3) in winter. Concentrations of SO_2 decrease more rapidly in spring than those of SO_4^{2-}.

Detailed studies[23,26,27] of particulate composition at Barrow, Alaska, in winter/spring 1986, provide further indirect evidence of SO_4^{2-} production from SO_2 oxidation. An absolute principal component analysis revealed that 52% of the fine particle SO_4^{2-} was associated with a component caused by SO_2 oxidation in the arctic troposphere. Gas phase oxidation of SO_2 in springtime arctic air is also implied by aircraft observations in 1983[28] that associated nucleation mode particles with SO_2 and hence SO_2 oxidation forming H_2SO_4 aerosols. In more than 80% of the cases in which nucleation mode aerosols were observed, SO_2 concentrations were greater than 2·85 $\mu g\, m^{-3}$ (i.e. 1 ppbv).

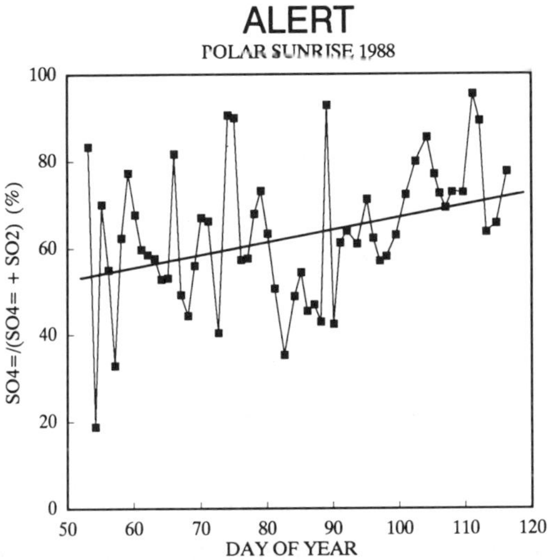

Figure 6.9. Changes in the fraction of total sulphur oxides that is oxidized to SO_4^{2-} at Alert in 1988 going from late February into April. Data from Bottenheim *et al.*[22]

6.3.2 Nitrogen compounds

While there is a wealth of information on the occurrence of sulphur compounds in the Arctic, particularly in the spring, much less is known about the nitrogen components. Early measurements during the 1970s and early 1980s were made with air filtering methods which are generally adequate for a determination of major sulphur pollution compounds but not for the predominantly gaseous nitrogen compounds. Only information on inorganic nitrate compounds (e.g. aerosol NO_3^- and gaseous HNO_3) can be obtained by this means.

Figure 6.10 shows the annual variation of total inorganic nitrate as observed at Alert since 1980. Concentrations of NO_3^- fall below 110 ng m^{-3} (i.e. <43 pptv). In general, on a molar basis, they are an order of magnitude lower than the concentrations of sulphur oxides (Fig. 6.3). Although a seasonal trend is apparent with a winter maximum and a summer minimum, a spring peak during the polluted arctic haze season such as that for SO_4^{2-} at Alert (Fig. 6.3) is conspicuously absent. Figure 6.10 also shows the complementary data for total inorganic nitrate at Cree Lake, Saskatchewan. The Alert samples were taken on Whatman 41 cellulose filters while those at Cree Lake are the total on a combination

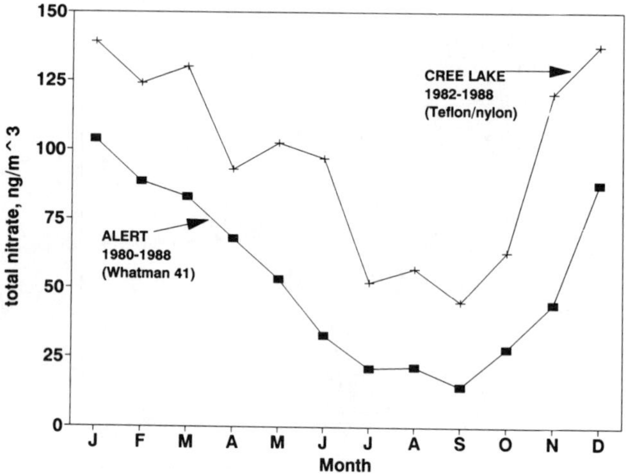

Figure 6.10. Average annual variation of NO_3^- concentrations obtained by filtering the air at Alert using Whatman 41 filters and at Cree Lake, Saskatchewan, using a Teflon–Nylon combination filter. Monthly medians of weekly and daily integrated samples are shown, respectively.

of Teflon and nylon filters in series. Comparisons show that on average, the Whatman 41 results are 68% of those from the Teflon/nylon.[21] Correcting for this would bring the observations in Fig. 6.10 even closer to one another. As in the case for SO_4^{2-}, seasonal variations of NO_3^- at the two stations are similar. The difference in seasonal variation between SO_4^{2-} and NO_3^- is that the spring maximum in SO_4^{2-} is absent for NO_3^-.

Early in arctic research, observations of low levels of inorganic nitrogen oxides led to the hypothesis that nitrogen oxides played a minor role in arctic atmospheric chemistry.[29] However, the situation has changed considerably since the first measurements of arctic NO_y and PAN were published in the mid-1980s. Dickerson[30] reported high mixing ratios of NO_y, typically about 700 pptv, observed during aircraft flights near Spitsbergen in the Norwegian Arctic in March 1983. In the lower atmosphere this was mostly due to NO_x, but at higher altitudes this was not the case. Further calculations suggested that compounds like PAN, HO_2NO_2 and alkyl-nitrates were the most likely nitrogen oxides in this case.

Surface observations of PAN in the Canadian high Arctic at Alert during a relatively low haze period in March 1985, showed an average mixing ratio of *c.* 200 pptv, about an order of magnitude higher than the inorganic nitrates.[31] Concurrent NO_2 observations showed mixing ratios of *c.* 50–100 pptv.

Further observations of nitrogen oxides have been made, but so far they have been limited to Alert and Barrow (see Fig. 6.3 for location). The data essentially confirm what has been discussed above. At Alert, PAN was measured in the spring of 1986,[21] and on a continuing basis since the fall of 1986. An annual profile of the mixing ratio for 1988 is shown in Fig. 6.11.[32] It shows a steady increase from very low summer levels (<20 pptv, the detection limit) to a spring maximum of 500–600 pptv. These data suggest a slow but steady rise of NO_y in the lower arctic troposphere from September until May. In Section 6.4.2 evidence is given supporting the conclusion that this NO_y is mostly from Eurasian anthropogenic sources.

As was discussed earlier, arctic spring is marked by the occasional occurrence of severe haze periods indicated by high concentrations of aerosol SO_4^{2-}. High concentrations of PAN in the spring as shown in Fig. 6.11 (>500 pptv in 1988) correlate very well with the occurrence of severe haze episodes (Fig. 6.12). These episodes represent more direct evidence for the pollution origin of PAN.

Observations of other nitrogen oxides at Alert are limited but instructive. In the spring of 1986 and 1988, attempts were made to measure NO_2 with a Scintrex luminol chemiluminescence instrument.[21,22] Part of the

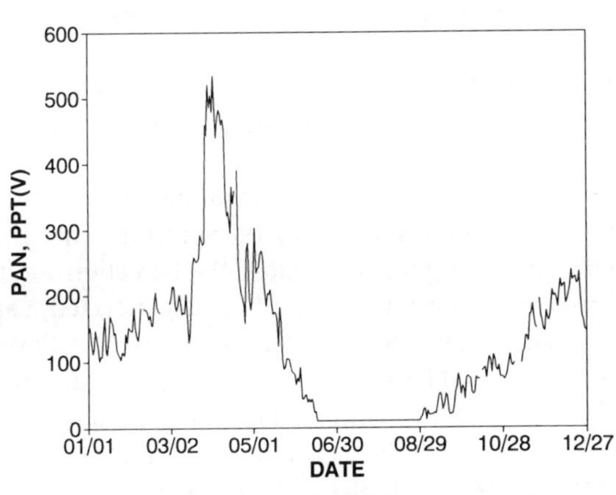

Figure 6.11. Annual variation of the mixing ratio of PAN for the year 1988 at Alert, N.W.T.[32] Between mid-June and late August the mixing ratio is below the detection limit of 15 pptv.

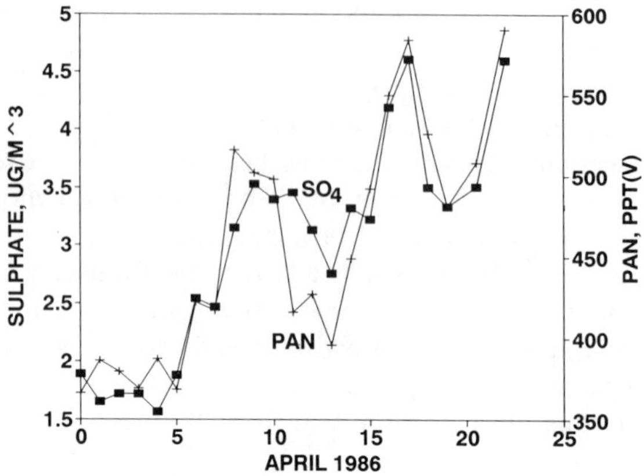

Figure 6.12. Time series for PAN (in pptv) and total sulphate (in $\mu g\, m^{-3}$) in April 1986 at Alert, N.W.T.[21]

observed NO_2 was possibly N_2O_5. Although instrumental uncertainties were large, the data suggest (i) the average level of NO_2 is less than 100 pptv (85 pptv during the spring of 1988) and (ii) during haze episodes the NO_2 mixing ratio is elevated above this average.

Several alkyl-nitrates were determined in 1988[33] at Alert. They were found to correlate reasonably well with PAN, and the total observed alkyl-nitrate mixing ratio was up to 50% of the PAN mixing ratio. No data for NO_y have been collected at Alert, but a summation of all measured nitrogen oxides (inorganic and organic nitrates, and NO_2) does strongly suggest that PAN is the dominant nitrogen oxide during the arctic spring.

At Barrow, Alaska, NO and NO_y measurements were made by Jaffe and co-workers, both in the summer of 1988, and the spring of 1989.[34,35] In the summer, NO was generally below the detection limit of 50 pptv. Periodically however, it could be quantitatively detected, usually during periods of apparent local pollution. The observed NO_y levels showed a similar trend as the PAN observations at Alert. Thus in clean background air, the median, mean, and standard deviation were 100, 120, and 60 pptv respectively.[34] Such levels are in general agreement with model predictions from Crutzen & Gidel.[36] At Alert, PAN is usually observed below 20 pptv in the summer (see above). Combining the Barrow and Alert data therefore suggests that the inorganic nitrate fraction of NO_y is somewhat higher during the summer than in other seasons.

Similarly, spring NO_y observations at Barrow are in agreement with the Alert data, being at 350–650 pptv.[34] The concurrent very low NO levels suggested very little NO_x ($NO+NO_2$) during that time. However, several episodes of substantially elevated NO_y concentrations were observed, reaching peaks of 3000–9800 pptv. Combined with the concurrent NO observations, it appeared that in this case NO_x was in fact the major part of the nitrogen oxides, indicating relatively recent NO_x emissions as the source. Trajectory analyses suggested that these episodes were due to emissions from the Prudhoe Bay industrial area some 300 km away from Barrow.

Filter measurements at Barrow during spring 1989 were conducted not only for the usual major ions (e.g. SO_4^{2-}, NO_3^-, etc.) but also for gaseous and particulate nitrite (NO_2^-).[37] The implications of these observations for atmospheric chemistry in the spring haze may be significant but have not been fully explored to date.

6.4 SOURCES AND FATE OF SULPHUR AND NITROGEN POLLUTION TO THE ARCTIC

6.4.1 Sulphur compounds

The primary source of anthropogenic sulphur in the Arctic is that released at mid-latitudes mainly as SO_2 from fossil fuel combustion and smelting of sulphide ores. The estimated spatial distribution of sulphur dioxide emissions in 1980 in the primary potential source regions in Europe, Asia and North America is shown in Fig. 6.13. Major sources in southeast Asia are not included since they tend to blow westward into the stormy Pacific rather than northward into the Arctic. They are therefore of little consequence to the North. Eurasian sources that are available to the Arctic in winter are more than double those in North America (54 versus 24 million tonnes SO_2 year^{-1}). The difference is even greater in the region north of 60° latitude (6·3 versus 0·013 million tonnes year^{-1}).

A quantitative assessment of the sources of sulphur to the arctic region for the year July 1979 to June 1980 using observed winds, precipitation, the emissions in Fig. 6.13 and a Lagrangian transport model[38] yielded the results in Figs 6.14 and 6.15. It showed that for this year 4·8% of the anthropogenic sulphur released in North America and Eurasia entered the Arctic. Furthermore, of this amount, 94% was from Eurasia (25% W. Europe, 27% E. Europe and 42% USSR) and only 4% from North

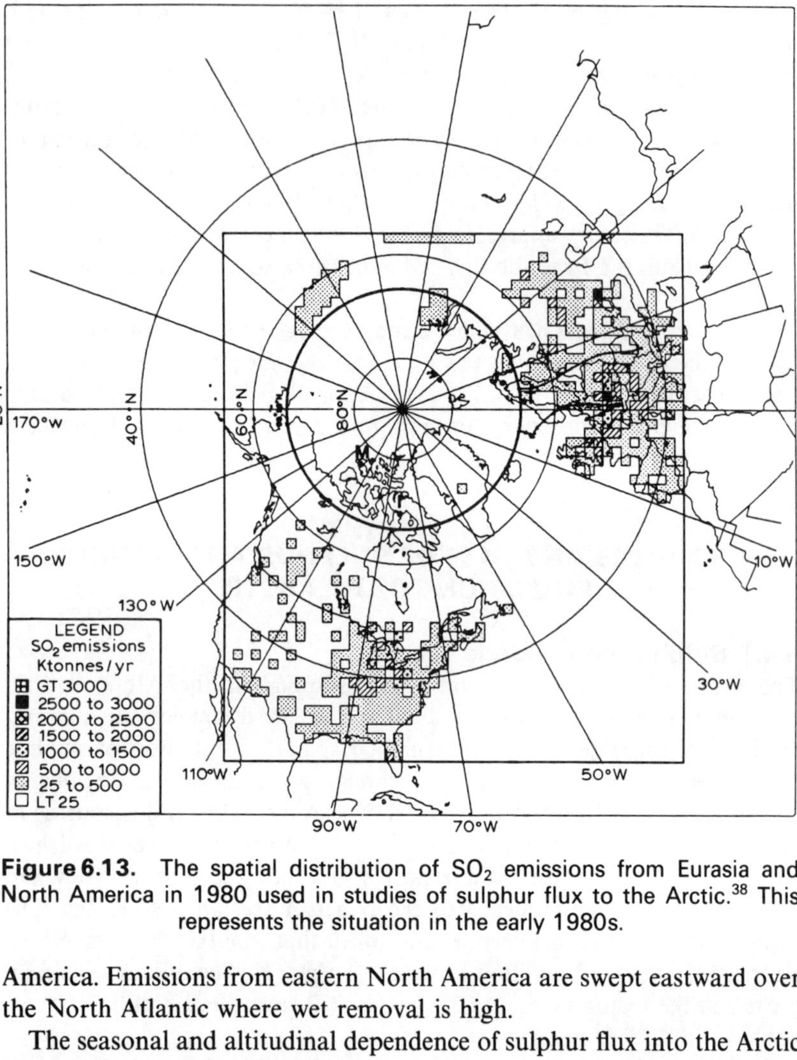

Figure 6.13. The spatial distribution of SO_2 emissions from Eurasia and North America in 1980 used in studies of sulphur flux to the Arctic.[38] This represents the situation in the early 1980s.

America. Emissions from eastern North America are swept eastward over the North Atlantic where wet removal is high.

The seasonal and altitudinal dependence of sulphur flux into the Arctic (Fig. 6.15) shows that the seasonal variation of observed concentrations of SO_4^{2-} in the Arctic is at least in part caused by seasonal variations in pollution inputs from Eurasia. Also most of the sulphur enters the Arctic in the lowest 2·5 km of the atmosphere. Based on limited observations of sulphur deposited in precipitation within the Arctic Circle, it is estimated that 12% of the anthropogenic sulphur reaching this region from mid-latitudes is deposited there.[39]

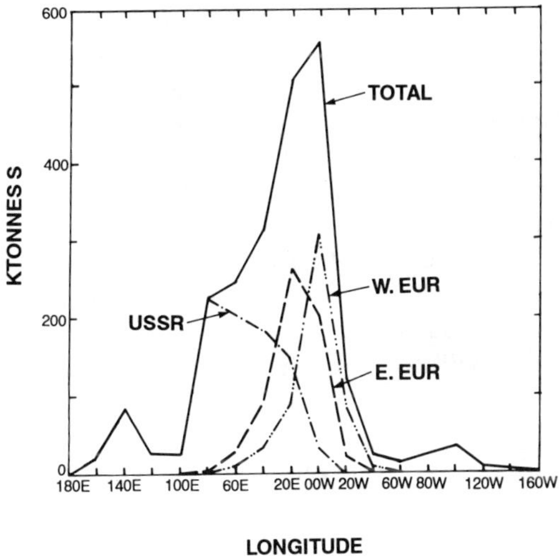

Figure 6.14. Estimated sulphur flux into the area bounded by the Arctic Circle as a function of longitude and source region for the year July 1979 to June 1980.[38]

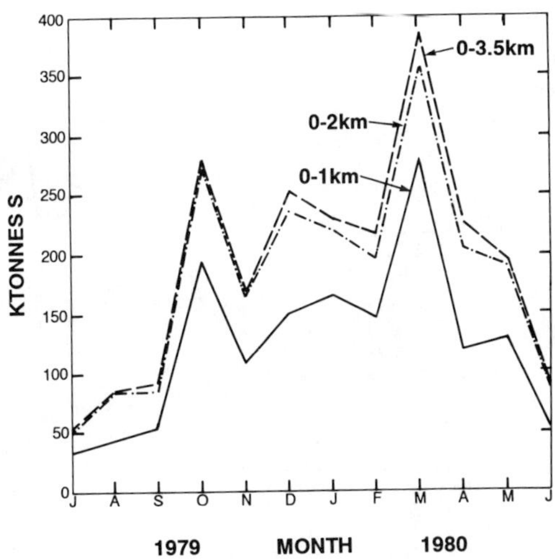

Figure 6.15. Estimated sulphur flux into the area bounded by the Arctic Circle as a function of time of year and altitude for the year July 1979 to June 1980.[38]

Another piece of evidence that supports qualitative meteorological arguments and quantitative chemical transport modelling results that strongly link arctic air pollution with Eurasian sources is the similarity of historical records of glacial ice acidity in winter in northern Canada and of SO_2 emissions from Eurasia in this century (Fig. 6.16).[40] The similarity with historical emissions from North America is not so strong. Aside from a peak in glacial ice acidity around 1912 that is explained by the eruption of Mt. Katmai in the Aleutian Islands, the ice acidity and Eurasian emission records are roughly constant in the first half of this century at levels of about half of present day observations. Since 1956, they have risen markedly as post-war economies expanded.

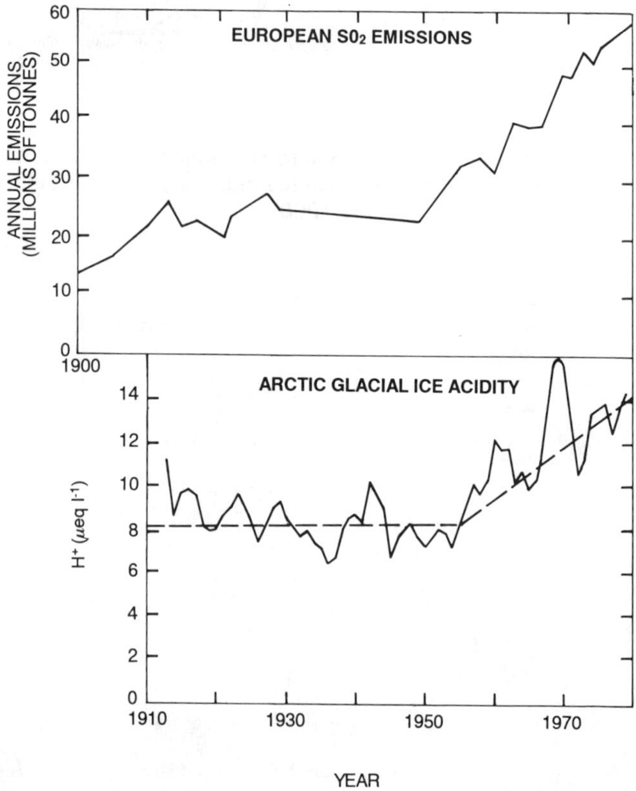

Figure 6.16. A comparison of historical records of glacial ice acidity (caused in large part by acidic sulphur compounds) on Agassiz glacier with SO_2 emissions in Europe estimated for this century.[48]

The trends in ice acidity in northern Canadian glaciers are similar to trends in glacial sulphates observed in Greenland[41] (see also Chapter 3). Between 1900 and 1910, ice levels rose to about half the current SO_4 concentrations in Greenland where they remained until about 1956. Thereafter, they increased steadily. In contrast, NO_3^- concentrations remained close to pre-1900 levels until about 1950 after which they increased, presumably as the use of automobiles rose.

6.4.2 Nitrogen compounds

Budget calculations by Logan[11] suggest that *c.* 40% of the global source of NO_x is due to the combustion of fossil fuels, 25% to biomass burning, and the balance to lightning and agricultural activity. Figure 6.17 shows the spatial distribution of combustion nitrogen emission to the atmosphere as derived by Levy & Moxim.[42] Similar to sulphur oxides, the three major potential source areas are Europe and western Siberia (Eurasia), East Asia (China and Japan), and North America.

Since biomass burning and lightning are predominantly tropical sources of nitrogen oxides, it can be safely stated that most tropospheric NO_y in the Arctic is of anthropogenic origin. As with the sulphur oxides, global circulation patterns are such that most of this NO_y originates from Europe and western Siberia. This is confirmed by a general circulation model estimate of the origin of NO_y in the Arctic during the winter.[42] It predicts that in the lower troposphere about 50% of NO_y originates from sources north of 60°. Such sources are mainly present in Eurasia. A further 30% comes from lower latitudes in Eurasia. At higher altitudes in the troposphere, North American sources are expected to contribute proportionally more than at low levels. However even then, European sources contribute more (>50% versus 30%). Thus for the whole troposphere, the model predicts that more than 90% of NO_y is from Eurasia.

If Eurasia dominates as the source of arctic NO_y, high atmospheric concentrations are expected in the Eurasian Arctic. In fact, the model of Levy & Moxim[42] predicts mixing ratios of greater than 1 ppbv NO_y in the winter and spring, in good agreement with observations.[30] Other atmospheric chemical transport model calculations[43] predict similar high levels of NO_y at 80°N latitude for the Norwegian Arctic in the winter. These calculations focused on the concentration of organic gases from anthropogenic sources. It predicted that a combination of these compounds and NO_x leads to substantial concentrations of PAN in the arctic

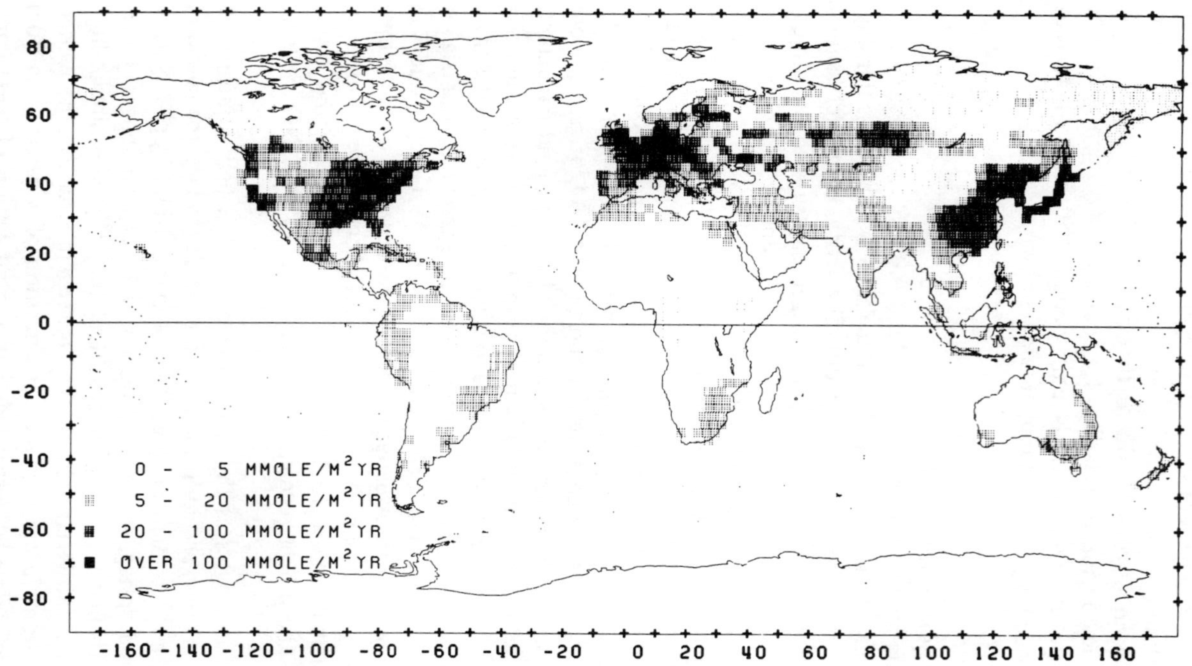

Figure 6.17 Annual total combustion nitrogen emissions in mmole m^{-2} $year^{-1}$. (From Levy and Maxim[42] (1989). Simulated global distribution and deposition of reactive nitrogen emitted by fossil fuel construction. Reprinted by permission from *Tellus* **41 B**).

troposphere. Both modelling studies calculate at least an order of magnitude less NO_y in the summer than in the winter.

A strong longitudinal gradient over the Arctic is calculated by Levy & Moxim.[42] However, a comparison of model predicted NO_y at Alert (150–200 pptv)[42] with observed levels (up to 600 pptv)[21] suggests that this gradient is actually weaker than calculated, and may be more like the gradient in total anthropogenic sulphur. This discrepancy is probably due to the fact that the model does not treat PAN explicitly. Under arctic conditions, the lifetime of NO_y is probably determined by compounds like PAN (which can be estimated to be over 300 days[44]). Because sulphur and nitrogen oxides are inefficiently removed from the arctic air mass in winter and spring, it is not surprising that on a molar basis the ratio of total-S/total-N observed in the spring at Alert (0·67) was not significantly different than the ratio of S/N in total Eurasian anthropogenic emissions.[21]

At present, little is known about the fate of the arctic nitrogen oxides. As previously stated, it is predominantly present in organic compounds that are stable in the Arctic. That is, they are neither removed via deposition processes nor effectively lost through decomposition. The most likely fate is transport out of the Arctic where higher temperatures would increase the decomposition rate of PAN into NO_2 followed by formation of more readily removed inorganic nitrates. It is well established that the concentrations of tropospheric ozone and PAN at mid-latitude rural locations peak in the spring.[45–47] It is an open question whether polluted arctic air reaching mid-latitudes contributes to spring maxima in O_3 and PAN concentrations.

6.5 SUMMARY AND CONCLUSION

From research carried out in the 1980s, it has become apparent that sulphur and nitrogen released from mid-latitudinal industrial sources is polluting the arctic air mass in the winter half of the year and thereby affecting large fractions of the northern hemisphere troposphere. The connection between Eurasian sulphur and nitrogen emissions and elevated background concentrations of their oxides in arctic air impinging on the North American airshed has been made. The full circle is completed since North American emissions enhance the sulphur content of air and precipitation of North Atlantic air entering Europe.[49]

Because of a peculiar twist of nature the northern Eurasian continent is essentially upwind of the Arctic in winter and spring. Thus, sulphur and nitrogen emissions from this region have the greatest effect on 'background' sulphur and nitrogen pollution levels in the northern region. It is therefore this region's environmental policies that will to a large extent determine future arctic pollution.

The arctic atmosphere is a unique laboratory for studying chemical transformations of sulphur and nitrogen oxides as well as of other photoreactive trace constituents. The abrupt change at polar sunrise from complete dark to complete light in a matter of weeks without an attendant shift in the meteorological situation is accompanied by chemical changes in the oxidation state of atmospheric sulphur from SO_2 to SO_4^{2-}. This leads to a maximum in arctic haze in March and April. Similar changes in nitrogen oxides bound up mainly in PAN are not evident in the seasonal cycle of tropospheric nitrate. This is consistent with the photostability of PAN. It is more sensitive to temperature than to sunlight. The temperature sensitivity of PAN stability may have broader implications. Building up to maximum concentrations in March and April, it may affect the oxidant chemistry of the northern hemispheric troposphere when it decomposes to nitrogen oxides upon reaching warmer southern climes during outbreaks of arctic air.

ACKNOWLEDGEMENT

The authors thank Dr. A. Semb of the Norwegian Institute for Air Research (NILU) for providing annual observations of SO_2 and SO_4^{2-} from the Norwegian Arctic and Dr. D. Jaffe for providing observations at Barrow, Alaska, before publication.

REFERENCES

1. Rahn, K. A. & McCaffrey, R. J., Long range transport of pollution aerosol to the Arctic. A problem without borders. *Proc. WMO Symp. on the Long Range Transport of Air Pollutants and Its Relation to General Circulation Including Stratospheric/Tropospheric Exchange Processes*, WMO No. 538, 1980, pp. 25–35.
2. Rahn, K. A. & McCaffrey, R. J., On the origin and transport of the winter arctic aerosol. *Ann. N. Y. Acad. Sci.*, **338** (1980) 486–503.
3. Barrie, L. A., Arctic air pollution: An overview of current knowledge, *Atmos. Environ.*, **20** (1986) 643–63.

4. Shaw, G. E. & Khalil, M. A. K., Arctic haze. In *The Handbook of Environmental Chemistry*, Volume 4, Part B, ed. O. Huntziger. Springer-Verlag, Berlin, 1989, pp. 69–111.
5. Greenaway, K. R., Experiences with arctic flying weather, *Journal of Canadian Royal Met. Soc.*, **1** (1950) 1–12.
6. Mitchell, M., Visual range in the polar regions with particular reference to the Alaskan Arctic, *J. Atmos. Terr. Physics*, (Special Supplement) (1956) 195–211.
7. Blanchet, J. P., Toward estimation of climatic impacts of arctic aerosols, *Atmos. Environ.*, **23** (1989) 2609–26.
8. Finlayson-Pitts, B. J. & Pitts, J. N., Jr., *Atmospheric Chemistry, Fundamentals and Experimental Techniques*. J. Wiley & Sons, New York, 1986.
9. Warneck, P., Chemistry of the natural atmosphere. *International Geophysics Series*, **41**, Academic Press, San Diego, 1988.
10. Bottger, A., Gravenhorst, G. & Ehhalt, D. H., Atmospherische Kreislaufe von Stickoxiden und Ammoniak, *Ber. Kernforschungsanlage Julich*, **1558** (1978).
11. Logan, J. A., Nitrogen oxides in the troposphere: Global and regional budgets, *J. Geophys. Res.*, **88C** (1983) 10785–807.
12. Atherton, C. S., Organic nitrates in remote marine environments: Evidence for long range transport, *Geophys. Res. Lett.*, **16** (1989) 1289–92.
13. Whitby, K.T., The physical characteristics of sulphur aerosols. *Atmos. Environ.*, **12** (1978) 135–59.
14. Bassett, M.E. & Seinfeld, J. H., Atmospheric equilibrium model of sulphate and nitrate aerosols—II. Particle size analysis. *Atmos. Environ.*, **18** (1984) 1163–70.
15. Hoff, R. M., Leaitch, W. R., Fellin, P. & Barrie, L. A., Mass–size distributions of chemical constituents of the winter arctic aerosol, *J. Geophys. Res.*, **88C** (1983) 10947–56.
16. Radke, L. F., Lyons, J. H., Hegg, D. A., Hobbs, P. V. & Bailey I. H., Airborne observations of arctic aerosols—I. Characteristics of arctic haze. *Geophys. Res. Lett.*, **11** (1984) 393–6.
17. Barrie, L. A., Bottenheim, J. W., Schnell, R. C., Crutzen, P. J. & Rasmussen, R. A., Ozone destruction and photochemical reactions at polar sunrise in the lower arctic atmosphere, *Nature*, **334** (1988) 131–41.
18. Finlayson-Pitts, B. J., Livingston, F. E. & Berko, H. N., Ozone destruction and bromine photochemistry at ground level in the arctic spring. *Nature*, **343** (1990) 622–5.
19. Sturges, W. T., Comment on the paper by Finlayson-Pitts & Johnson, *Atmos. Environ.*, **22** (1988) 1107; *Atmos. Environ.*, **23** (1989) 1167–8.
20. Bryson, R. A., Air masses, streamlines and the boreal forest. *Geographical Bull.*, **8** (1966) 228–67.
21. Barrie, L. A., den Hartog, G., Bottenheim, J. W. & Landsberger, S., Anthropogenic aerosols and gases in the lower troposphere at Alert Canada in April 1986. *J. Atmos. Chem.*, **9** (1989) 101–27.
22. Bottenhim, J. W., Barrie, L. A., Atlas, E., Heidt, L. E., Niki, H., Rasmussen, R. A. & Shepson, P. B., Depletion of lower tropospheric ozone during arctic spring: The Polar Sunrise Experiment. *J. Geophys. Res. D*, (in press).
23. Li, S.-M. & Winchester, J. W., Geochemistry of organic and inorganic ions of late winter arctic aerosols. *Atmos. Environ.*, **23** (1989), 2401–15.

24. Barrie, L. A. & Hoff, R. M., Five years of air chemistry observations in the Canadian Arctic. *Atmos. Environ.*, **19** (1985) 1995–2010.
25. Barrie, L. A. & Hoff, R. M., The oxidation rate and residence time of sulphur dioxide in the arctic atmosphere. *Atmos. Environ.*, **18** (1984) 2711–22.
26. Li, S.-M. & Winchester, J. W., Resolution of ionic components of late winter arctic aerosols. *Atmos. Environ.*, **23** (1989), 2387–99.
27. Li, S.-M. & Winchester, J. W., Haze and other aerosol components in late winter arctic Alaska, 1986. *J. Geophys. Res.*, **95D** (1990) 1797–1810.
28. Radke, L. F., Lyons, J. H., Hegg, D. A., Hobbs, P. V. & Bailey I. H., Airborne observations of arctic aerosols—I. Characteristics of arctic haze. *Geophys. Res. Lett.*, **11** (1984) 393–6.
29. Barrie, L. A., Hoff, R. M. & Daggupaty, S. M., The influence of mid-latitudinal pollution sources on haze in the Canadian Arctic. *Atmos. Environ.*, **15** (1981) 1407–19.
30. Dickerson, R. R., Reactive nitrogen compounds in the Arctic. *J. Geophys. Res.*, **90D** (1985) 10739–43.
31. Bottenheim, J. W., Gallant, A. J. & Brice, K. A., Measurements of NO_y species and O_3 at 82°N latitude. *Geophys. Res. Lett.*, **11** (1986) 113–16.
32. Bottenheim, J. W. (unpublished observations).
33. Bottenheim, J. W., Barrie, L. A. & Atlas, E., The partitioning of NO_y in the lower arctic troposphere during spring 1988. Abstract to the 7th International Symposium of the CACGP on the Chemistry of the Global Atmosphere, Chamrousse, France, 1990.
34. Jaffe, D. A., Honrath, R. E., Herring, J. A., Li, S. M. & Kahl, J. D., Measurement of nitrogen oxides at Barrow, Alaska during Spring: evidence for regional and northern hemispheric sources of pollution. *J. Geophys. Res.* **96** (1991) 7395–406.
35. Honrath, R. E. & Jaffe, D. A., Measurements of NO and NO_y during springtime at Barrow Alaska. Abstract to the Annual Fall Meeting of the AGU, San Francisco, 1989.
36. Crutzen, P. J. & Gidel, L. T., A two-dimensional photochemical model of the atmosphere. 2: The tropospheric budgets of anthropogenic chlorocarbons, CO, CH_4, CH_3Cl and the effect of various NO_x sources on tropospheric ozone. *J. Geophys. Res.*, **88C** (1983) 6641–61.
37. Li, S.-M., Particulate and snow nitrite in the spring arctic troposphere, *Atmos. Environ.* (submitted).
38. Barrie, L. A., Olson, M. P. & Oikawa, K. K., The flux of anthropogenic sulphur into the Arctic from mid-latitudes in 1979/80. *Atmos. Environ.*, **23** (1989) 2505–12.
39. Jickells, T. D., Arimoto, R., Barrie, L. A., Church, T. M., Dehairs, F., Dulac, F., Mart, L., Sturges, W. T. & Zoller, W. H., The long range transport of trace elements: Four case studies. In *The Long Range Atmospheric Transport of Natural and Contaminant Substances*, Chapter 9, ed. A. H. Knap. NATO ASI Series 297, 1988, pp. 177–96.
40. Barrie, L. A., Fisher, D. & Koerner, R. M., Twentieth century trends in arctic air pollution revealed by conductivity and acidity observations in snow and ice in the Canadian high Arctic. *Atmos. Environ.*, **19** (1985) 2055–63.
41. Mayewski, P. A., Lyons, W. B., Spencer, J. J., Twickler, M., Dansgaard, G.,

Koci, B., Davidson, C. I. & Honrath, R. E., Sulphate and nitrate concentrations from a south Greenland ice core. *Science*, **232** (1986) 975–7.
42. Levy II, H. & Moxim, W. J., Simulated global distribution and deposition of reactive nitrogen emitted by fossil fuel combustion. *Tellus*, **41B** (1989) 256–71.
43. Isaksen, I. S. A., Hov, Φ, Penkett, S. A. & Semb, A., Model analysis of the measured concentration of organic gases in the Norwegian Arctic. *J. Atmos., Chem.*, **3** (1985) 3–27.
44. Singh, H. B., Reactive nitrogen in the atmosphere. *Environ. Sci. Technol.*, **21** (1987) 320–7.
45. Oltmans, S. J. & Komhyr, W. D., Surface ozone distributions and variations from 1973–1984 measurements at the NOAA Geophysical Monitoring for Climate Change baseline observatories. *J. Geophys., Res.*, **91D** (1986) 5229–36.
46. Penkett, S. A. & Brice, K. A., The spring maximum in photooxidants in the Northern Hemisphere troposphere. *Nature*, **319** (1986) 655–7.
47. Bottenheim, J. W., Fricke, W. & Brice, K. A., Measurements of PAN, a highly sensitive tracer for photochemical oxidant chemistry. *Proceedings of the 80th APCA Annual Meeting*, New York, Air Pollution Control Association, Pittsburg, Pennsylvania, 1987.
48. Emissions of SO_2 in Europe and the Western Soviet Union from Fjeld, B., Forbruk av fossilt brensel i Europa og utslip av SO_2 i perioden 1900–1972. *NILU Teknisk Notat* Nr 1/76, 6pp. Updated to 1980 by A. Semb, pers. comm.
49. Whelpdale, D. M., Eliassen, A., Galloway, J. N., Dovland, H. & Miller J. M., The transatlantic transport of sulphur, *Tellus*, **40B** (1988) 1–15.

Chapter 7

Arctic Ozone Chemistry

SAMUEL J. OLTMANS

National Oceanic and Atmospheric Administration
Climate Monitoring and Diagnostics Laboratory
325 Broadway, Boulder, Colorado 80303, USA

7.1 INTRODUCTION

The depletion of ozone in the stratosphere and the production of ozone as a secondary pollutant in the lower atmosphere highlight the impacts encompassed by human activities on ozone behavior in the atmosphere. The striking effect of chlorofluoromethanes on the springtime polar stratosphere and the importation of pollutants in the troposphere during springtime arctic haze events point out the particular susceptibility of the arctic ozone regime to modification. In addition, the unique meteorology and chemistry of the Arctic produce patterns of ozone behavior that provide difficult tests for our ability to model such behavior. The focus here will be on the unique features of the arctic ozone distribution and some aspects of this distribution subject to human modification.

Most of the available information on arctic ozone, especially in the troposphere, is from North America and Europe. Data from the vast areas of the Arctic encompassed by the USSR are not readily available.

The division of the topic of arctic ozone chemistry into discussions about the stratosphere and troposphere is mostly arbitrary and a matter of convenience. Research efforts have usually focused on one area or the other. Inclusion of a section on stratosphere/troposphere exchange makes it obvious that although these portions of the atmosphere have identifiable characteristics, their interrelationship is equally important.

Processes outside the Arctic also influence the ozone distribution within the Arctic, both in the stratosphere and troposphere. For example, most of the ozone in the arctic stratosphere has its origin at low latitudes and is transported into the Arctic.

7.1.1 Characteristics of the ozone distribution in the arctic stratosphere

The formation of ozone (O_3) in the stratosphere (the layer between about 10 and 50 km) begins with the photolytic separation of O_2 molecules by solar ultraviolet (UV) radiation through the reaction

$$O_2 + h\nu \rightarrow O + O \tag{7.1}$$

Subsequently the oxygen atoms combine with O_2 molecules in a three body collision to form O_3 according to

$$O + O_2 + M \rightarrow O_3 + M \tag{7.2}$$

where M is usually N_2 or O_2. Ozone is also destroyed by dissociation by UV radiation and recombination with O_2 represented by

$$O_3 + h\nu \rightarrow O + O_2 \tag{7.3}$$

$$O + O_3 \rightarrow 2O_2 \tag{7.4}$$

Net reaction:

$$2O_3 + h\nu \rightarrow 3O_2 \tag{7.5}$$

Reactions of ozone with compounds of nitrogen, hydrogen and chlorine in a number of catalytic cycles are also a major contributor to the chemical balance of ozone in the stratosphere. An important example is the set of gaseous reactions involving the oxides of nitrogen

$$NO + O_3 \rightarrow NO_2 + O_2 \tag{7.6}$$

$$O_3 + h\nu \rightarrow O + O_2 \tag*{(7.3)}$$

$$NO_2 + O \rightarrow NO + O_2 \tag{7.7}$$

Net reaction:

$$2O_3 + h\nu \rightarrow 3O_2 \tag*{(7.5)}$$

In the perturbed arctic stratosphere a similar set of reactions involving chlorine monoxide (ClO) is important as we will see later. The production of ozone from the UV photolysis of O_2 suggests that in the tropics where sunlight is abundant, ozone also should be most plentiful. Observations, on the other hand, show that the largest ozone amounts are found at higher latitudes.

The appearance of high ozone amounts at middle and high latitudes reflects the role of atmospheric circulation in redistributing air in the stratosphere. During winter and spring the tendency of air to flow around the Earth's poles (the polar vortex) is disrupted by waves of large scale.[1] These planetary waves originating in the troposphere can amplify in the stratosphere redistributing ozone in two ways. First, air moving horizontally transports ozone from its source region at lower latitudes, poleward. Because of the reduced exposure to sunlight as the air reaches higher latitudes, the photochemical lifetime of ozone increases. Planetary waves may also enhance ozone abundances by moving air vertically. Air moving poleward tends to descend under the action of the waves. This increases the ozone density and total column amounts as the air is compressed.

The marked difference between the north and south polar stratosphere follows from the weaker orographic forcing of planetary waves in the Southern Hemisphere. In the Arctic, major midwinter warmings occur frequently leading to the breakdown of the circumpolar winter circulation.[2] This breakdown is accompanied by strong transport of ozone-rich air from middle to polar latitudes. Maximum ozone concentrations may thus be seen as early as January in the Arctic. This also means that large year-to-year differences may occur in the time of the ozone maximum.[2] The seasonal trend of ozone in the stratosphere can be seen in the ozonesonde measurements (see Fig. 7.1) at the 50 mb level (~20 km) over Resolute, North West Territories (75°N). Ozone amounts, on average, rise throughout the winter reaching a maximum in March and then

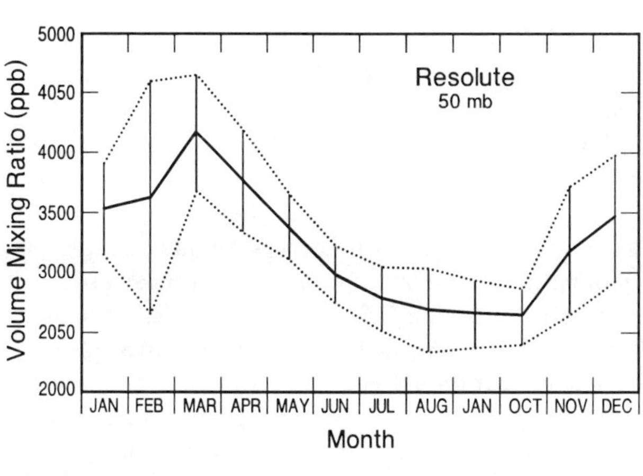

Figure 7.1. The monthly ozone mixing ratio at 50 mb (~20 km) over Resolute, NWT. The data are from balloon-borne ozonesondes and cover the period 1980–89. The solid line connects the monthly means. The dashed lines encompass two standard deviations (± 1 S.D.).

decline throughout the spring. The strong relationship between circulation patterns and ozone in the lower stratosphere can be seen in Fig. 7.2. Total ozone amounts from the Solar Backscatter Ultraviolet (SBUV) instrument on the Nimbus 7 satellite are shown alongside the average temperature and height contours of the 70 mb surface for January 1979–1986.[3] Near the pole an area of low ozone corresponding to the lowest heights and temperatures (the polar vortex) is surrounded by an area of high total ozone amounts. It is in this vortex where heterogeneous chemical processes may be responsible for ozone depletion. More will be said about this in a later section.

7.1.2 Characteristics of the ozone distribution in the arctic troposphere

As in other parts of the global atmosphere, the troposphere in the Arctic generally has much lower ozone concentrations than the overlying stratosphere. Within the arctic troposphere there are important temporal and spatial variations. These include synoptic scale variations, a prominent annual cycle, important differences in the annual cycle in the boundary layer and free troposphere and long-term changes. In addition, there appear to be significant differences in the seasonal cycle at higher arctic latitudes from that seen elsewhere in the Arctic. This will be discussed further in the section dealing with the spring ozone depletion in the boundary layer.

7.1.2.1 Synoptic scale disturbances

A feature of the monthly ozone record at Barrow, Alaska (see Fig. 7.3) is the longer period multi-day cycle. These multi-day or synoptic scale cycles are irregular in form and while they represent what might be expected from different air mass influences, we have not yet developed a satisfactory quantitative explanation involving obvious source areas. On a qualitative basis, changing influences from different source and sink areas such as could be generated by continental areas, areas of stratospheric exchange or meridional gradients could explain these cycles.

The synoptic features seem to be less pronounced, however, in the Arctic than at middle or even tropical latitudes.[4] This may result from the narrower range of air mass types affecting this high latitude location. In particular, the lack of tropical air masses where strong sink mechanisms are at work seem to preclude the large day-to-day variability due to synoptic changes often seen at mid-latitudes. Large day-to-day variations

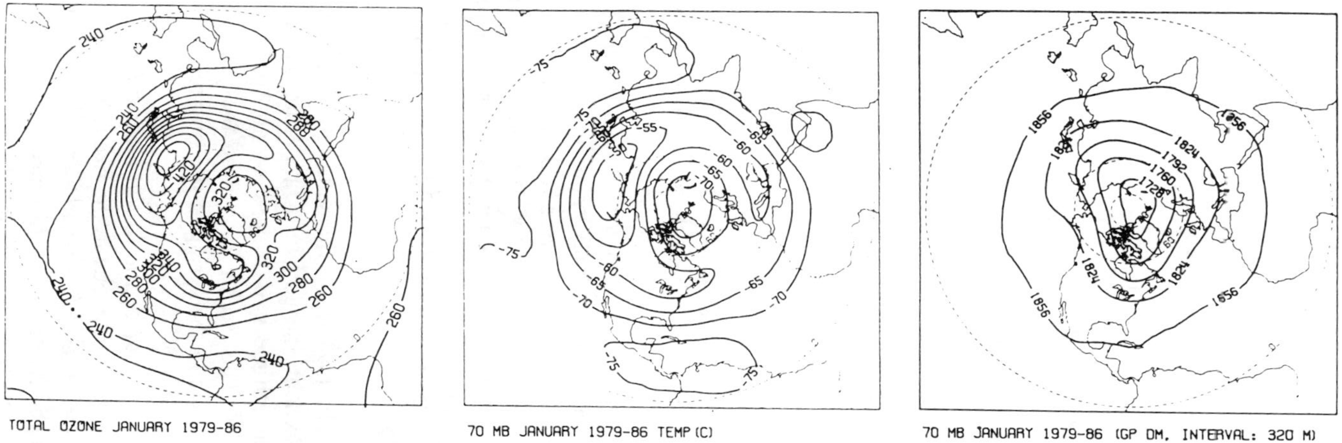

Figure 7.2. Average geopotential height and temperature of the 70 mb surface and total ozone for the month of January. Data covers the period 1979–86. Total ozone is from the Solar Backscatter Ultraviolet (SBUV) instrument on the NIMBUS 7 satellite. (Adapted from Ref. 3.)

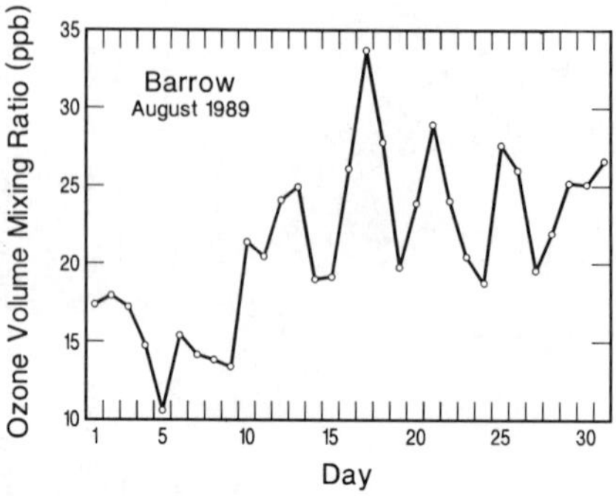

Figure 7.3. Daily mean ozone at the surface from Barrow, Alaska, for August 1989 illustrating the day-to-day variability in the record.

are seen in the spring, but this involves other processes to be discussed later.

7.1.2.2 Seasonal variation at the surface

In the non-arctic Northern Hemisphere the seasonal variation in ozone near the surface is characterized by a spring or summer maximum and an autumn minimum.[5,6] Throughout much of the Southern Hemisphere including polar latitudes a winter maximum and summer-to-autumn minimum are typical. The average behavior at the surface at a high arctic location like Barrow, Alaska (71°N, 157°W), on the other hand, is characterized by a late autumn and winter maximum and spring minimum[7] (see Fig. 7.4(a)). Surface data from ozonesonde measurements at Resolute, NWT (75°N, 95°W) and T-3 on Fletcher Ice Island near the geographic North Pole (85–86°N) show a similar pattern (see Figs 7.4(b) and (c)).

At Churchill, Manitoba (59°N), on the southern edge of the Arctic, a spring maximum is seen at the surface. Data from Tromso, Norway (69·5°N), near the same latitude as Barrow, show an April maximum and September minimum.[8,9] This seems to reflect the differences in the processes controlling ozone in the spring at the surface.

7.1.2.3 Vertical distribution of ozone in the troposphere

Vertical profile measurements of ozone in the Arctic are limited. The only location with a lengthy series of measurements is Resolute. On the southern periphery of the Arctic, Churchill also has an extensive set of ozonesonde soundings. At Barrow a much more limited set of data is

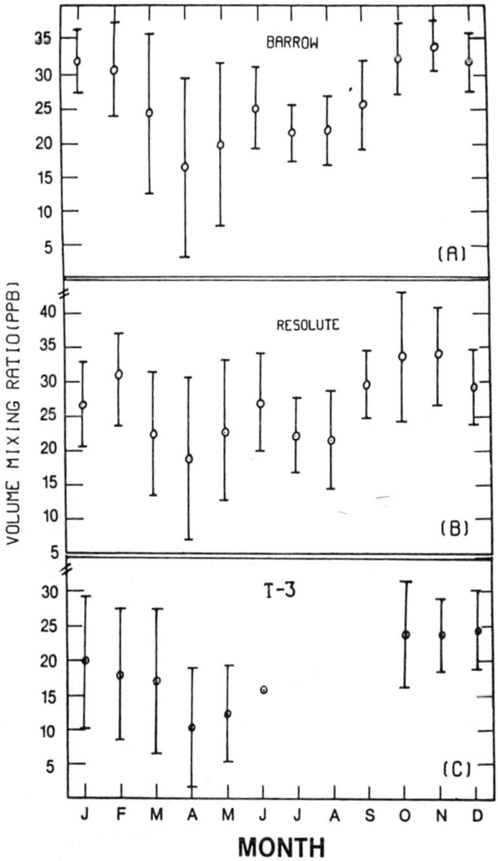

Figure 7.4. Monthly average mixing ratio of ozone at the surface in parts per billion. (A) At Barrow, Alaska, during the period 1976–86. Averages are from hourly means, vertical bars represent plus and minus one standard deviation. (B) At Resolute, NWT, during the period 1979–86. Averages are from the surface value for individual ozonesonde profiles. Vertical bars are plus and minus one standard deviation for the approximately 15 soundings each month. (C) At T-3 ice island during the period 1968–71. The number of soundings per month was variable from 35 in February and two in June. (From Ref. 7.)

available, as is the case at T-3. All of these data come from the North American sector which we have seen from the surface data may be somewhat different from other regions of the Arctic. Outside the boundary layer it is likely, however, that features such as the seasonal cycle are similar in much of the Arctic.

Examination of the seasonal ozone pattern at various levels at Resolute (see Fig. 7.5) shows that between 700 and 300 mb (~3–9 km) the peak occurs in May and seems primarily to reflect a lag in the lower stratosphere spring maximum. This is not true in the boundary layer as

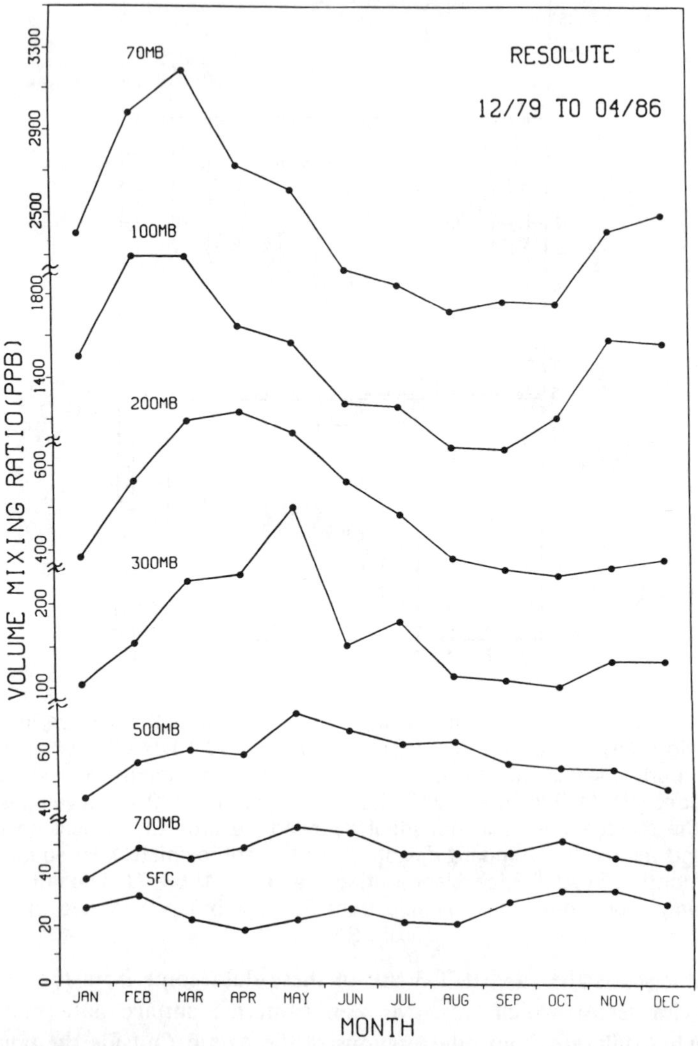

Figure 7.5. Monthly average ozone mixing ratio at Resolute for various levels in the troposphere and stratosphere during the period 1979–86. (From Ref. 7.)

noted earlier. The more limited data at Barrow and T-3 are in agreement with this pattern seen at Resolute. Even at Churchill the free tropospheric cycle is similar to that seen at Resolute. This suggests that in much of the arctic troposphere, transport from the stratosphere plays a major role in determining the seasonal pattern.

7.2 OZONE DEPLETION IN THE ARCTIC STRATOSPHERE

While natural processes such as the effect of planetary waves described earlier can lead to large changes in the ozone distribution, it is the possibility of human alteration of the ozone layer that is the source of rising concern about the polar stratosphere. An analysis of observations from the ground-based Dobson total column ozone observing network found that ozone decreased by about 5% at latitudes 53°N–64°N during winter months for the period 1969–1986.[10] This decrease was in addition to changes that could be attributed to known natural variations. Such a decrease is much larger than any predicted to have occurred to this point from the growth of atmospheric concentrations of ozone-destroying gases such as man-made chlorofluorocarbons based on standard photochemical modeling using gas phase chemistry alone. It now appears clear that in the Antarctic the dramatic springtime ozone depletion is a result of increasing levels of man-made chlorine in the atmosphere and the special chemistry resulting from the unique conditions of the winter polar stratosphere.[11] In the Arctic the conditions are markedly different from those in the Antarctic. The less vigorous wave activity in the Southern Hemisphere results in a strongly developed polar vortex. This not only prevents the importation of ozone into the polar region until the vortex weakens, but also leads to much colder temperatures since the transport of heat is also inhibited. The colder temperatures of the Antarctic lead to the widespread formation of polar stratospheric clouds (PSCs) which play a crucial role in ozone depletion.[12]

7.2.1 Polar stratospheric clouds—structure and composition

Although PSCs are not as widespread or as persistent in the Arctic as in the Antarctic, they are routinely observed in the Arctic.[12,13] This suggests that the potential does indeed exist for ozone depletion in the arctic stratosphere. The importance of PSCs in polar ozone depletion lies in their twofold role as liberators of reactive chlorine compounds and removers of nitrogen compounds.

The term polar stratospheric cloud was coined by McCormick and co-workers[12] to describe the large aerosol extinctions seen in the stratosphere during the polar winter from satellite observations. These clouds appear when the air temperature falls below about 197K. Airborne lidar measurements in the Arctic[14,15] show these clouds to be layered with a vertical extent of about 1 km or greater centered around an altitude of 20 km but extending between about 15–25 km (see Fig. 7.6). From these observations it was determined that the PSCs form an extensive cloud bank which covers the north polar region surrounded by the 188K isotherm. Fifty percent occurrence is found within the 193K isotherm.

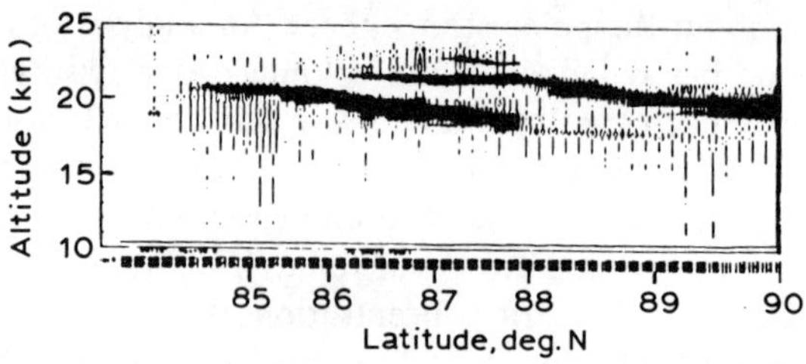

Figure 7.6. Airborne lidar data from 83°N to 90°N at 60°W longitude on 24 January 1984 depicting PSCs. The darkness of the display is proportional to the backscattered lidar signal. (Reproduced with permission from Ref. 14.)

Our knowledge of the composition and formation of PSC particles has rapidly expanded as a result of recent observational and modeling efforts. Originally thought to be composed primarily of water and sulfuric acid[16] more recent studies[17-20] conclude that nitric acid condensation is of major importance. In the stratosphere, water vapor is about 300 times more abundant than nitric acid. Water vapor concentrations are typically about 4·5 parts per million by volume while nitric acid volume mixing ratios are about 15 parts per billion. During the winter, as temperatures fall in the polar stratosphere, between an altitude of about 15–25 km a temperature is reached where nitric acid vapor begins to condense. This occurs at a temperature about 2–6 K above that at which water vapor condenses (the frost-point). Observational[21] and laboratory[22] evidence indicate strongly that the particles that are formed are made of nitric acid trihydrate ($HNO_3 \cdot 3H_2O$) in crystalline form consisting of 54% nitric acid. An amorphous form of 54% nitric acid ($HNO_3 + 3H_2O$)[18] may also be suggested by observations.[19] The widely observed nitric acid PSCs are referred to as 'type I'. The formation of type I PSCs only removes a small proportion of the much more plentiful water vapor.

If, however, temperatures continue to fall below the frost point (around 190 K) water vapor will begin to condense. Because of the greater abundance of water, these particles should be able to grow to larger sizes ($\sim 10\,\mu m$ in radius) than the nitric acid particles ($< 1\,\mu m$). These larger ice crystal particles have been designated 'type II'. The smaller type I particles probably serve as condensation nuclei for the larger type II particles. In the Arctic the type II PSCs seem to be associated with the uplifting and adiabatic cooling of stratospheric air that results from large scale synoptic events in the troposphere and mountain lee waves.

Lidar measurements of the optical properties of PSCs made in the Arctic during the winter of 1989 suggest that nitric acid particles may attain different sizes, shapes and number densities due to the different rates of cooling of the air parcels.[15,23,24] Rapid cooling leads to relatively numerous, small particles while slower cooling rates allow selective growth to larger sizes.[25] The attainment of larger particle sizes may be significant in the removal of nitrogen as discussed later.

7.2.2 Chemistry of ozone depletion

Under the conditions normally prevailing in the stratosphere, chlorine is found primarily in the form of hydrochloric acid (HCl) and chlorine nitrate ($ClONO_2$). On the surfaces of PSCs in which both gaseous and liquid or solid phases are involved, conversion of the chlorine reservoir gases (HCl and $ClONO_2$) to active forms such as Cl_2 and HOCl takes place in reactions such as[11]

$$ClONO_2(g) + HCl(s) \rightarrow Cl_2(g) + HNO_3(s) \qquad (7.8)$$

$$ClONO_2(g) + H_2O(s) \rightarrow HOCl(g) + HNO_3(s) \qquad (7.9)$$

Ozone destruction takes place when parcels of air processed during the polar night are exposed to sunlight. These parcels containing active forms of chlorine react, for example, according to

$$Cl_2 + h\nu \rightarrow Cl + Cl \qquad (7.10)$$

$$2 \times (Cl + O_3 \rightarrow ClO + O_2) \qquad (7.11)$$

The reaction of ClO with itself, involving the ClO dimer, is now considered to be the primary catalytic cycle responsible for ozone depletion in the Antarctic[26,27]

$$ClO + ClO + M \rightarrow Cl_2O_2 + M \quad (7.12)$$

$$Cl_2O_2 + h\nu \rightarrow Cl + ClO_2 \quad (7.13)$$

$$ClO_2 + M \rightarrow Cl + O_2 + M \quad (7.14)$$

$$2 \times (Cl + O_3 \rightarrow ClO + O_2) \quad (7.11)$$

In the Arctic this process also appears to be important[28] along with a cycle involving HOCl[29] and a synergistic cycle involving ClO and BrO.[30] Reactions (7.8) and (7.9) sequester almost all of the NO_2 in the form of HNO_3 which is crucial to keeping chlorine in its active form.[29] Also if the HNO_3 is photolyzed, it will reform NO_2 thus short circuiting the ozone destruction process. This leads to the other important role of PSCs—the irreversible removal of nitrogen. It is suspected that this happens through the incorporation of smaller nitric acid particles into somewhat larger particles that can settle out of the photochemical region. Evidence from Antarctica[21] suggests that this sedimentation process takes place as the temperature falls sufficiently to form water ice particles (type II) that incorporate smaller HNO_3 particles (type I) which form at temperatures several degrees warmer than the water frost-point temperature. This conclusion was reached on the basis of the observation that low reactive nitrogen values (NO_y) were accompanied by low water vapor amounts (see Fig. 7.7). More recent work in the Arctic by the same investigators[31] found removal of nitrogen in air processed by PSCs but water vapor values not markedly depressed (see Fig. 7.7). Based on the relationship between N_2O (an inert tracer in this region of the atmosphere) and NO_y over a broad range of conditions, it was possible to determine that the region shown by the hatching in Fig. 7.7 represents a loss of reactive nitrogen (primarily HNO_3) from this part of the atmosphere. Although denitrification was the rule in Antarctica, the geographical extent of it was more limited in the Arctic.[32]

One method suggested for dry nitrogen removal is the selective growth of nitric acid trihydrate particles.[21,33,34] Instead of every particle growing uniformly, growth occurs only on particles larger than about 0·3 μm[33] with a fraction of the particles becoming large enough to fall out of the active ozone depletion region. Larger particle sizes are achieved, as mentioned earlier, when the air cools at a slower rate. A fuller understanding of the denitrification process remains as a significant step in predicting ozone loss.

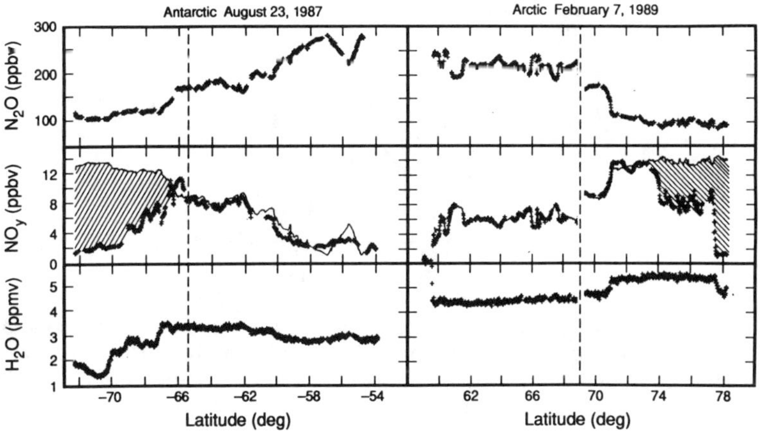

Figure 7.7. Data from aircraft flights in the Antarctic (23 August 1987) and Arctic (7 February 1989). The dashed vertical line represents the boundary of the polar vortex. The thin solid line represents the unperturbed NO_y concentration based on the relationship between NO_2 and NO_y outside of the vortex boundary. The shaded area highlights the difference between the expected gas phase relationship between NO_2 and NO_y and the measured NO_y. (Reprinted by permission from *Nature*, Vol. 344, p. 322. Copyright © 1990 Macmillan Magazines Limited.)

In determining whether the arctic winter stratosphere is susceptible to ozone loss from perturbed chemistry, two required chemical conditions are the presence of elevated amounts of reactive chlorine (e.g. ClO) and depressed levels of reactive nitrogen. Both of these conditions were found in aircraft measurements in the Arctic in January and February 1989.[31,35,36] The dramatic nature of the perturbation to the chlorine chemistry is shown by the extremely high values of ClO found within the polar vortex as illustrated by the data from 10 February 1989 (see Fig. 7.8). Outside the vortex amounts were of the order of 50 pptv and jump to 400 pptv upon entering the vortex (~1115 UT) and eventually rise to over 1000 pptv. The dip at 1245 UT corresponds to aircraft descent below about 15 km. The remainder of the measurements were made betwen 18 and 20 km. Although the 10 February flight represents the highest levels observed, mixing ratios above 500 pptv were found within the vortex on all five flights beginning on 30 January and continuing through to 10 February. These high levels of ClO, and the relatively rare occurrence of type II PSCs, imply that the conversion of chlorine from the inactive

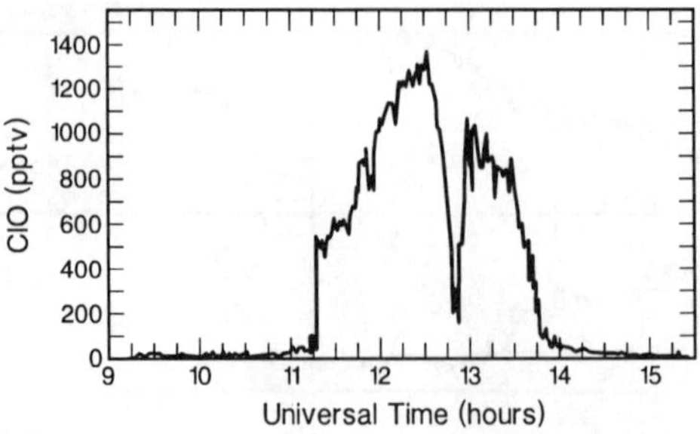

Figure 7.8. Mixing ratio of ClO observed from an aircraft on 10 February 1989 flight into the arctic polar vortex at altitudes of 18–20 km. The dip at 1245 UT is associated with aircraft descent below 15 km. (From Ref. 35, reproduced by permission of W. H. Brune *et al.*)

reservoir species HCl and $ClONO_2$ to active forms such as ClO occurs very efficiently on type I PSCs.[32] These high levels of reactive chlorine can only be produced by heterogeneous chemistry on PSCs,[35] and are very close to those found in Antarctica.[37,38]

Low values of NO_2 in the polar vortex indicate that NO_2 is being suppressed in favor of HNO_3 and that the formation of $ClONO_2$ will be inhibited. Reduced amounts of HNO_3 in turn show that reactive nitrogen is being removed from a portion of the atmosphere. Low column abundances of NO_2 have been found during the winter in the arctic polar vortex both from ground-based and aircraft measurements. Values in both cases were as low or lower than those seen in the Antarctic in late winter and early spring. It is clear from the measurements in the arctic stratosphere, that a highly perturbed chemistry indicative of heterogeneous processes in PSCs takes place in the wintertime polar regions of both hemispheres.

7.2.3 Observations of ozone depletion

The detection of ozone depletion in an antarctic ozone vertical profile is unmistakable. Comparison of September profiles in Antarctica made in recent years with those made prior to 1975 show the remarkable changes.[11,39] With conditions in the Arctic less conducive to ozone depletion because of generally warmer vortex temperatures and the tendency for the vortex to breakdown before sufficient sunlight reaches high arctic latitudes, a more subtle signature in the profiles is expected. Such a signature does indeed appear to have been detected both in the Scandinavian Arctic[40] and in the Canadian Arctic.[41]

At Kiruna, Sweden, the beginning of ozone depletion was seen between 22 and 26 km during January 1989 as shown in Fig. 7.9.[40] This represents a local reduction of about 25% on 23 January and 10% on 30 January. From the viewpoint of total column abundances of ozone the notch in the 23 January sounding, however, represents about a 3% reduction. The region of diminished ozone appears to be just above the region of coldest temperatures and largest aerosol amounts indicative of PSCs.[40] With the

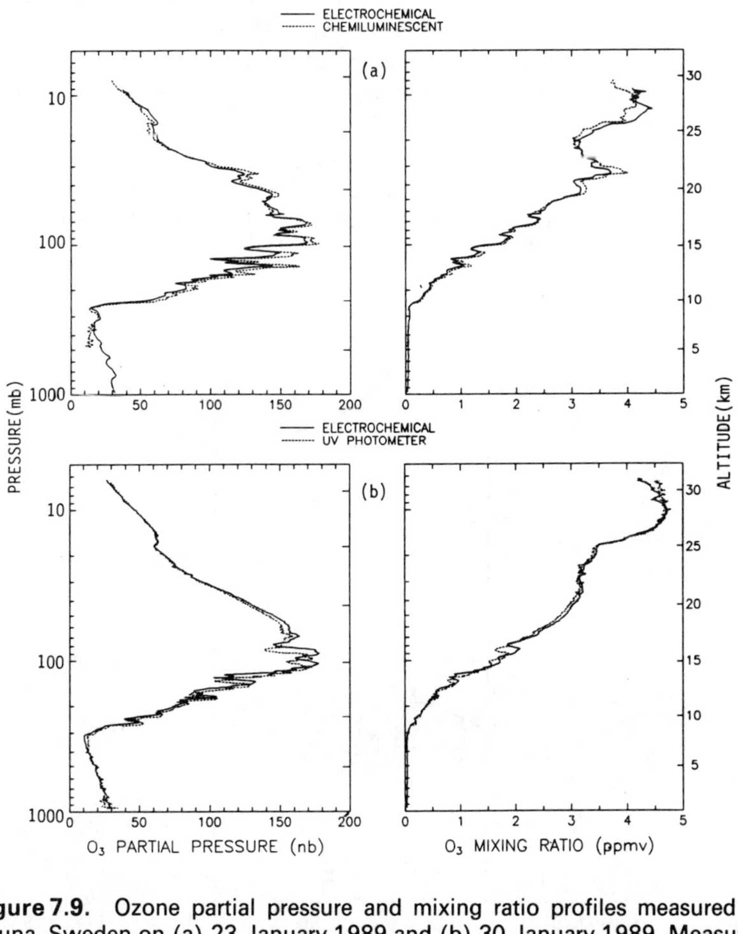

Figure 7.9. Ozone partial pressure and mixing ratio profiles measured at Kiruna, Sweden on (a) 23 January 1989 and (b) 30 January 1989. Measurements by different instruments (electrochemical, chemiluminescence and UV photometer) are shown. (Reproduced with permission from Ref. 40.)

breakdown of the polar vortex in early February and the ensuing warming, PSC formation would cease and reactive forms of nitrogen are again able to sequester the ozone-destroying forms of chlorine.[40]

Over Alert, NWT, a series of ozonesonde soundings from 27 January to 22 February 1989 showed the progressive decrease in the ozone mixing ratio at 21 km.[41] This culminated in the sounding of 22 February shown in Fig. 7.10 with a prominent notch in the ozone partial pressure between 18 and 24 km. Although PSC measurements were not made during the time of ozone decrease, this was the region where persistent PSCs were observed during January at Alert.[13] At both Kiruna and Alert the similarity between these profiles (see Figs 7.9 and 7.10) and those seen over Antarctica are strong indicators of similar depletion mechanisms taking place in both polar regions.[40,41]

Since the depletion seen in the arctic profiles occurs a month or more sooner than the corresponding season in Antarctica, it is necessary to account for the earlier exposure to sunlight implied by the arctic case.

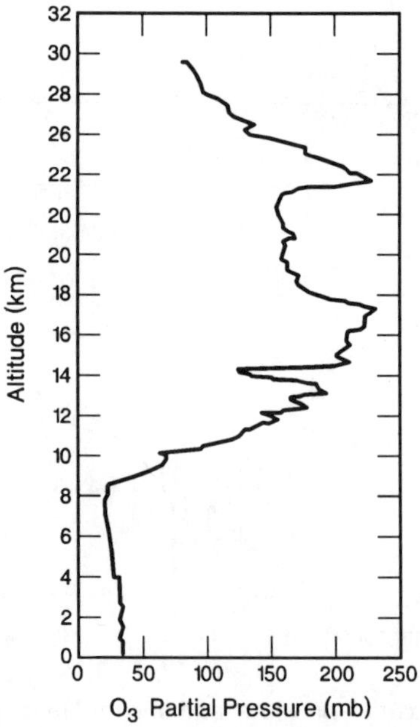

Figure 7.10. Ozone partial pressure profile measured at Alert, NWT, on 22 February 1989. (Reproduced with permission from Ref. 41.)

Kiruna is relatively close to the Arctic Circle (68°N) and by the end of January is receiving more than eight hours of sunlight in the region of coldest temperatures. Alert, on the other hand, at 82°N is not sunlit at this time. Air in the stratosphere over Alert did, however, pass increasingly through sunlit areas in its traverse around the vortex as the measurement period progressed. Part of the ozone reduction might also be attributed to mixing of air from nearer the edge of the vortex, where chemical loss had already taken place, with air deeper within the vortex.

A large research campaign, the Airborne Arctic Stratospheric Expedition (AASE), in January and February 1989 which studied possible ozone depletion and its causes was patterned after a similar campaign in Antarctica a year and a half earlier.[42] Two NASA instrumented aircraft flying from Norway were able to observe a number of the features of the processes leading to ozone destruction in the Arctic.[32] In addition to the observation of PSCs, the detection of highly elevated levels of ClO, and significant denitrification, an airborne lidar system capable of measuring ozone and particles found regions in which ozone was reduced by more than 15%.[43] The flight into the polar vortex on 9 February in which the region of lower ozone was found was located near the edge of the vortex, with higher concentrations found further poleward. Analysis of meteorological data showed no dynamical processes that could account for this lower ozone region. In situ measurements from a high flying research aircraft made on the same day near the bottom of the region of reduced ozone confirmed the presence of a chemically perturbed region including high levels of ClO.[35] Earlier flights into this region found PSCs, indicating that the air had had an opportunity for processing by heterogeneous chemical reactions within the PSCs to produce the observed ozone depletion.

Although ozone reductions were detected at particular altitudes at Kiruna and Alert as well as from the aircraft, this probably underestimates the extent to which ozone is being destroyed chemically in the region of the arctic polar vortex.[44] This is because during the winter there is a constant resupply of air from outside the vortex. Based on the relationship between ozone and the conservative tracer, N_2O, inside and outside the vortex, the ozone loss in the vortex was found to be 12–35% in the region studied.[44] This translates into a reduction in the column ozone amount of 5–7% which is consistent with the reported long-term decreases noted earlier.[10]

The picture emerging of the arctic stratosphere is that winter depletions of ozone, although much less severe than observed in Antarctica during spring, are in fact occurring. This more modest thinning of the ozone layer in the Arctic is in keeping with the expected effects of PSCs to catalyze chlorine into its active forms and remove reactive nitrogen. The ameliorating effects of warmer arctic stratospheric temperature (and hence fewer PSCs), the less confining nature of the north polar vortex and the earlier breakdown of the vortex appear to combine to limit the ozone destruction in the Arctic compared to the Antarctic. Gaps in our knowledge of such processes as the removal of active nitrogen, the role of a cooling stratosphere because of the influence of the build-up of greenhouse gases,[45] and increasing amounts of stratospheric water vapor resulting from increasing atmospheric methane[46] make future projections of the state of the arctic stratosphere uncertain. However, given the important role played by PSCs in polar ozone depletion, a cooling of the stratosphere and increase in water vapor which would lead to more numerous and extensive occurrences of PSCs is likely to enhance the potential for even greater losses, especially since it will be a number of years before any reduction of the chlorine content of the stratosphere can be realized even with reductions in the emissions of long-lived, man-made chlorine containing compounds.

7.3 SPRINGTIME OZONE DEPLETION IN THE ARCTIC BOUNDARY LAYER

It was noted earlier that in the high Arctic the seasonal variation shows a springtime minimum which is in contrast to the usual spring and summer maxima seen elsewhere in the troposphere. An examination of hourly or even daily ozone concentrations at the surface during this time of the year (March through May) reveals that there are periods when the ozone concentration is essentially zero. At Barrow these very low values alternate with the highest amounts seen during the course of the year (see Fig. 7.11). A vertical profile of ozone at Barrow (see Fig. 7.12) shows that even during the times of very low ozone near the surface there is abundant ozone above the boundary layer as defined by the temperature inversion. This has also been observed at Alert.[47]

7.3.1 Bromine in the Arctic

At Alert, NWT, Barrie and co-workers[48] found that the sudden springtime disappearances of ozone at the ground were strongly anticorrelated with bromine collected on filters (see Fig. 7.13). This filterable bromine

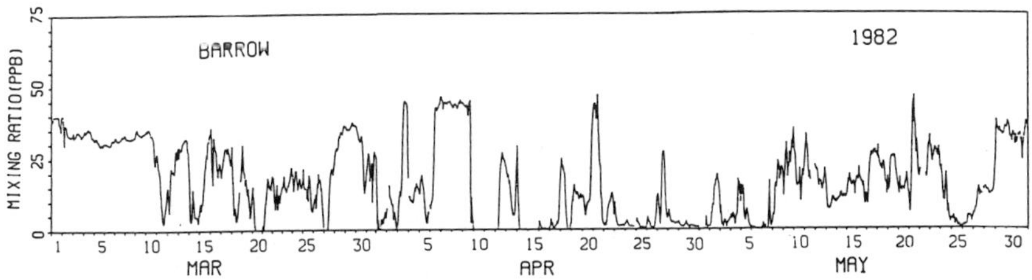

Figure 7.11. Hourly ozone mixing ratios in ppb at Barrow, Alaska, during spring 1982. (From Ref. 7.)

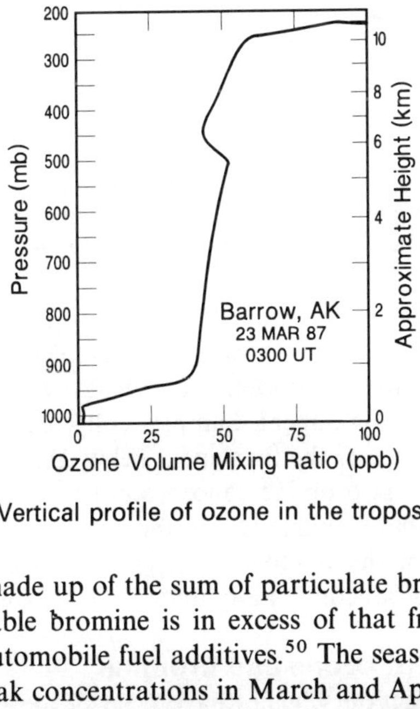

Figure 7.12. Vertical profile of ozone in the troposphere at Barrow.

was found to be made up of the sum of particulate bromine and gaseous HBr.[49] The filterable bromine is in excess of that from sea-salt, wind-blown dust and automobile fuel additives.[50] The seasonal cycle in excess bromine shows peak concentrations in March and April (see Fig. 7.14) as sunlight returns to high arctic latitudes. This pattern has been observed at several locations in the North American Arctic.[50,51] With the appearance of the sun, organic bromine compounds such as bromoform ($CHBr_3$) can be converted photolytically to bromine compounds captured on filters. Bromoform builds up during the winter months and reaches a peak prior to sunrise then drops to much lower values through the spring[52] (see Fig. 7.14). The likely source of the bromoform is the Arctic Ocean where several types of algae are thought to produce large amounts of organic bromine gases (see references in Ref. 7).

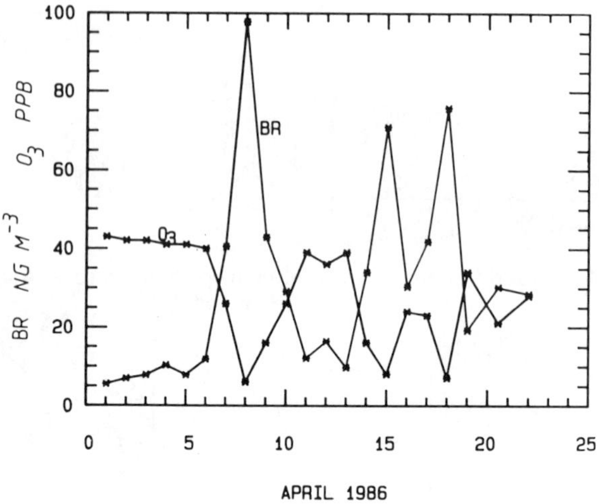

Figure 7.13. Time series of daily averaged ozone mixing ratio in ppb and of filterable bromine concentration in ng m^{-3} at ground level at Alert. (Reproduced with permission from Ref. 49.)

Filter samples of excess bromine obtained during the aircraft flights of AGASP I and II consistently showed that samples collected in the free troposphere or over the open ocean water had low filterable bromine in excess of the marine component. Samples collected over pack ice, with even a small percentage of the time spent below the top of the inversion, showed elevated excess bromine concentrations.[7] These data show that the excess bromine has a source within the boundary layer and that it is found over much of the ice-covered Arctic. They also show that the filterable bromine production is a boundary layer phenomenon.

7.3.2 Chemistry of ozone and bromine

Several chemical reaction schemes have been proposed involving ozone destruction by bromine.[48,53,54] The presence of high concentrations of bromoform in the arctic spring suggests that reactions

$$CHBr_3 + h\nu \rightarrow CHBr_2 + Br \tag{7.15}$$

$$Br + O_3 \rightarrow BrO + O_2 \tag{7.16}$$

should be efficient in removing ozone. In addition, atomic bromine can be reformed from

$$BrO + BrO \rightarrow 2Br + O_2 \tag{7.17}$$

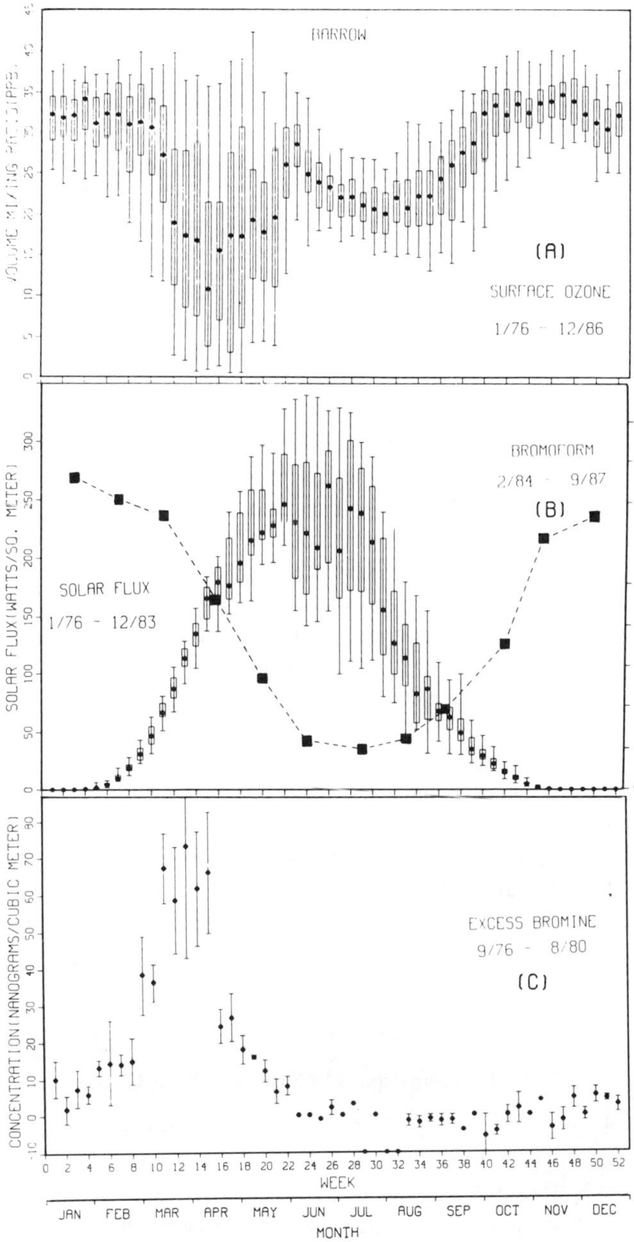

Figure 7.14. (A) The seasonal ozone variation at Barrow for 1976–86. The median (dot), the upper and lower quartiles (ends of the boxes) and the upper and lower fifth percentiles (ends of lines) are shown. (B) Total global flux of solar radiation at Barrow for 1976–83 (E. Dutton, pers. comm.) symbols as in (A), and mean monthly bromoform mole fraction[52] (squares). (C) Weekly mean and standard deviations of the excess filterable bromine at Barrow.[51]

Other reactions involving nitrogen dioxide (NO_2) and hydrogen peroxide (H_2O_2) can also result in the formation of atomic bromine.[48] Photolysis of $CHBr_2$ may also produce additional bromine radicals. Further along the reaction chain, bromine gases such as HBr and particles that may be captured on filters are produced. A number of uncertainties in reaction rates and levels of constituents such as NO_x ($NO+NO_2$) and HBr make the mechanism of ozone destruction by gas phase reactions a matter of further research. In particular the rate at which ozone is destroyed in the air by the mechanisms described may not be fast enough to account for the rapid plunges in ozone that are observed in the surface ozone record. Evidence for rapid removal of ozone, although somewhat preliminary at this point, seems to be suggested by several recently reported investigations. The low level temperature inversion, under which the severely depleted ozone layer is located, has been found to erode frequently during late winter to early spring allowing warmer air to be mixed downward to the surface.[55] During March and April along the Alaskan arctic coast there is only about a 25% probability that an inversion will persist a day or more,[55] which limits the lifetime of an ozone-depleted air mass. It has also been found that ozone loss in the spring at Barrow occurred with air transported from open ice leads often less than one day's travel time upwind from Barrow.[56] In aircraft flights in the Norwegian Arctic as part of the Arctic Gas and Aerosol Experiment (AGASP III)[57] in spring 1989, it was observed that in the boundary layer areas of full sunlight were depleted of ozone relative to adjacent cloud-covered areas.[57]

Another mechanism that has been suggested involves the release of bromine from the compound sodium bromide (NaBr) which is found in sea-salt particles.[53,54] The reactions

$$2NO_2(g)+NaBr(s)\rightarrow BrNO(g)+NaNO_3(s) \tag{7.18}$$

$$N_2O_5(g)+NaBr(s)\rightarrow BrNO_2(g)+NaNO_3(s) \tag{7.19}$$

can form bromine atoms as BrNO or $BrNO_2$ photolyze. The atomic bromine will in turn react with ozone as in reaction (7.16). At the low levels of NO_2 thought to exist in the Arctic the first of these reactions appears to be too slow to produce the magnitude of the ozone depletion observed. The levels of N_2O_5 are not known for the Arctic, but it appears that based on estimates of NO_2 and NO_3 from which it is formed, significant ozone reductions may be possible.

Although a chemical link between bromine and the springtime ozone depletion appears to be established, the high degree of anticorrelation between ozone and filterable bromine may partly result from changing air flow patterns. At Alert, for example, air from off the ocean below the inversion, which is high in filterable bromine and low in ozone, alternates with air from the free troposphere mixed down to the surface by winds blowing over the rougher and higher terrain of Ellesmere Island.[48] At Barrow where the regimes are less well-defined, the relationship between ozone and bromine is not quite as clear on a daily basis although the seasonal relationship is strong.[7]

7.4 OZONE TRENDS IN THE ARCTIC TROPOSPHERE

Long-term ozone observations are limited in the Arctic to the surface measurements at Barrow, Alaska, and the ozonesonde record at Resolute, NWT. At Barrow there has been a large increase in ozone of nearly 1% year^{-1} over the 17 years of the observational record (see Fig. 7.15). This increase has taken place during the summer and autumn months (June–November) with the largest increase during August and September of

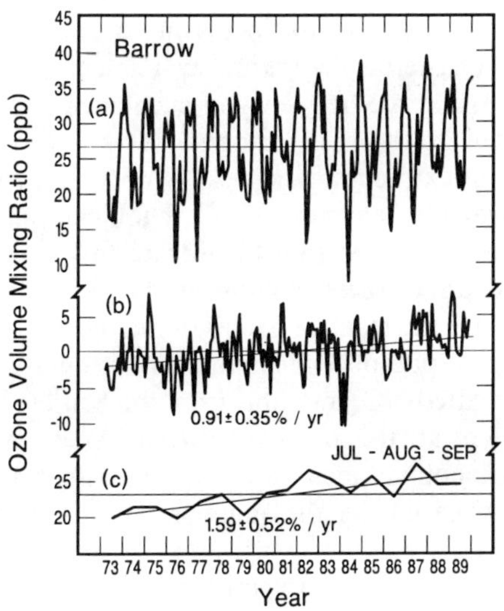

Figure 7.15. (a) Monthly mean surface ozone amounts at Barrow, Alaska, for 1973–89, and (b) the monthly ozone anomalies fitted with a linear least-squares trend line. (c) Seasonal means for July–August–September with the linear fit to the data.

1·6% year^{-1}. This means that during these two months, the average concentration has increased from about 20 ppb in 1973 when measurements began to 27 ppb in 1989. This represents a significant change in the summertime ozone distribution in the Arctic.

The ozonesonde record at Resolute goes back to the mid-1960s. The type of instrument used for making the ozone profile measurements was changed in 1979 and the results from each period of observation should probably be considered separately. An analysis of the data from 1966–1979[5] shows increases in the troposphere below 5 km of 1–2% per year with the largest increase during the summer. For the period 1980–1989 there is little evidence of a continuing increase in ozone amounts in the troposphere. If anything a decrease is indicated throughout the year.

Winter and spring months are recognized as those during which the Arctic receives pollutants from mid-latitudes of the Northern Hemisphere. It is not clear whether during the summer, there is a large anthropogenic contribution from sources at lower latitudes. Stratospheric intrusions, which will be discussed later, can contribute to the tropospheric ozone burden during the summer months, but there is no reason to expect these to increase preferentially the amount of ozone in the lower and middle troposphere while not increasing ozone in the upper troposphere even if a long-term change in such intrusion events had been observed. No change in the frequency of such events has been reported.

Barrow is located on the North Slope of Alaska which is the primary region of petroleum extraction in Alaska. Substantial oil production at the Prudhoe Bay facility began in 1977.[58] Significant amounts of natural gas are used for power generation to operate this facility. The primary pollutants emitted under these circumstances with potential impact on ozone production are the reactive oxides of nitrogen ($NO_x = NO + NO_2$). Simulations using a photochemical model with natural hydrocarbon emissions and estimated NO_x amounts from the Prudhoe Bay facility give ozone enhancement at Barrow of between three and six parts per billion.[58] Such an enhancement appears consistent with the changes that have taken place during the summer. This is discussed further in Chapter 9.

In the Arctic Boundary Layer Experiment (ABLE-3A) conducted in the summer of 1989, several cases of photochemically produced ozone were observed.[59] Enhanced ozone was found in conjunction with increased aerosol scattering. These layers were traced to regions of biomass burning (forest fires). The enhancement was more than 50% over the average ozone concentration. Such results suggest that the distribution of ozone in the arctic troposphere may be significantly modified by biomass burning as it is in the tropics.[60]

7.5 STRATOSPHERE–TROPOSPHERE EXCHANGE OF OZONE IN THE ARCTIC

At mid-latitudes, tropopause folds associated with the polar front have been studied extensively.[61–64] Such folding events in the Arctic have only recently received similar attention.[65–67] In the spring at mid-latitudes stratospheric ozone intrusions due to tropopause folding events occur nearly twice as often as during other times of the year due to the greater temperature contrasts of the atmosphere.[62] In the Arctic these events occur at colder temperatures and lower altitudes.[67] The lower altitude of occurrence suggests that the potential for injecting ozone lower into the atmosphere is higher for such events in the Arctic, and thus probably has a significant impact on the seasonal cycle in the arctic troposphere.[68] This can be seen in Fig. 7.16 which shows the cross-section of ozone in the troposphere from mid-February through March at station T-3 near the North Pole. A particularly dramatic event early in March brought enhanced ozone amounts to near the surface and produced high ozone amounts at mid-tropospheric levels for more than a week.

It also appears that lower stratospheric air may subside into the middle troposphere under the influence of anticyclonic conditions producing noticeable but less dramatic increases in tropospheric ozone. Aircraft observations made during the summer months as part of ABLE-3A also showed that stratospheric intrusions strongly influenced the ozone distribution down to the top of the mixed layer.[59] Air originating from large scale stratospheric intrusions increased the ozone amounts from average concentrations of 20–30 ppb in the region below 2 km to over 40 ppb. The air was identified as being of stratospheric origin by the low aerosol scattering.

7.6 CONCLUSIONS

The recent discovery that ozone in the winter arctic stratosphere is being lost due to the effects of long-lived, man-made chlorine compounds highlights both the importance and susceptibility of the arctic atmos-

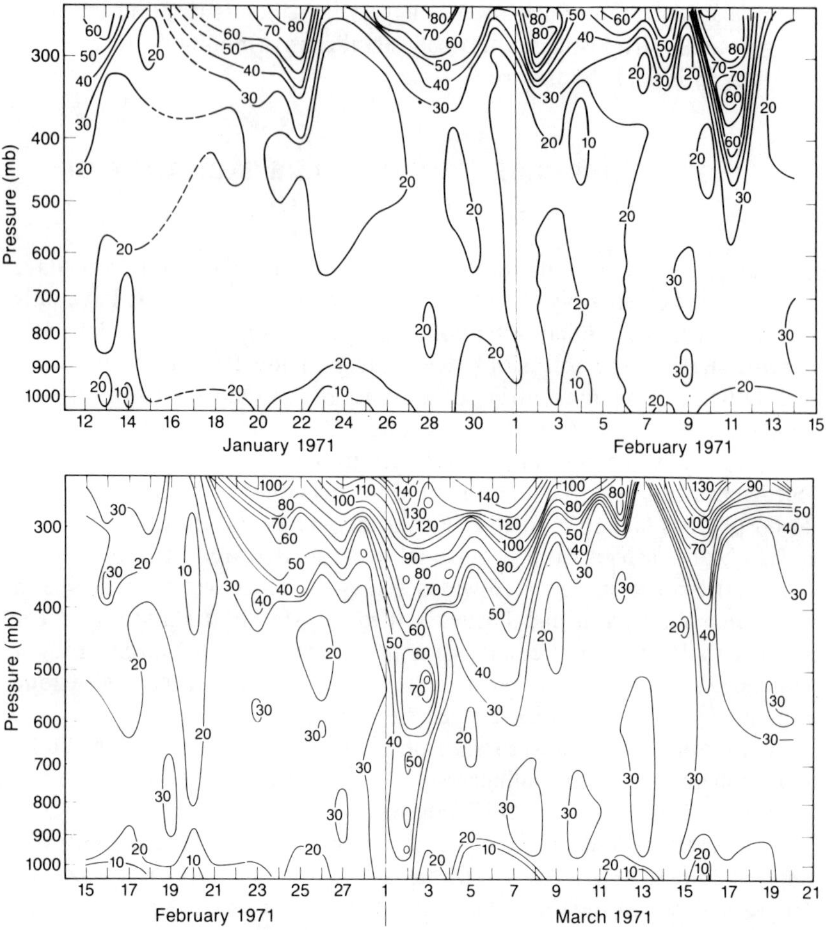

Figure 7.16. Time–height cross-section of ozone partial pressure (nanobars) at Station T-3 on Fletcher's Ice Island for 12 January 1971–20 March 1971. (From Ref. 61.)

phere. The key roles of PSCs in liberating chlorine into its ozone-destroying forms and removing reactive nitrogen from the stratosphere are much better understood in the light of recent, intensive investigations in the Arctic. While the processes at work in the arctic stratosphere appear to be similar in many ways to those responsible for the springtime antarctic ozone 'hole' the conditions in the Arctic appear to mitigate against an exact duplication of the antarctic situation. In particular the effects of warmer stratospheric temperatures (thus fewer PSCs), the less confining nature of the north polar vortex, and the earlier breakdown of the vortex which shortens the period of exposure of reactive chlorine to sunlight, combine to limit the extent of ozone destruction. Trends in several constituents of the stratosphere could, however, lead to an intensification of the thinning of the arctic ozone layer that is already taking place.

In the troposphere the spring depletion of ozone in the boundary also appears to be related to halogen chemistry, but in this case bromine is the active molecule. The sudden springtime disappearances of ozone at the ground are strongly anticorrelated with bromine collected on filters. Organic bromine gases such as bromoform, which are thought to be produced by marine microalgae, build up during the winter. With the appearance of the sun, the organic bromine compounds are photolyzed with Br in turn reacting with ozone and further along the reaction chain the bromine captured on filters is produced. Recent findings that the destruction of ozone may be a fairly rapid process suggest that heterogeneous reactions (gas–particle interactions) may play a prominent role.

The summertime increase in ozone at the surface at Barrow, Alaska, may be a result of gas and oil extraction activities on the north slope of Alaska with the associated increases in nitrogen oxide emissions. Biomass burning, primarily forest fires, may also lead to enhanced summer ozone production.

Transfer of ozone from the stratosphere into the troposphere through exchange events such as tropopause foldings is an important contributor to the ozone budget of the troposphere. This process will need to be better understood and quantified to properly assess the role of human impacts on the ozone distribution of the troposphere.

It is clear that human activities both within the Arctic and outside influence the ozone distribution of the arctic atmosphere and that this influence will continue in the future.

REFERENCES

1. Salby, M. L. & Garcia, R. R., Dynamical perturbations to the ozone layer. *Physics Today*, **43** (1990) 38–46.
2. Labitzke, K., The lower stratosphere over the polar regions in winter and spring: Relation between meteorological parameters and total ozone. *Annales Geophysicae*, **5A** (1987) 95–102.

3. Nagatani, R. M., Miller, A. J., Johnson, K. W. & Gelman, M. E., An eight-year climatology of meteorological and SBUV ozone data. NOAA Technical Report NWS 40, 1988, 125 pp.
4. Oltmans, S. J., Surface ozone measurements in clean air. *J. Geophys. Res.*, **86** (1981) 1174–80.
5. Logan, J. A., Tropospheric ozone: Seasonal behavior, trends and anthropogenic influence. *J. Geophys. Res.*, **90** (1985) 10463–82.
6. Oltmans, S. J. & Komhyr, W. D., Surface ozone distributions and variations from 1973–1984 measurements at the NOAA Geophysical Monitoring for Climatic Change baseline observatories. *J. Geophys. Res.*, **91** (1986) 5229–36.
7. Oltmans, S. J., Schnell, R. C., Sheridan, P. J., Peterson, R. E., Li, S. M., Winchester, J. W., Tans, P. P., Sturges, W. T., Kahl, J. D. & Barrie, L. A., Seasonal surface ozone and filterable bromine relationship in the high Arctic. *Atmos. Environ.*, **23** (1989) 2431–41.
8. Fabian, P. & Pruchniewicz, P. G., Meridional distribution of ozone in the troposphere and its seasonal variations. *J. Geophys. Res.*, **82** (1977) 2063–73.
9. Schmidt, M., Further results from the MPAE—project 'Tropospheric Ozone'. In *Proceedings of the Quadrennial International Ozone Symposium*, Boulder, CO, 1980, ed. J. London. National Center for Atmospheric Research, pp. 454–61.
10. Watson, R. T., Prather, M. J. & Kurylo, M. J., *Present State of Knowledge of the Upper Atmosphere 1988: An Assessment Report*. NASA Reference Publication 1208, 1988, 200 pp.
11. Solomon, S., Mystery of the antarctic ozone 'hole'. *Rev. Geophys.*, **26** (1988) 131–48.
12. McCormick, M.P., Steele, H. M., Hamill, P., Chu, W. P. & Swissler, T. J., Polar stratospheric cloud sightings by SAM II. *J. Atmos. Sci.*, **39** (1982) 1387–97.
13. Rosen, J. M., Oltmans, S. J. & Evans, W. J. F., Balloon borne observations of PSCs, frost point, ozone and nitric acid in the north polar vortex. *Geophys. Res. Lett.*, **16** (1989) 791–4.
14. McCormick, M.P., Hamill, P. & Farrukh, U. O., Characteristics of polar stratospheric clouds as observed by SAM II, SAGE, and LIMS. *J. Meteor. Soc. Japan*, **63** (1985) 267–76.
15. Browell, E. V., Butler, C. F., Ismail, S., Robinette, S., Carter, A. F., Higdon, N. S., Toon, O. B., Schoeberl, M. R. & Tuck, A. F., Airborne lidar observations in the wintertime arctic stratosphere: Polar stratospheric clouds. *Geophys. Res. Lett.*, **17** (1990) 385–8.
16. Steele, H. M., Hamill, P., McCormick, M.P. & Swissler, T. J., The formation of polar stratospheric clouds. *J. Atmos. Sci.*, **40** (1983) 2055–67.
17. Crutzen, P. J. & Arnold, F., Odd nitrogen incorporation in polar stratospheric clouds: A possible major cause for springtime ozone decay in Antarctica. *Nature*, **324** (1986) 651–5.
18. Toon, O. B., Hamill, P., Turco, R. P. & Pinto, J., Condensation of HNO_3 and HCl in the winter polar stratosphere. *Geophys. Res. Lett.*, **13** (1986) 1284–7.
19. Poole, L. R. & McCormick, M. P., Polar stratospheric clouds and the Antarctic ozone hole. *J. Geophys. Res.*, **93** (1988) 8423–30.
20. Toon, O. B., Turco, R. P., Jordan, J., Goodman, J. & Perry, G., Physical

processes in polar stratospheric ice clouds. *J. Geophys. Res.*, **94** (1989) 11359–80.
21. Fahey, D. W., Kelly, K. K., Ferry, G. V., Poole, L. R., Wilson, J. C., Murphy, D. M., Loewenstein, M. & Chan, K. R., In situ measurements of total reactive nitrogen, total water, and aerosol in a polar stratospheric cloud in the Antarctic. *J. Geophys. Res.*, **94** (1989) 11299–316.
22. Hanson, D. & Mauersberger, K., Laboratory studies of the nitric acid trihydrate: Implications for the south polar stratosphere. *Geophys. Res. Lett.*, **15** (1988) 855–8.
23. Poole, L. R., Kent, G. S., McCormick, M. P., Hunt, W. H., Osburn, M. T., Schaffner, S. & Pitts, M. C., Dual-polarization airborne lidar observations of polar stratospheric cloud evolution. *Geophys. Res. Lett.*, **17** (1990) 389–92.
24. Toon, O. B., Browell, E. V., Kinne, S. & Jordan, J., An analysis of lidar observations of polar stratospheric clouds. *Geophys. Res. Lett.*, **17** (1990) 393–6.
25. Salawitch, R. J., Gobi, G. P., Wofsy, S. C. & McElroy, M. B., Denitrification in the Antarctic stratosphere. *Nature*, **339** (1989) 525–7.
26. Molina, L. T. & Molina, M. J., Production of Cl_2O_2 by the self reaction of the ClO radical. *J. Phys. Chem.*, **91** (1987) 433–6.
27. Barrett, J. W., Solomon, P. M., deZafra, R. L., Jaramillo, M., Emmons, L. & Parrish, A., Formation of the ozone hole by the ClO dimer mechanism. *Nature*, **336** (1988) 455–8.
28. Chipperfield, M. P. & Pyle, J. A., Two-dimensional modeling of the northern hemisphere high latitude lower stratosphere. *J. Geophys. Res.*, **95** (1990) 11865–74.
29. Solomon, S., Garcia, R. R., Rowland, F. S. & Wuebbles, D. J., On the depletion of Antarctic ozone. *Nature*, **321** (1986) 755–8.
30. McElroy, M. B., Salawitch, R. J., Wofsy, S. C. & Logan, J. A., Antarctic ozone: Reductions due to synergistic interactions of chlorine and bromine. *Nature*, **321** (1986) 759–62.
31. Fahey, D. W., Kelly, K. K., Kawa, S. R., Tuck, A. F., Loewenstein, M., Chan, K. R. & Heidt, L. E., Observations of denitrification and dehydration in the winter polar stratospheres. *Nature*, **344** (1990) 321–4.
32. Turco, R., Plumb, A. & Condon, E., The Airborne Arctic Stratosphere Expedition: Prologue. *Geophys. Res. Lett.*, **17** (1990) 313–16.
33. Hofmann, D. J., Rosen, J. M., Harder, J. W. & Hereford, J. V., Balloon-borne measurements of aerosol, condensation nuclei and cloud particles in the stratosphere at McMurdo Station, Antarctica during the spring of 1987. *J. Geophys. Res.*, **94** (1989) 11253–69.
34. Toon, O. B., Turco, R. P. & Hamill, P., Denitrification mechanisms in the polar stratospheres. *Geophys. Res. Lett.*, **17** (1990) 445–8.
35. Brune, W. H., Toohey, D. W., Anderson, J. G. & Chan, K. R., In situ observations of ClO in the arctic stratosphere; ER-2 aircraft results from 59°N to 80°N latitude. *Geophys. Res. Lett.*, **17** (1990) 505–8.
36. Wahner, A., Callies, J., Dorn, H. -P., Platt, U. & Schiller, C., Near UV atmospheric absorption measurements of column abundances during Airborne Arctic Stratospheric Expedition, January–February 1989: 1. Technique and NO_2 observations. *Geophys. Res. Lett.*, **17** (1990) 497–500.

37. Brune, W. H., Anderson, J. G. & Chang, K. R., In situ observations of ClO in the Antarctic: ER-2 aircraft results from 54°S to 72°S latitude. *J. Geophys. Res.*, **94** (1989) 16649–63.
38. Mount, G. H., Solomon, S., Sanders, R. W., Jakoubek, R. O. & Schmeltekopf, A. L., Observations of stratospheric NO_2 and O_3 at Thule, Greenland. *Science*, **242** (1988) 555–8.
39. Komhyr, W. D., Oltmans, S. J. & Grass, R. D., Atmospheric ozone at South Pole, Antarctica in 1986. *J. Geophys. Res.*, **93** (1988) 5167–84.
40. Hofmann, D. J., Deshler, T. L., Aimedieu, P., Matthews, W. A., Johnston, P. V., Kondo, Y., Sheldon, W. R., Byrne, G. J. & Benbrook, J. R., Stratospheric clouds and ozone depletion in the Arctic during January 1989. *Nature*, **340** (1989) 117–21.
41. Evans, W. J. F., Ozone depletion in the arctic vortex at Alert during February 1989. *Geophys. Res. Lett.*, **17** (1990) 167–70.
42. Tuck, A. F., Watson, R. T., Condon, E. P., Margitan, J. J. & Toon, O. B., The planning and execution of ER-2 and DC-8 aircraft flights over Antarctica, August and September 1987. *J. Geophys. Res.*, **94** (1989) 11181–222.
43. Browell, E. V., Butler, C. F., Ismail, S., Fenn, M. A., Kooi, S. A., Carter, A. F., Tuck, A. F., Toon, O. B., Proffitt, M. H., Loewenstein, M., Schoeberl, M. R., Isaksen, I. & Braathen, G., Airborne lidar observations in the wintertime arctic stratosphere: Ozone. *Geophys. Res. Lett.*, **17** (1990) 325–8.
44. Proffitt, M. H., Margitan, J. J., Kelly, K. K., Loewenstein, M., Podolske, J. R. & Chan, K. R., Ozone loss in the Arctic polar vortex inferred from high-altitude aircraft measurements. *Nature*, **347** (1990) 31–6.
45. Schlesinger, M. E. & Mitchell, J. F. B., Climate model simulations of the equilibrium climatic response to increased carbon dioxide. *Rev. Geophys.*, **25** (1987) 760–98.
46. Blake, D. R. & Rowland, F. W., Continuing worldwide increase in tropospheric methane, 1978 to 1987. *Science*, **239** (1988) 1129–31.
47. Mickle, R. E., Bottenheim, J. W., Leaitch, W. R. & Evans, W., Boundary layer ozone depletion during AGASP-II. *Atmos. Environ.*, **23** (1989) 2443–9.
48. Barrie, L. A., Bottenheim, J. W., Schnell, R. C., Crutzen, P. J. & Ramussen, R. A., Ozone destruction and photochemical reactions at polar sunrise in the lower Arctic atmosphere. *Nature*, **334** (1988) 138–41.
49. Barrie, L. A., den Hartog, G., Bottenheim, J. W. & Landsberger, S., Anthropogenic aerosols and gases in the lower troposphere at Alert, Canada in April 1986. *J. Atmos. Chem.*, **9** (1989) 101–27.
50. Sturges, W. T. & Barrie, L. A., Chlorine, bromine and iodine in Arctic aerosols. *Atmos. Environ.*, **22** (1988) 1179–94.
51. Berg, W. W., Sperry, P. D., Rahn, K. A. & Gladney, E. W., Atmospheric bromine in the Arctic. *J. Geophys. Res.*, **88** (1983) 6719–36.
52. Cicerone, R. J., Heidt, L. E. & Pollack, W. H., Measurements of atmospheric methyl bromide and bromoform. *J. Geophys. Res.*, **93** (1988) 3745–9.
53. Sturges, W. T., The reaction of NO_2 with NaBr: Possible source of BrNO in polluted marine atmospheres. *Atmos. Environ.*, **23** (1989) 1167–8.
54. Finlayson-Pitts, B. J., Livingston, F. E. & Berko, H. H., Ozone destruction and bromine photochemistry at ground level in the Arctic spring. *Nature*, **343** (1990) 622–5.

55. Kahl, J. D., Characteristics of the low level temperature inversion along the Alaskan arctic coast. *International J. Climatology*, **10** (1990) 537–48.
56. Sturges, W. T., Schnell, R. C., Landsberger, S., Harris, J. M. & Li, S. M., Chemical and meteorological influences on surface ozone destruction at Barrow, Alaska during spring 1989. (Submitted to *Atmos. Environ.*)
57. Schnell, R. C. & Sheridan, P. J., Pers. Comm.
58. Jaffee, D. A. & Honrath, R. E., Ozone in the arctic troposphere. *EOS*, **69** (1988) 1078.
59. Browell, E. V., Lidar measurements of ozone, water vapor and aerosols. In *Optical Remote Sensing of the Atmosphere*. Summaries of papers presented at the topical meeting, Incline Village, Nevada, 12–15 February 1990. Optical Society of America, Washington D.C., pp. 570–3.
60. Browell, E. V., Gregory, G. L., Harriss, R. C. & Kirchhoff, V. W. J. H., Tropospheric ozone and aerosol distributions across the Amazon Basin. *J. Geophys. Res.*, **93** (1988) 1431–51.
61. Danielsen, E. F., Stratospheric–tropospheric exchange based on radioactivity, ozone and potential vorticity. *J. Atmos. Sci.*, **25** (1968) 502–18.
62. Danielsen, E. F. & Mohnen, V. A., Project Duststorm report: Ozone transport, in situ measurements and meteorological analyses of tropopause folding. *J. Geophys. Res.*, **82**, (1977) 5867–77.
63. Shapiro, M. A., Turbulent mixing within tropopause folds as a mechanism for the exchange of chemical constituents between the stratosphere and troposphere. *J. Atmos. Sci.*, **37** (1980) 994–1004.
64. Johnson, W. B. & Viezee, W., Stratospheric ozone in the lower troposphere—I. Presentation and interpretation of aircraft measurements. *Atmos. Environ.*, **15** (1981) 1309–23.
65. Shapiro, M. A., Schnell, R. C., Parungo, F., Oltmans, S. J. & Bodhaine, B. A., El Chichon volcanic debris in an Arctic tropopause fold. *Geophys. Res. Lett.*, **11** (1984) 421–44.
66. Raatz, W. E., Schnell, R. C., Shapiro, M. A., Oltmans, S. J. & Bodhaine, B. A., Intrusions of stratospheric air into Alaska's troposphere, March 1983. *Atmos. Environ.*, **19** (1985) 2153–8.
67. Shapiro, M. A., Hampel, T. & Krueger, A. J., The arctic tropopause fold. *Mon. Wea. Rev.*, **115** (1987) 444–54.
68. Oltmans, S. J., Raatz, W. E. & Komhyr, W. D., On the transfer of stratospheric ozone into the troposphere near the North Pole. *J. Atmos. Chem.*, **9** (1989) 245–53.

Chapter 8

Trace Organic Chemicals in the Arctic Environment: Atmospheric Transport and Deposition

D. J. GREGOR

Lakes Research Branch, National Water Research Institute, Environment Canada, P.O. Box 5050, Burlington, Ontario, Canada, L7R 4A6

8.1 INTRODUCTION

As noted elsewhere in this volume, scientists have studied the glacial record and the atmosphere of the Arctic for several decades to investigate the residuals of anthropogenic activities in the most remote parts of the world. The phenomenon of arctic haze is a well known manifestation of hemispheric pollution in the North. Publications have generally considered increased acidity, increased sulphate and nitrate depositions, increased metals and the increase and subsequent decline of radionuclides from atmospheric bomb tests. However, there has been little interest in potential long term effects of these and other contaminants in the arctic environment.

Since the 1970s, evidence has been presented that long-lived trace organic compounds are reaching the Arctic and accumulating in the ecosystem. Through meaurement of body burdens of trace organic compounds in marine mammals and fish; it has become all too evident that these important biological food chains contain relatively elevated concentrations of organic chemicals. Local or regional sources could not possibly explain the presence of many of the compounds, especially the organochlorine pesticides which are for all intents and purposes foreign, in terms of use, to the Arctic. Consequently the atmosphere has to be an important pathway for these contaminants.

The first report of organic contaminants in the Arctic was by Holden as early as 1972 concerning concentrations of dichlorodiphenyl-trichlorethane (DDT) and polychlorinated biphenyls (PCBs) in seals.[1] Presently, there is an extensive data base of body burdens in arctic marine mammals and fish for DDT, chlordane related compounds, hexachlorocyclohexanes, toxaphene, chlorobenzenes, mirex, chlorinated dibenzo-*p*-dioxins and dibenzofurans and PCBs.[2–5] The major organochlorine residue in Canadian arctic marine invertebrates and fish is toxaphene with arctic cod having toxaphene residues five to ten-fold higher than DDT or PCB residues.[6] Chlordane is present at levels similar to that of DDT in zooplankton and amphipods[7] and in ringed seal blubber at two east/central arctic locations.[2] In polar bears, chlordane is a major residue along with PCBs.[3]

Concentrations of *p,p'*-DDT plus *p,p'*-DDE and PCBs in ringed seal, which are relatively sedentary, were higher in seals from Spitzbergen[8] than in the Canadian Arctic.[2] Arctic marine fish levels of ΣPCB (sum of individual congeners), ΣDDT (*p,p'*-DDE plus *p,p'*-DDD plus *p,p'*-DDT) and chlordane are two to five times lower than levels of these contaminants in fish from the North Atlantic, Baltic and Pacific northwest.[9]

It is only recently that efforts have been directed at understanding more fully the sources, transport pathways and controlling processes of organic contaminants in the Arctic. This has come about largely as a result of human health concerns. Northern natives of Canada have a much higher dependence on 'country foods' (i.e. fish, game and marine mammals) as a dietary staple, than most other Canadian residents. The estimated annual per capita consumption of country foods by northern natives from the eastern Arctic of Canada ranges between 108 and 597 kg with a mean of 267 kg.[10] The residents of a single community in this region have been studied in some detail with respect to dietary patterns and PCB concentrations in blood and breast milk.[11] In northern Quebec, the PCB concentrations in milk fat of Inuit women was almost five times higher than Caucasian women and Dewailly *et al.*[12] suggest that PCBs could play a role in the impairment of immunity and in the high occurrence of infection amongst Inuit children.

This chapter reviews the relatively limited information available quantifying long-range atmospheric transport of trace organic contaminants to the Arctic. In order to focus the content, only organochlorine pesticides, polychlorinated biphenyls and polycyclic aromatic hydrocarbons are considered. These classes of compounds are important environmentally as they tend to be long lived in the environment, may bio-accumulate and are frequently toxic and/or carcinogenic at elevated concentrations. Emphasis is placed on work that demonstrates a pathway between the atmosphere and deposition through measuring the occurrence of these compounds in the atmosphere and the abiotic arctic aquatic environment. The marine environment will not be discussed further as it is not possible to differentiate between long-range oceanic transport and atmospheric sources to the Arctic Ocean. Finally, brief mention is made of the limited information regarding trends of these compounds in the Arctic. However, before discussing the actual data, it is necessary to summarily review the physicochemical properties of these compounds, the processes controlling their emission and transport to the Arctic and their source regions.

A note of caution is appropriate at the outset of this paper due to the fact that data are compared from numerous published sources. That is, inclusion here does not imply that the author has either investigated or concluded that the data are compatible with respect to sampling and analytical procedures and quantification methods. This is particularly problematic for complex compounds like PCBs for which different analytical methods have been used by different investigators over time and different reporting methods exist. Thus comparison of results from different studies must be accepted in this context regardless of whether or not this is explicitly stated in the related discussion. In an effort to reduce confusion with respect to PCBs, the term ΣPCB will be used to express total PCB concentrations expressed as a sum of congeners. Note, however, that the same number of congeners may not be used by all authors. The term Ttl. PCB will be used to signify concentrations determined as equivalents of a single Aroclor or a 1:1:1 by volume mixture of Aroclor 1242, 1254 and 1260. Muir *et al.*[9] indicated that the use of Aroclor equivalents overestimates PCB levels in marine mammals because of the extensive transformation of congeners. Although it is desirable to conduct congener specific PCB analyses, it is not possible to replace the Aroclor equivalent data.

The term ΣDDT can be equally confusing as different authors have included different components. The term ΣDDT will be used here to represent the summation of *p,p′*-DDT, the derivative *p,p′*-DDD and the metabolite *p,p′*-DDE. In order to minimize confusion when different definitions apply, the term DDT alone will be used and the compounds included will be stated.

8.2 SOURCES, EMISSION AND ATMOSPHERIC EXCHANGE

8.2.1 Potential sources

Two classes of chemicals, here called chlorinated industrial organics and pesticides (OCs) and polycyclic aromatic hydrocarbons (PAHs) are considered. OCs relevant to the Arctic are those produced in large quantities, released in part to the environment, long lived and widely dispersed. These include PCBs, hexachlorocyclohexane (HCH), chlordane, DDT, dieldrin, heptachlor and heptachlor epoxide, endosulfan, methoxychlor, hexachlorobenzene (HCB) and polychlorinated camphenes (PCCs) or toxaphene. Production and use statistics for the OCs are difficult to uncover but a recent summary for some of the compounds is provided in Barrie *et al.*[13] using information from a variety of sources (Table 8.1). Current regulations regarding these compounds have also been summarized for selected countries by these authors while more detailed information is maintained by the International Registry of Potentially Toxic Chemicals (IRPTC) of the United Nations Environmental Programme (UNEP).

PCBs are a concern as most of the 209 congeners, especially those with five or more chlorines, are biomagnified.[2,14,15] PCBs have been used worldwide for a variety of purposes with as much as 370×10^3 kg estimated to have been released into the environment.[16] PCBs are found in the North as a result of local use in transformers, capacitors and other equipment, and the disposal of wastes at industrial and military sites. There are, for example, 21 abandoned early warning radar stations and one abandoned US Coast Guard site across the Canadian Arctic. Recently, these sites were cleaned up with respect to intact containers of contaminated oils although the possibility of buried equipment and contaminated buildings and soils at a number of these sites remains. Nonetheless, site specific contamination of this nature can not account for the widespread presence of PCBs throughout the Arctic.[17]

Organochlorine pesticides are a concern in the Arctic because of their long half lives and proclivity to bio-accumulate in aquatic and terrestrial food chains. Because pesticides are applied intentionally over large areas of land, their sources and pathways to the Arctic may differ from PCBs, other industrial organics and PAHs.

Table 8.1 Global production and use estimates of major organochlorines (modified from Barrie *et al.*[13])

Compound	*Use*	*Period of manufacture*	*Estimated global production (megatonnes)*			*Ref.*
			Global	*USA*	*Other*	
PCBs	Dielectric and hydraulic fluids	1930–present	1·2	0·6	0·4	19
DDT	Insecticide	1940–present	1·5	≈1	≈0·2[a]	28
Toxaphene	Insecticide	1947–present	—	0·5	≈0·2[b]	19, 28
Chlordane	Insecticide	1945–present	—	≈0·25	—	25
Hexachlorobenzene (HCB)	Fungicides and inadvertent production	1915–present	—	≈0·1	—	30
Hexachloro-cyclohexane (HCH)	Insecticide	1945–present	—	≈0·4	≈1[c]	22, 23
Aldrin/dieldrin	Insecticide	1950–present	—	0·11	0·04[d]	19

[a] Based on reported annual usage in Mexico, South America and India.[19]
[b] Based on use in Mexico and Central America.[19]
[c] Estimated from reported annual use of 0·06 Mtonnes in India and China during the 1970s and 1980s.
[d] Estimated from reported annual use of 0·001 Mtonnes of aldrin in Central America and Mexico in the period 1978–1988.

HCH can theoretically exist as eight different stereo isomers of which only five are found in the technical product. The dominant isomers are α-HCH at 70% and lindane at 13% of the total.[18] The insecticidal isomer lindane is currently used in pure form in North America and western Europe.[19,20] Canadian usage in the early 1980s was estimated to be 100–150 tonnes per annum of lindane.[21] Technical HCH is widely used in southern Asia and China with estimated annual use in the late 1970s of about 6×10^4 t.[22-24]

Technical chlordane, a multi-component product, was used as an insecticide but has more recently been used only for subterranean application in Canada and the US.[21,25] Since heptachlor is more insecticidal than chlordane, efforts have been made to purify the technical product to more than 70% heptachlor. Heptachlor is mainly used as a soil insecticide and for seed dressing, and has been reported to bioaccumulate. Heptachlor epoxide, a metabolite, is common in the environment, reportedly more resistant to chemical and biological changes in the aquatic environment than heptachlor, and has a predicted half life of several years.[21]

DDT has been banned or restricted in Canada, the USA and western Europe since the early 1970s. Its use in Eastern Europe as recently as the mid 1980s is suggested from pine needle studies[26] and it continues to be manufactured and used in southern Asia, Africa and in Central and South America.[19] Dieldrin is considered to be persistent, volatile, and bioaccumulative.[21] This product has also been restricted in Canada and the USA since the mid 1970s although it is still produced and used in a number of western European counries.[20] Endrin has properties similar to dieldrin and has been applied widely to cotton and rice.[27] Its use is presently restricted or banned in many developed countries.[20]

Endosulfan, consisting primarily of the α-isomer (70%) photolyzes in the gaseous state and, although it is bioconcentrated, it appears to be rapidly broken down in the organism or is eliminated, and there is little danger of accumulation.[21,27] It is widely used on cotton and orchards, the latter because it is not toxic to bees.[21] Methoxychlor is not stored in fatty tissues of higher organisms[21,27] and is still approved for agricultural use in Canada, particularly for the control of blackfly larvae. Use has decreased considerably since the 1970s.[21] It is also approved for use in the USA, USSR and many European countries.[20]

Toxaphene was the most heavily used insecticide in the USA during the 1960s and 1970s. There it was used mainly for controlling weevil on cotton,[19,28] while it has been used only in a very restricted way in

Canada.[21] Use in Mexico was estimated to be about 2200 t in 1978 and projected as 2000 t in 1988.[19] The USSR produces toxaphene derived from turpentine rather than camphene, for use as an agricultural pesticide,[29] but production and use figures are not available.

HCB, representative of a number of chlorinated benzenes, has been used variously as a fungicide, insecticide and as a wood preservative, and can be a by-product of many industrial activities as well as the chlorination of wastewaters. As a result it is ubiquitous in the aquatic and atmospheric environments and is a concern because it is relatively resistant to biotic and abiotic degradation. Chlorobenzenes tend to bioaccumulate.[21,30]

PAHs occur naturally in the environment as a result of forest and prairie grass fires. However, the quantity of PAHs in the environment has increased remarkably as a result of incomplete combustion of fossil fuels. A number of the PAHs are potentially carcinogenic.[21] Studies of PAHs in sediments of lakes in the Adirondacks of the eastern USA show that atmospheric emissions there peaked in the 1950s corresponding to the decline in the use of coal and increased use of oil for space heating.[31] Since the combustion of fossil fuels for space heating increases during the winter, it is to be expected that PAHs will be present in the Arctic during the winter season.

The sources of the compounds of concern in the Arctic are potentially global in nature and may be emitted to the atmosphere either directly through use (e.g. agricultural pesticides) or indirectly and unintentionally as part of industrial, combustion and other processes (e.g. PCBs and PAHs). Once introduced into the atmosphere, these compounds may circulate in the atmosphere and hydrosphere for long periods of time prior to going to a permanent sink or being broken down or bound up such that they are effectively removed from the system. Clearly, for all these compounds, the major sources are outside the arctic region although local combustion sources of PAHs will exist and improper disposal of dielectric fluids and PCB-contaminated wastes has resulted in very localized contamination.

The information regarding sources and transport of trace organic contaminants to the arctic environment is inadequate at present. However, it is possible that the more reliable information for sulphur sources can be instructive. Barrie *et al.*[32] estimated that 96% of the sulphur input to the Arctic in 1979/80 was from Eurasia while only 4% was from North America (see Chapter 6). The Eurasian inputs were almost equally split between the USSR and Europe (east and west). These authors further noted that most of the sulphur was transported into the region in the lower 2·5 km of the atmosphere. Consequently, future research with respect to sources and atmospheric pathways and processes for trace organic contaminants should emphasize Eurasia and transpolar circulation.

8.2.2 Emission and atmospheric exchange

If a specific organic compound is chemically stable with respect to, for example, photodegradation and chemical transformation, and is semi-volatile, this compound will tend to migrate over long distances through repeated deposition and volatilization. Many organic compounds are likely to be transported in this way.[33] The vapor pressure, organic content and the concentration, size and surface area of total suspended particles control the distribution or partitioning between vapor and particles. Vapor–particle partitioning, in turn, will influence the distance travelled by these contaminants by determining whether the compound is scavenged by rain and snow, by dry particle deposition, or by direct vapor exchange across the air–water interface. Overviews of these atmospheric processes are available.[33-36] It is sufficient here to note that compounds with saturation vapor pressures greater than $1{\cdot}3 \times 10^{-2}$ Pa should exist almost entirely in the vapor phase while those having vapor pressures less than $1{\cdot}3 \times 10^{-6}$ Pa should exist almost entirely in the particulate phase.[34] Most high molecular weight organics have vapour pressures between these extremes, as shown in Table 8.2, and their atmospheric lifetimes depend on the nature and concentration of particles in the air mass.

PCBs have been shown to be between 80 and 100% in the vapor phase in urban atmospheres and in the area of the Laurentian Great Lakes.[34,37,38] Duinker & Bouchertall[39] investigated the partitioning of several PCB congeners and concluded that although only a very small fraction of PCBs were found in aerosols in air samples, more than 99% of the PCBs in rain were particulate associated. More highly chlorinated congeners were particle associated while the lower chlorinated congeners remained essentially in the vapor phase. Thus the method of deposition can affect the toxicity of the PCBs and can influence how readily they can be revolatilized from the Earth's surface.

Particle deposition processes in polar regions have been reviewed by Barrie.[40] They depend mainly on physical factors including particle size distribution, surface roughness and wind speed. Vapor phase exchange with the surface, which may be water (fresh or saline), snow or soil, depends on the physicochemical properties of the compound as discussed above, which are in turn temperature dependent.[41] Limited summer

Table 8.2 Physicochemical properties of selected organochlorine compounds and PAHs[a]

Compound	*Henry's Law Constant pa(mol m^{-3})*		*Vapor pressure over liquid*	*Water solubility*[b]
	freshwater (20°C)	*seawater (−2°C)*	*(20°C)(pa)*	*(ug litre^{-1} @ 25°C)*
Organochlorines				
Lindane	0·20	0·03	$3{\cdot}9 \times 10^{-2}$	2 000–12 000
α-HCH	0·68	0·10	$1{\cdot}4 \times 10^{-1}$	1 200–2 000
Toxaphene	0·61	0·09	$6{\cdot}1 \times 10^{-2}$	300
Hexachlorobenzene	7·1	1·1	$7{\cdot}8 \times 10^{-2}$	<20
α-endosulfan	0·44	0·07	$3{\cdot}5 \times 10^{-3}$	150–600
Dieldrin	1·1	0·16	$5{\cdot}6 \times 10^{-3}$	186–200
Chlordane	9·0	1·3	$2{\cdot}7 \times 10^{-3}$	56–1 850
DDT	2·4	0·36	$2{\cdot}6 \times 10^{-4}$	<1–25
DDE	8·0	1·2	$1{\cdot}8 \times 10^{-3}$	1–140
PCB (IUPAC #) (#Cl)				
33 (3)	22·6	3·3	$1{\cdot}2 \times 10^{-2}$	
52 (4)	24·1	3·6	$9{\cdot}0 \times 10^{-3}$	
101 (5)	18·1	2·7	$1{\cdot}5 \times 10^{-3}$	
153 (6)	10·0	1·5	$2{\cdot}5 \times 10^{-4}$	
PAHs				
Pyrene	0·78	0·12	$8{\cdot}2 \times 10^{-3}$	
Phenanthrene	2·6	0·38	$6{\cdot}0 \times 10^{-2}$	1 000–1 800[c]
Napthalene	28·0	4·1	17·5	26 000[c]

[a] Modified from Barrie *et al.*[13] unless indicated otherwise.
[b] From Arctic Laboratories[67].
[c] From CCREM[21].

arctic air and seawater measurements show a wide range of atmospheric saturation indices averaging about 60% and as high as 90% for α-HCH, 40% for lindane[7,41,42] and 30% for toxaphene.[7,41] Although published data do not appear to be available to confirm this, the cold dry arctic winter conditions should increase the proportion of dry deposition of aerosol associated contaminants to total contaminant deposition.

Exchange processes and rates between water or soil and the atmosphere have been researched, but usually not at arctic temperatures.[38,43–46] Only recently, have exchange processes between the surface snow and the arctic atmosphere been considered for volatile organic compounds. Results of replicate sampling of the 1985/86 snow layer on the Agassiz Ice Cap, Ellesmere Island, Canada (80°40′N, 73°30′W) in May of both 1986 and 1987 are presented in Table 8.3.[13,47] Except for Ttl. PCBs, the

Table 8.3 Comparison of 1985/86 season concentrations for selected chlorohydrocarbons in Agassiz Ice Cap as meaured in 1986 and again in 1987

Compound	*Concentration* ($pg\ litre^{-1}$) (*mean ± standard deviation*)		
	1986 (*n*=5)	*1987* (*n*=3)	*residue* (%)
Lindane	4 080±909	128±43	3
α-HCH	6 576±1 441	497±90	8
Heptachlor epoxide	358±55	47±15	13
Chlordane (*cis*+*trans*)	760±89	40±26	5
Dieldrin	1 346±238	213±75	16
α-Endosulfan	1 094±156	ND	nil
Ttl. PCB	972±344	648±118	67

ND = not detectable.

concentration of the OCs measured in this layer in 1986 was significantly greater than the concentration measured in the same layer in the following year. Sampling was also repeated over the two years in annual firn layers with good agreement between measured concentrations indicating that this was not an analytical artifact. Comparison of the snow accumulation layer for the winter of 1986/87 sampled in May of 1987 and again in May of 1989 shows residual concentrations of OC pesticides of about 1% in the glacier relative to the depositional concentration measured in 1987 (Gregor, unpublished data). Since melting on the glacier was limited, as indicated by the preserved specific conductance profile presented in Fig. 8.1, and the snow layers below showed consistent low concentrations, it must be concluded that large portions of the OCs, with the exception of the PCBs, revolatilized to the atmosphere during the summer season as the air and snow temperatures rose. It is conceivable that, although a large portion of the PCBs stay, it is the lower chlorinated congeners that are lost from the snow.

Supporting evidence for revolatilization from the snow in the Arctic comes from Hargrave *et al.*[48] who detected HCB in snow collected in May–June, 1986 (33 pg $litre^{-1}$) but not in snow from the same area in August–September. HCB has the highest Henry's Law Constant and the lowest water solubility of the compounds listed in Table 8.2 and thus it is readily volatilized from the snow. At 0°C, more than 99% of HCB was

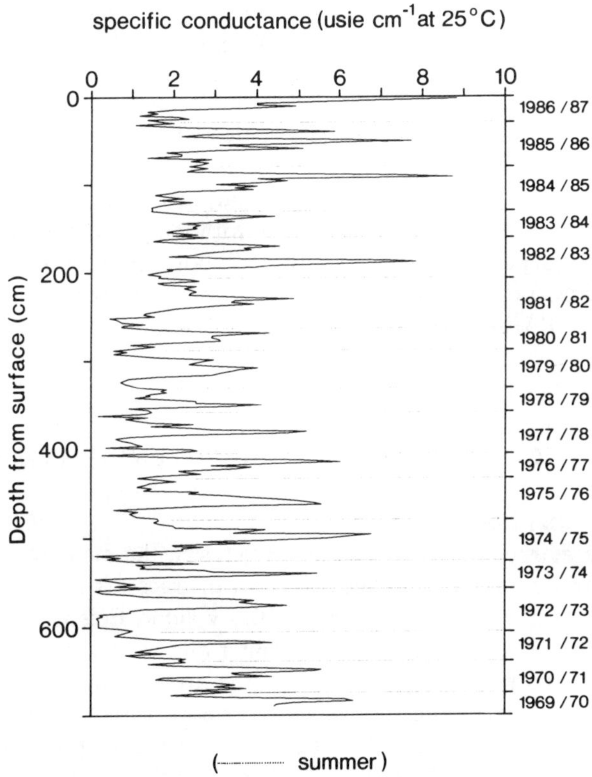

Figure 8.1. Plot of specific conductance versus depth for the Agassiz Ice Cap from the snow surface in the spring of 1987 to a depth of 6·9 m.

still found in the vapor phase by Bidleman *et al.*[49] Similarly, chlordane was found at 91 pg $litre^{-1}$ in the snow sampled in May–June while it was undetectable (<1 pg $litre^{-1}$) in August.[48]

The knowledge base for OCs remains severely inadequate and in order to understand the processes and fate of organochlorines in the Arctic, more research is essential in polar and sub-polar conditions.

Since the combustion of fossil fuels in the Northern Hemisphere increases during the winter season and transport from the mid-latitudes to the Arctic is more efficient in winter, it is to be expected that the products of combustion would occur in remote arctic regions. Local sources may also be important. The sampling and analysis of PAHs is considerably more difficult than for OCs, especially in the Arctic, where local conditions and the equipment associated with sampling (e.g. aircraft) can themselves be sources. Edwards[50] notes that the extraction and recovery of PAHs is problematic, resulting in underestimates and considerable sample variability.

Tuominen *et al.*[51] demonstrated that the largest proportion of PAHs in the atmosphere in Finland was associated with particulate matter and that precipitation, specifically snowfall, was an efficient scavenger of particulate associated PAH. Bidleman[33] argued that the proportion of aerosol bound PAHs increased inversely with temperature and thus both wet and dry deposition were likely to be important at ambient northern temperatures. Losses of PAHs from the arctic snowpack and air–water exchange of PAHs have apparently not been addressed in the published literature.

8.3 OCCURRENCE IN THE ARCTIC ENVIRONMENT

8.3.1 Atmosphere

There is a marked seasonal distinction between the meteorologic conditions controlling the long-range transport of trace organic pollutants to the Arctic, as discussed in Chapter 2 of this volume. Episodes of pollution are much more prevalent in the late winter and early spring.

Much of the published OC data for the arctic atmosphere has come from Scandinavian countries which are more directly affected at all latitudes by both summer and winter atmospheric transport from sources to the south and east. Selected data from Scandinavia are compared to the relatively recent and geographically limited Canadian data and data from the Bering and Chukchi Sea region in Table 8.4. HCH has

Table [illegible].4 Selected arctic (north of 60°N) air concentration data (ranges except where noted as arithmetic means) from published literature (n = number of samples, α-HCH = α-hexachlorocyclohexane, HCB = hexachlorobenzene, ΣPCB = total PCB as reported by individual references, ΣDDT = DDT + DDE, ND = not detectable)

Location	Time	n	α-HCH ($pg\ m^{-3}$)	Lindane ($pg\ m^{-3}$)	HCB ($pg\ m^{-3}$)	ΣPCB ($pg\ m^{-3}$)	ΣDDT ($pg\ m^{-3}$)	Ref.
Norway								
Bear Island	Fall '80	5	217–516	15–42	29–45			52
	Winter '81	5	185–270	12–16	59–188			
Spitzbergen	Fall '80	4	708–1 290	25–50	50–85			52
	Winter '81	5	161–274	13–17	71–95			
Hopen	Fall '82	11	250–1 500	ND–6	100–230			52, 53
	Winter '83	11	200–750	20–90	90–210			
Norway								
Ny Alesund	Fall '82/83	[a]	407–1 416	0·1–67	75–227			54
	Winter/spring '83/84	[a]	121–787	12–102	29–389			
	Summer '84	[a]	260–774	24–100	20–201			
Bjornoya	Fall '82/83	[a]	277–1 550	0·1–32	78–200			54
	Winter/spring '83	[a]	110–469	23–80	87–201			
	Summer '84	[a]	38–305	5–41	42–149			
Sweden	'84/85	[b]				≈20 – 140	≈10–120	55
Canada								
Ice Island	Spring '86	6	451[c]	70[c]	73[c]	<2[c]	2[c]	48
(81°N, 97°W)	Summer '86	4	253[c]	17[c]	63[c]	<9[c]	<1[c]	
	Summer '86	7	417–731	24–37	131–233	9–20	0·5–1·4	41
	Summer '87	6	283–377	36–53	44–155	15–28	0·8–4·3	
Alert	Feb–April '88	5	100–270	15–31	100–140	29–60	0·6–3·5	56
Bering & Chukchi Sea	'88	17	140–410	52–84	120–270			57

[a] Number of samples is relatively large and is stated to be based on 48- or 72-hour sampling over sampling periods each year of about 30 days.
[b] Sampled over a period of one year but number of samples not indicated.
[c] Mean of the indicated number of samples.

consistently been found to be the dominant atmospheric OC in the Arctic with α-HCH exceeding lindane. It has been observed[52,53] that concentrations of α-HCH in wintertime are lower than in summer while higher concentrations of lindane are observed in the winter season. Air masses in the Arctic show α-HCH to lindane ratios of 10:1 in winter and as much as 50:1 and 500:1 in summer[52] and fall,[53] respectively. The higher summer ratio is thought to be a result of the more rapid photodegradation of lindane during the summer[52] and to preferential scavenging of lindane from the atmosphere.

HCB is frequently the next most dominant OC compound in the arctic atmosphere (see Table 8.4). Concentrations are similar from around the world and show little seasonal variation.[54] PCBs and DDT are detected throughout the Arctic but at concentrations much lower than the HCHs. Interestingly, Larsson & Okla[55] have been able to distinguish a decrease in concentration of DDT (*p,p′*-DDT plus *p,p′*-DDE) along a south–north gradient in Sweden. A gradient of this type is expected with increasing distance from the source region. However, these authors also show that the levels of PCBs and DDT in the lower atmosphere are positively related to the mean temperatures of the sampling station. Since the amount of gas phase pollutant in the atmosphere is dependent on vapor pressure, which in turn is a function of temperature, it is conceivable that as temperature decreases northward, a decreasing amount of a pollutant is able to remain in the gas phase and thus adsorption or deposition occurs.

Chlordane-related compounds were first observed in the Canadian arctic air at Mould Bay (76°15′N, 119°16′W) in July, 1984[58] and more recently on the Canadian Ice Island (approximately located at 81°N, 97°W) in 1986 and 1987 at 2–10 pg m^{-3} (total chlordane related compounds).[41,48] Between February and April 1988, total chlordane concentrations in air averaged 3·3 pg m^{-3} at Alert (82°30′N, 62°18′W).[13,56] Concentrations of α-chlordane alone have been reported generally below 5 pg m^{-3} for arctic sites in Norway.[52–54] These concentrations were comparable to southern Sweden but much lower than sites in the continental USA in the mid 1980s.[49]

Dieldrin has been detected in arctic air at between 0·1 and 5 pg m^{-3} at the Canadian Ice Island in 1986 and 1987.[41] Only trace quantities were reported by Hargrave *et al.*[48] but with higher detection limits. The most abundant OC after HCH and HCB at this site was toxaphene at concentrations between 36 and 44 pg m^{-3} in the summers of 1986 and 1987.[7,41]

PCCs (toxaphene) were the most abundant pesticide residues in Canadian arctic fish and marine mammals.[2,3,6] Concentrations of PCCs in the arctic atmosphere at the Canadian Ice Island were reported for the first time at 44 pg m^{-3} for August 1986 (mean of seven samples) and at 36 pg m^{-3} in June 1987 (mean of six samples).[7,41] Average concentrations of toxaphene at Alert in February to April 1988 were nearly identical to the earlier summer measurements of 35 pg m^{-3}.[56] Although other published measurements are not available for north of 60°N, these measurements are comparable to atmospheric concentrations in Sweden in 1983–1985.[49]

Most studies have shown that the organochlorine compounds are concentrated in the gas phase, although the use of relative concentrations on filters (particulate associated) and absorbents (gas phase) in large volume air samplers may not be precise. Specifically, adsorption and desorption to and from particles may occur under varying conditions (e.g. temperature changes) over the course of the sampling period for large volume air samples or due to physical desorption as a result of pulling air through the filter.[33,43] Larsson & Okla[43] determined that relatively high, near source particulate PCB concentrations ranged from 0·4 to 7·3% of the total PCBs. Oehme & Mano[52] reported more than 98% of all OC compounds in the vapor phase while Bidleman *et al.*[49] reported less than 10% of OCs were particle associated at 20°C. These authors noted that the percentage of particle associated *p,p'*-DDT increased to as much as 74% at 0°C. Other OCs increased less dramatically. HCH and HCB remained at less than 1% particle associated even at 0°C. The absence of quantifiable concentrations of OCs in aerosols was noted at the Canadian Ice Island by Hargrave *et al.*[48] Further understanding of vapor–particle partitioning of trace organic contaminants at sub-zero temperatures is essential to understanding processes controlling air–surface exchanges in the Arctic.

Relatively few measurements of PAHs in the arctic atmosphere are available. Data from Alaska and Canada have been summarized in Table 8.5. Winter concentrations at remote sites (Alert and Narwhal Island) are similar to summer concentrations at Barrow, somewhat less than winter concentrations at Barrow and a great deal less than winter concentrations at Fairbanks. Daisey *et al.*[59] concluded that much of the PAH at Barrow during the winter comes from long-range transport as local sources are minimal due to the use of natural gas as a heating fuel. However, the summer concentrations appear to be influenced by local emissions as suggested by the predominance of coronene and ben-

Table 8.5 Concentrations of selected PAHs in the arctic atmosphere expressed as ranges for the period of samples (pg m^{-3})

	Canada	*Alaska*			
	Alert	*Barrow*		*Narwahl I.*	*Fairbanks*
	Feb–April 1988	*March 1979*	*August 1979*	*April 1979*	*Nov 1976 to Apr 1977*
n	5	3	5	2	21
Phenanthrene	6·6–130	80–170	4–30	4–420	100–21 600
2-Methyl-phenanthrene	0·16–3·2	NA	NA	NA	NA
Fluoranthene	7·7–74	120–520	8–55	23–95	500–29 000
Pyrene	1·9–50	50–540	5–72	150	400–39 800
Benzo(a)-anthracene	0·75–8·3	NA	NA	NA	NA
Chrysene	4·0–24	40–170	ND–15	ND	600–44 700
Benzo(b) fluoranthene	6·4–23	NA	NA	NA	NA
Benzo(e) pyrene	1·8–8·0	120–370	ND	ND	300–20 700[a]
Benzo(a) pyrene	1·0–3·4	7–44	ND–35	ND	
Benzo(ghi) perylene	1·2–8·6	70–74	ND–40	ND	2 900–154 700
Reference	56	59	59	59	61

[a]Sum of benzo(e)pyrene and benzo(a)pyrene.
NA = not analysed. ND = not detected.

zo(ghi)perylene which are more typical of automobile and fuel oil emissions. These authors further observed that the mean PAH distribution for March samples appears to be closely related to those observed in Norway and Sweden between December 1976 and April 1977 for periods when air masses circulate from Europe.[60] In contrast, the much higher concentrations in Fairbanks are thought to be due to vehicular exhaust combined with coal-fired power plant emissions.[61] Clearly, a better understanding of PAHs is required.

8.3.2 Concentrations in the arctic freshwater system

This section considers two compartments of the aquatic system. First, measurements of contaminants in snow are discussed as direct evidence of long-range atmospheric transport and deposition to the arctic aquatic environment. Second, the limited data for river and lake systems are presented to link deposition to the aquatic system. In both cases, comparisons will be made with temperate regions.

The earliest known reported work on organochlorine compounds in the Canadian Arctic was undertaken in 1970 as part of the Ice Field Ranges Research Project (IRRP) in the Yukon Territory.[62] Sampling for subsequent analyses of DDT was undertaken at an elevation of 5364 m on Mount Logan. Nineteen samples were taken in snow at depths of 1 to 15 m with DDT not detected in any of these samples at a detection limit of 5 ng litre^{-1}.

The earliest measurements in northern Sweden were conducted in 1972 and 1973 using a silicone oil impregnated nylon mesh to measure airborne fallout.[55] These data (see Table 8.6) were compared with results from a similar study conducted in 1984–1985 and it was concluded that the overall fallout of DDT (*p,p'*-DDT plus *p,p'*-DDE) decreased during the intervening period. The ratios of DDT to DDE were found to have

Table 8.6 Comparison of published contaminant deposition for northern Sweden and Moscow with Canadian Arctic data (μg m^{-2} year^{-1})

Compound	*Moscow*[a] (*1978–79*)	*Sweden* (*1972–73*)[b]	*Sweden* (*1974–76*)[c]	*Sweden* (*1984–85*)[b]	*Arctic*[d] (*1986–87*)
α-HCH	14				0·02–9·1
Lindane	12				0·01–2·0
DDT	52[e]	≈1·2–13·2[f]		≈0·1–8·79[f]	0·01–3·0[g]
Ttl. PCB		≈1·2–16·6	24–60	≈24–154	0·04–0·6

[a] From the Moscow area as reported by Bobobnikova & Dibtseva.[68]
[b] Estimated from figures published by Larsson and Okla[55] for stations north of 60°N only. Actual data were reported for a monthly time period and these data have been multiplied by 12 to provide a comparable annual value.
[c] From Lunde *et al.*[63] only for stations north of 60°N.
[d] From Barrie *et al.*[13] and determined from annual snow pack samples representing September to May and here assumed to represent the annual deposition.
[e] Sum of *p,p'*-DDT, *o,p'*-DDT, *p,p'*-DDD and *p,p'*-DDE.
[f] Sum of *p,p'*-DDT and *p,p'*-DDE.
[g] Includes only 1987 data, sum of *p,p'*-DDE, *p,p'*-DDD, *o,p'*-DDT and *p,p'*-DDT.

decreased from 4·10 ($n=17$) in 1972–1973 to 2·64 ($n=52$) in the more recent study. Since DDT degrades to DDE in ultraviolet light in the atmosphere and by the metabolism of organisms, it was concluded that the restrictions on DDT use in Sweden in 1971 had been effective and that the DDT reaching Sweden presently has 'aged' during long-range transport from distant source regions. The trend of PCB concentrations over this time period was more complicated having decreased slightly in the northern part of Sweden.[55]

Although still limited, recent work has been concentrated in the Canadian Arctic. The Finnish Huure Expedition to the North Pole in 1984 collected 1 litre of water equivalent snow samples for analyses of a broad range of chlorinated pesticides and chlorophenols.[64] None of the target compounds were detected in these snow samples. Based on the work of Gregor & Gummer[65] it would have been reasonable to expect to see at least lindane (α-HCH was not one of the target compounds) at the Finnish study's reported detection limit of 0·5 ng litre^{-1}. The most comparable station from the 1987 study is the Arctic Ocean site which had Ttl. PCB and lindane concentrations of approximately 1·0 and 0·9 ng litre^{-1}, respectively. Annual variation may well be such that these compounds in 1984 were below the level of detection of the Finnish study.

In 1979, McNeely & Gummer[66] undertook snow sampling in the area of east-central Ellesmere Island. Sampling included snow pack at five sites, fresh snow at three sites and rime frost from two sites between 24 May and 19 June 1979, yielding 19 samples with replicates and repeat analyses for a total of 28 analyses. The results for organochlorine pesticides are summarized and compared with selected data from the much more extensive study of Gregor & Gummer[65] in Table 8.7. The two stations selected for comparison were Alexandra Fiord sampled in 1986, the only station overlapping between the two studies, and the surface snow for both 1986 and 1987 from the Agassiz Ice Cap, approximately 175 km to the north. Although the results for Alexandra Fiord for 1986 were below the level of detection available to McNeely & Gummer, the observations at the Agassiz Ice Cap support HCH concentrations of the magnitude found in this earlier study.

Concurrent with the 1987 snow chemistry survey of Gregor & Gummer, Arctic Laboratories Limited undertook work at the Canadian Ice Island during 1986 and 1987.[67] Data from these two studies are compared in Table 8.8 and show good agreement given the very low concentrations seen in these samples. HCH data for 1987 Ice Island snow samples are compared for three different laboratories using three different

Table 8.7 Summary of pesticide snow chemistry data collected in 1979 by McNeely & Gummer[66] from east-central Ellesmere Island compared with selected data from Gregor & Gummer[65] for Alexandra Fiord (1986), and the surface snow layers from Agassiz Ice Cap in 1986 and 1987

Pesticide	*1979 range* ($ng\ litre^{-1}$)	*Confirmed detections*[a]	*Unconfirmed detections*[b]	*Alexandra Fiord*[c] ($ng\ litre^{-1}$)	*Agassiz L−1*[d] ($ng\ litre^{-1}$)	*Agassiz L+1*[c] ($ng\ litre^{-1}$)
α-HCH	<1–18	21	5	0·43	6·7	19·1
Lindane	<1– 8	10	5	0·22	4·4	6·8
Chlordane	<1– 4	3	7	0·17	0·74	0·23
DDT[e]	<1– 6	4	10	0·07	0·01	0·13
α-Endosulfan	<1– 2	1	2	0.24	1.1	1.1
Dieldrin	<2– 4	5	12	0·83	1·4	1·0

[a] Includes both detections above the detection limit and below the detection limit (i.e. not quantified) but confirmed to be present by MID-GC/MS.
[b] Indicates instrument responses below the detection limit but of insufficient concentration to permit confirmation by MID-GC/MS.
[c] Single sample.
[d] Represents the mean of five samples from this layer.
[e] Sum of *p,p′*-DDD, *p,p′*-DDE, *o,p′*-DDT and *p,p′*-DDT.

sampling methods in Table 8.9. The University of South Carolina results are relatively high and lindane is dominant, which does not agree with the other analyses and is unusual for the Arctic. The difficulty in measuring the very low concentrations in the arctic aquatic system, as illustrated here, demonstrates the need for more extensive inter-laboratory evaluations to permit data pooling and comparisons.

Toxaphene measurements in the arctic snowpack are restricted to four samples. Two of these were collected from the Canadian Ice Island in May 1986 and two others were collected from the Devon Ice Cap on Devon Island (75°6′N, 82°45′W) and from the vicinity of Resolute Bay on Cornwallis Island (74°42′N, 94°54′W) in May 1987. Concentrations for the two Ice Island samples were 85 and 307 $pg\ litre^{-1}$ while the Resolute Bay and Devon Ice Cap samples were 600 and 1700 $pg\ litre^{-1}$, respectively.[7] The wide range in concentration between Resolute Bay and Devon Ice Cap was not surprising given the distance between these two stations, the difference in elevation (nearly 1700 m) and different snow accumulation rates. There are no known published measurements of toxaphene concentrations in arctic surface fresh waters.

Table 8.8 Comparison of 1987 Ice Island samples for organochlorine residues as reported by Arctic Laboratories[67] with results from Gregor & Gummer[65] based on three replicates for each laboratory

Compound	*Concentration (pg litre^{-1}) in melted snow*	
	Gregor & Gummer (mean ± 1SD)	*Arctic Laboratories (mean ± 1SD)*
Aroclor (1254)[a]	129 ± 37	185 ± 160
Hexachlorobenzene	11 ± 6	12 ± 3
Heptachlor	< 3[b]	6 ± 3
Aldrin	Int	0·5 ± 0·5
o,p'-DDE	ND	2·8 ± 2·4
p,p'-DDE	ND	20 ± 14
Mirex	ND	< 4·0
trans-Chlordane	8 ± 1	20 ± 15
cis-Chlordane	< 12[c]	25 ± 18
p,p'-DDT	23 ± 19	< 47 ± 26
Heptachlor epoxide	30 ± 11	46 ± 26
α-Endosulfan	21 ± 1	97 ± 58
Dieldrin	54 ± 5	< 140 ± 55
Endrin	Spike	27 ± 11
Methoxychlor	55 ± 35	37 ± 26

[a] For this comparison, the PCB data from Gregor and Gummer are reported as Aroclor 1254 whereas in all other Gregor and Gummer references in this report PCBs are reported as the 1:1:1 mixture of Aroclor 1242, 1254 and 1260.
[b] One of the three samples was ND. Therefore the concentration is reported as less than the maximum concentration observed with no standard deviation.
[c] Two of the three samples were ND. Therefore the concentration is reported as less than the concentration of the single value with no standard deviation.
SD = standard deviation. Int = interference on chromatograms. ND = not detectable.

Table 8.9 Comparison of 1987 Ice Island snow samples for hexachlorcyclohexane (HCH) residues among Gregor & Gummer[65], Arctic Laboratories[67] and University of South Carolina (USC) as reported by Arctic Laboratories

Compound	*Concentration (pg litre^{-1}) in melted snow*		
	Gregor & Gummer[a] (mean ± 1SD)	*Arctic Labs[b] (mean ± 1SD)*	*USC[c] (mean ± 1SD)*
α-HCH	143 ± 13	290 ± 130	315 ± 25
Lindane	83 ± 9	192 ± 96	365 ± 15

[a] Based on three samples collected 6 June 1988.
[b] Based on three samples collected 8, 9 and 15 June 1988.
[c] Based on two samples collected 7 and 9 June 1988.

The most northerly data from the USSR found to date, although not within the definition of the Arctic used here, are the results of precipitation samples collected in the Moscow region from January 1978 to September 1979.[68] Although Bobovnikova & Dibtseva[68] did not indicate specifically which monthly samples were snow, it can be reasonably assumed for this latitude that samples from December through March were primarily snowfall. Based on these four months, it can be estimated that DDT (sum of *p,p'*-DDT, *o,p'*-DDT, *p,p'*-DDD and *p,p'*-DDE) ranged from 76 to 110 ng $litre^{-1}$; α-HCH ranged from 'trace' to 25 ng $litre^{-1}$; and lindane ranged from 'trace' to 30 ng $litre^{-1}$. The detection limit, apparently indicated by the term 'trace', for any of these compounds was not stated. The concentrations of α-HCH were similar to those noted for the Canadian Arctic[65] while lindane tends to be somewhat higher, likely reflecting proximity to the source region. Similarly, DDT was an order of magnitude higher in the Moscow snow samples than the arctic snow samples collected by Gregor & Gummer.[65]

There do not seem to be published data for organochlorine concentrations in arctic rivers and lakes except as reported by Erickson & Fowler[69] for the Mackenzie River and by Gregor[70] and Lockhart *et al.*[71] for Sophia and Amituk Lakes on Cornwallis Island and for Lake Hazen on northern Ellesmere Island. Data from these sites for α-HCH, lindane and HCB are compared with measurements from rivers in northern Ontario[72] and northern Quebec[73] and from Lakes Superior and Ontario[74] in central Canada in Table 8.10. Although the number of samples in the Canadian Arctic are few, the concentrations of HCHs and HCB are within the same ranges as those in rivers in northern Quebec and Ontario. These in turn are generally comparable to the concentrations of the two Laurentian Great Lakes reported here. The very high concentration of HCH in the surface waters of Sophia Lake is perhaps attributable to the fact that this is a meromictic lake with a very shallow surface layer resulting in trapping and minimal dilution of surface runoff. α-HCH was found above the detection limit in four of five lakes in Labrador, Canada, while lindane was detected in only one of these lakes.[75]

Lindane, in daylight, can be transformed to the α-HCH isomer with a half life, dependent upon temperature and other conditions, of the order of one year.[53] Although degradation rates are likely to be substantially different in a polar climate than in a temperate one, this ratio may be informative with respect to possible age or length of time of HCH transport. Also, chemical formulations differ. North America uses pure

Table 8.10 Comparison of selected organochlorine pesticide concentrations (pg litre^{-1}) in arctic lakes and the Mackenzie River with rivers from northern Ontario and Quebec and Lakes Superior and Ontario in central Canada

	Sample depth (m)	*Number of samples*	*α-HCH*	*Lindane*	*HCB*	*α-HCH/ lindane*	*Reference*
L. Hazen	2	3	590	170	90	3·5	70, 71
L. Hazen	35	3	340	100	120	3·4	70, 71
Sophia L.	2	2	13 620	1 740	140	7·8	70, 71
Amituk L.	2	1	3 270	450	50	7·3	70, 71
Amituk L.	20	1	3 350	420	70	8·0	70, 71
Amituk L.	39	1	2 220	210	30	10·6	70, 71
Mackenzie R. at delta	NA		3 000–4 900	400–500	NA		67
N. Ontario Rivers	NA	60	5 000–8 400	400–1 500	100–600	7·1[a]	72
N. Quebec rivers Hudson Bay	NA	87	8 000	1 100	ND	7·6[a]	73
N. Quebec rivers Ungava Bay	NA	87	5 800	800	ND	7·3[a]	73
L. Superior	1	19	5 700–11 000	800–1 430	20–40	7·4[a]	74
L. Ontario	1	31	1 170–5 920	310–2 280	20–110	3·1[a]	74

[a] Ratio has been determined from means as reported by respective authors.
NA = not applicable. ND = not detected.

lindane with aerosol ratios reported at 1:7 in the Great Lakes region[34] while the HCH mixtures utilized in Europe produce ratios of 1:3–4.[52] As shown in Table 8.2, the water solubility of lindane is one to two orders of magnitude higher than α-HCH. This could result in preferential scavenging of lindane by precipitation which will, in turn, affect the ratios of the residual compounds measured in the arctic atmosphere.

Alpha/gamma-HCH ratios in the air at the Canadian Ice Island were 6:1 and 15:1 in the spring and summer, respectively, of 1986 (see Table 8.4).[48] Based on Table 8.3, the ratio for surface snow from the Agassiz Ice Cap in 1986 was 2:1 but 4:1 a year later The average ratio for all surface snow samples collected by Gregor & Gummer[65] in 1986 and 1987 was 2·5:1. These low ratios in the snow relative to the atmosphere, may indicate preferential scavenging of lindane by precipitation. Whether the difference in solubility is maintained at sub-freezing temperatures is unknown. Another explanation could be that the snow pack ratios represent deposition during pollution events which carry relatively fresh HCH to the Arctic whereas the air measurements possibly represent a background ratio of older, recycled HCH.

The ratios of α-HCH to lindane for various water bodies are also shown in Table 8.10. Although by no means conclusive, it is noteworthy that Hazen Lake has HCH ratios similar to Lake Ontario while Sophia Lake and Amituk Lake have ratios comparable to Lake Superior and northern Ontario and Quebec rivers. Further speculation on the significance of HCH ratios requires a better understanding of degradation processes in the Arctic as well as a more extensive data base.

Arctic snow samples collected in 1987 have been analyzed by Gregor (unpublished data) for 17 PAHs by capillary gas chromatography with electron impact, multiple ion monitoring mass spectrometric detection (MIM–GC/MSD). ΣPAH concentrations were highly variable ranging from non-detectable to 125 ng litre^{-1}. The seven PAHs detected consistently were phenanthrene, 1- and 2-methylnapthalene, 1,2,3,4-tetrahydronaphthalene and indene with generally minor quantities of fluorene and/or acenaphthene.

All seven of these PAHs are coal tar derivatives, practically insoluble or having low water solubilities (although having higher solubilities than many of the OCs) and of relatively low molecular weights for PAHs (Table 8.2). These PAHs are mildly toxic with some having a tendency to bio-accumulate, to be slightly carcinogenic and even to be mutagenic. Some, such as acenaphthene and 1,2,3,4-tetrahydro-naphthalene have commercial uses either alone or in conjunction with other products.[76]

Compounds commonly associated with internal combustion exhausts (e.g. fluoranthene, phenanthrene and pyrene) were either not detected or were detected only occasionally or were believed to be sample contamination as evidenced in method blanks. Relative to other PAHs, it appears that it is the low molecular weight, high vapor pressure, more water soluble PAHs that are being found in the arctic snow.

PAHs were not detected above the limit of detection for any of the arctic snow samples from 1979.[66] PAHs have been measured in fresh snow samples in Bavaria, Germany,[77] and in Switzerland[78] at concentrations ranging from about 0·2 to 2 μg litre^{-1}, an order of magnitude higher than seen in the Arctic. Phenanthrene, fluoranthene and pyrene were dominant in the Swiss samples of which only phenanthrene was detected in the Arctic. Evidently, the PAHs transported to the Arctic generally have higher Henry's Law Constants than the PAHs commonly found in these temperate areas.

8.3.3 Aerial deposition rates

The aerial deposition rates of contaminants to the land surface from precipitation is a more appropriate value than concentrations for determining spatial patterns and temporal trends. Median winter deposition rates for selected organochlorines based on concentrations in Canadian Arctic snowpack in the spring of 1986 and 1987 are presented in Table 8.11.[13,70] Although these are shown as winter deposition rates, it is reasonable to assume that these represent, for all intents and purposes, annual deposition in the Arctic. There are two facts supporting this assumption. First, approximately 80% of the total precipitation in the high Arctic arrives as snowfall.[79] Second, the period of snow accumulation at these arctic stations generally ranges from September to May and this encompasses the period during which the arctic aerosol has been shown to be most heavily contaminated with anthropogenic pollutants.[32,80,81]

These deposition estimates are limited by the fact that snow samples are commonly collected from remote sites and total annual snowfall has to be assumed to be equal to that recorded at the nearest weather station which may be several hundreds of kilometres distant. In addition, total snow storage in a basin generally exceeds weather station records by as much as from 130 to 300%.[79,82] It must also be remembered that deposition to the snow pack does not necessarily mean a direct transfer of this quantity of contaminant to the aquatic system following snow melt. As discussed in Section 8.2.2, the majority of the measured pesticides

Table 8.11 Median and range of concentrations and corresponding winter season deposition of trace organic compounds from Canadian Arctic snow chemistry surveys in 1986 and 1987[13,70]

	1986 (n=12)		*1987 (n=15)*	
Compound	*Concentration (pg litre^{-1})*	*Deposition (ng m^{-2} year^{-1})*	*Concentration (pg litre^{-1})*	*Deposition (ng m^{-2} year^{-1})*
Ttl. PCB	745	73	907	95
	(20–1 760)		(257–1 773)	
α-HCH	3 185	285	910	95
	(430–8 120)		(143–42 700)	
Lindane	1 160	104	700	73
	(220–4 446)		(83–10 500)	
Heptachlor epoxide	97	10	117	16
	(ND–362)		(ND–1 600)	
α-Endosulfan	345	37	175	24
	(100–1 094)		(ND–4 880)	
Dieldrin	540	58	400	56
	(205–1 402)		(ND–4 420)	
HCB	28	3	16	2
	(10–70)		(ND–104)	
DDT[a]	18	2	93	8
	(10–50)		(ND–1 380)	
CBZ[b]	NA	NA	240	30
			(ND–915)	
PAH[c]	NA	NA	19 000	4 246
			(190–149 000)	

[a] DDT is the sum of *p,p'*-DDD, *p,p'*-DDE, *o,p'*-DDT and *p,p'*-DDT.
[b] Sum of all detected chlorobenzenes excluding HCB.
[c] Sum of all detected PAHs.
NA = not analysed. ND = not detectable. HCB = hexachlorobenzene.

disappear from the surface layer of Agassiz Ice Cap over the course of a year without substantial melting.[47] Attempts to confirm that this phenomenon occurs in snow that totally melts annually were undertaken in the vicinity of Resolute Bay in 1989 but with inconclusive results (Gregor, unpublished data). Nonetheless, the concentrations of HCHs present in several arctic lakes supports the conclusion that large quantities of trace organic contaminants are transported during snowmelt.

Loading estimates of organochlorine compounds have been made for the Great Lakes area[34,83] and for stations from across Canada.[84] The samplers in the latter study collected only rainfall, using lids to preclude dry deposition. The Canada-wide data, adjusted by Strachan[84] to represent annual precipitation, are compared to ranges for the Canadian Arctic data in Table 8.12. Interestingly, the annual deposition of pesticides to the Canadian Arctic was estimated to be nearly equal to, or only marginally less than, all regions sampled by Strachan except for the British Columbia (B.C.) station which received much more rainfall than the other sites. Total PCB deposition to the Arctic, on the other hand, was lower than all other sites. Highest PCB deposition, reported for the Lake Superior region of Canada, was approximately 10 times more than the Arctic. The high PCB deposition to the Upper Laurentian Great Lakes region is supported by the work of Murphy & Rzeszutko[85] and Swain.[86]

The most northerly published data for the USSR appear to be limited to the annual fluxes in rainfall and snowfall provided by Bobovnikova & Dibtseva[68] for the Moscow area. Fluxes are estimated to be about 52 $\mu g\ m^{-2}\ year^{-1}$ for DDT (*p,p'*-DDT, *o,p'*-DDT, *p,p'*-DDD and *p,p'* DDE) and 14 and 12 $\mu g\ m^{-2}\ year^{-1}$ for α-HCH and lindane, respectively, for the 1978 calendar year. The flux of α-HCH and lindane is about an

Table 8.12 Comparison of annual contaminant deposition in precipitation in non-arctic Canada[84] with Canadian arctic (September to May)[70] deposition estimated from concentrations in snowpack ($\mu g\ m^{-2}\ year^{-1}$)

Compound	*Lake Superior*[b]	*Central Ontario (1984)*	*North Saskatchewan*[c]	*South Alberta (1985)*	*South B.C.*[d]	*Arctic*[e]
α-HCH	3·9–11	14	2·1–6·5	3·3	22–36	0·01–2·3
Lindane	2·0–3·7	7·3	0·39–1·9	1·3	4·8–6·2	<0·01–1·0
Dieldrin	0·15–0·50	0·30	0·01–0·12	0·022	0·03–0·34	0·0–0·25
DDT[a]	0·08	0·02	0·002–0·01	0·007	0·03	<0·01–0·2[f]
Ttl. PCBs	1·2–4·9	1·6	1·5–1·7	1·2	1·0	0·01–0·3
HCB	0·04–0·05	0·08	0·003		0·05	<0·01

[a] Strachan[84] reported only *p,p'*-DDE as opposed to *p,p'*-DDD, *p,p'*-DDE, *o,p'*-DDT plus *p,p'*-DDT for the Arctic.
[b] The Lake Superior data are from 1981, 1983 and 1984.
[c] The station located at Cree Lake in north Saskatchewan included data collected in 1984 and 1985.
[d] The station at Kanaka Creek in southern B.C. included data from 1985 and 1986.
[e] The arctic data are ranges for the two years, 1986 and 1987 for all sites except Alexandra Fiord.
[f] Represents the range of loadings for 1987 sites only.

order of magnitude greater and DDT is more than two orders of magnitude greater in the Moscow region than in the Canadian Arctic. As suggested above, these data likely show much closer proximity to source regions. However, the Moscow data precede the arctic data by nearly 10 years and conditions could have changed considerably during this time due to different pesticide formulations, application procedures and quantities used.

In the Antarctic, DDT (*o,p'*-DDT plus *p,p'*-DDT) residues of the order of 100–2000 pg $litre^{-1}$ have been reported in snow for samples representing the period of snowfall between 1965 and 1969.[87] These results question those of Peterle[88] which were considerably higher.

Rapaport *et al.*[89] analysed snow and rain samples collected at Minneapolis, Minnesota, USA from 1981 through 1983 for *p,p'*-DDT. Mean annual fluxes of *p,p'*-DDT for snow reported by these authors averaged 0·2, 0·4 and 0·13 $\mu g\,m^{-2}\,year^{-1}$ for 1981/82, 1982/83 and 1983/84, respectively. These fluxes were 10 to 100 times higher than the DDT (*p,p'* – DDE, *p,p'*-DDD, *o,p'*-DDT and *p,p'*-DDT) fluxes observed for the artic sites in which the technical product of *p,p'*-DDT commonly constituted more than 75%.

8.3.4 Deposition trends

To date there has been limited conclusive research regarding trends of trace organic contaminant deposition in the Arctic despite the extensive work conducted on glacial records for other anthropogenic contaminants (see Chapter 3). Periodic measurements in air and fresh snow have been too restricted temporally to provide any time series information with the exception of the Swedish study of Larsson & Okla[55] noted in Section 8.3.2. Stengle *et al.*[62] sampled at depth on Mount Logan but could not quantify DDT. Thus the only work intended specifically to determine temporal trends of OC deposition is from the Agassiz Ice Cap on north-central Ellesmere Island.

The Agassiz Ice Cap has been sampled at a remote site at an elevation of approximately 2000 m asl in both 1986 and 1987 to include 17 years of samples between 1969/70 and 1986/87.[13,47] Intuitively, this cold glacier should provide a credible historical record of anthropogenic pollutant deposition because (i) average summer melt affects only about 3% of the winter snow layer resulting in negligible redistribution of ions between adjacent snow layers[81] and (ii) annual snow accumulation at this latitude and elevation represents, on average, more than 75% of the total annual precipitation.[79] The seasonal specific conductance record of the firn (Fig.

8.1) is clearly preserved and is consistent with that reported by Barrie *et al.*[81] As noted by these authors, this profile reflects the trace constituent composition of the atmospheric aerosol which undergoes a strong seasonal variation with maximum ion concentrations, and therefore specific conductance, during the arctic winter.[90,91]

The large volume sample required for trace organic chemistry restricts the profiling for these compounds to annual layers. At first glance, it appears that annual quantities of chlorohydrocarbons in the firn at the Agassiz Ice Cap decreased from the winter of 1969/70 through to the winter of 1986/87 as follows:

- α-HCH from >300 to $70\ \text{ng m}^{-2}$
- lindane from 90 to $<20\ \text{ng m}^{-2}$
- heptachlor epoxide from 20 to $5\ \text{ng m}^{-2}$
- dieldrin from >50 to $20\ \text{ng m}^{-2}$
- chlordane from 20 to $<5\ \text{ng m}^{-2}$
- Ttl. PCB from >300 to $100\ \text{ng m}^{-2}$

The consistent decline amongst all compounds is quite remarkable. However, as noted in Section 8.2.2, the firn does not seem to be an effective permanent trap of the volatile trace organic compounds with the possible exception of PCBs. The residues in the firn therefore are not quantitatively representative of the annual flux of these pesticides to the glacier. It must then be concluded that the trends seen in the glacier for lindane, α-HCH, heptachlor epoxide, dieldrin and chlordane are not entirely the result of a trend in annual loadings, but rather are being controlled, to some degree, by annual meteorological conditions influencing the snow/atmosphere exchange.

Until the processes controlling deposition and revolatilization are better understood, it appears that only Ttl. PCBs demonstrate a real trend in annual deposition to the ice cap. The residue masses of Ttl. PCBs in the glacier will be influenced somewhat by the annual variability of summertime losses; nevertheless, it is conceivable that the actual annual deposition is dominant. Since meteorologic conditions that control atmospheric transport to the Arctic and deposition to the snow vary from year to year, it is necessary to consider this variability and to attempt to adjust for it. Gregor *et al.*[47] normalized the Ttl. PCB trend for annual meteorologic variability with ^{210}Pb flux measurements which were also determined for the glacier. The radioactive isotope ^{219}Pb is derived from the decay of ^{222}Rn which emanates from the Earth's surface where it is produced as a member of the ^{238}U decay series. ^{222}Rn decays in the atmosphere to ^{210}Pb with a half life of 3·8 days. The charged daughter ions produced are chemically reactive and soon become irreversibly associated with aerosol particles.[92] Since the source of ^{210}Pb (i.e. the Earth's surface) remains more or less constant with time, it can be hypothesized that the deposition of ^{210}Pb to the glacier would be uniform from year to year if the air mass movements and the complex interaction of factors controlling transport and deposition to the glacier remain constant. Thus, the differences of measured decay corrected fluxes in the snow layers can be assumed to represent, simplistically, the year to year differences in atmospheric transport and deposition of terrestrially derived anthropogenic pollutants. Under these assumptions, any trends in these adjusted trace organic substance flux data should therefore be real changes in the flux of contaminants to the glacier.

The adjusted Ttl. PCB residues in the annual layers of the Agassiz Ice Cap are compared to measured quantities in Fig. 8.2. These results imply that the tripling of the aerial mass of PCBs in the glacier beginning about 1975/76 is largely an artifact of variable meteorologic conditions affecting transport and deposition. The quantity of PCBs in the glacier has evidently remained stable since 1976/77 at a level approximately one half of that which prevailed between 1970/71 and 1975/76.

Although this decline in the glacier agrees in general with trends in PCB body burdens in ringed seals from the western Arctic between 1972 and 1981[93] and for the Lancaster Sound area between 1975–76 and 1984[2] it is possible that correspondence among these limited data is merely fortuitous and a great deal more work is required to thoroughly evaluate trends in contaminant deposition in the Arctic.

8.4 SUMMARY AND CONCLUSIONS

The arctic atmosphere and polar glaciers have received considerable research over the past several decades for the purpose of measuring the extent of anthropogenic contamination of remote regions. But only relatively recent research has addressed the transport, deposition and accumulation of trace organic contaminants. Extensive data on contaminant body burdens of marine fish and seals have been compiled. However, information on the sources, atmospheric pathways, present deposition rates, deposition trends and processes controlling the fate of these contaminants in the Arctic is sparse.

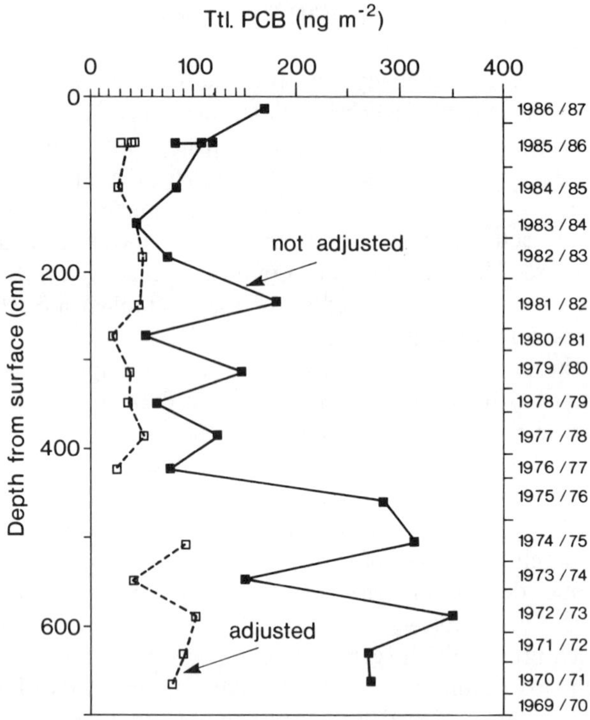

Figure 8.2. Ttl. PCB residues in annual layers of the Agassiz Ice Cap from the winter of 1970/71 to the winter of 1986/87 compared to the residues adjusted in accordance with the ^{210}Pb normalization procedure to compensate for annual meteorological variability.

If the source areas and atmospheric pathways of OCs and PAHs are similar to those for sulphates, then it is conceivable that Eurasia is a major contributor to the concentrations seen in the Arctic. This, however, needs to be confirmed. Local sources in the Canadian Arctic are not significant at the regional scale. Significant quantities of OCs and PAHs are found in the arctic atmosphere during the winter season indicating that large-scale, long-range transport does occur. In addition, some evidence exists that large percentages of the high vapor pressure compounds may be revolatilized to the atmosphere from the snow as temperatures rise. These processes warrant more research under arctic conditions.

Evidence of deposition and incorporation into the arctic aquatic system is provided by the rather limited data on fluxes and river/lake concentrations. Annual deposition of OCs in the Canadian Arctic are of the same order as for southern Canada. PAHs, which are more variable and more readily influenced by local activities, are more difficult to assess. Overall, there are no great differences between the concentrations of HCHs and HCB in arctic aquatic systems than more southerly rivers and lakes of Canada. However, some of the less volatile OCs do not seem to be present in northern waters.

Despite extensive work utilizing glacial ice cores for monitoring pollutant trends for acid deposition and climate change, there has been only limited research on the glacial record of OC deposition. The only glacier in the northern hemisphere for which published data exist is the Agassiz Ice Cap on Ellesmere Island. This research, although limited to 17 years of record at this time, indicates primarily that the glacial contaminant residues are not necessarily indicative of actual deposition. Only the PCB record appears relatively intact and these data, after correction for annual meteorological differences, suggest that Ttl. PCB deposition has decreased by about a factor of two since 1970.

Clearly, more comprehensive reseach is required to expand the information base for trace organic contaminants in the Arctic. This research must seek a broader geographical and temporal coverage. Equally as important, though, is the reseach directed at understanding the processes and fate of these contaminants in a polar environment. This must come from both laboratory and field based research. Only with this information, combined with ecological effects assessments, will an understanding of the potential of trace organic contaminants to impair the delicate arctic ecosystem be attained. Protection of the balance of this ecosystem, which is so intertwined with the support systems of the indigenous peoples, may depend on rapid acquisition of essential information to demonstrate the need for more extensive, hemispheric restrictions of the contaminants considered here.

ACKNOWLEDGEMENTS

I wish to express my appreciation to Dr. T. Bidleman, University of South Carolina and Dr. M. Oehme, Norwegian Air Research Institute, who provided helpful review comments on an earlier draft as well as providing additional information for this chapter. Finally, my own reseach in the Arctic would not have been possible without the support of Environment Canada, Energy, Mines and Resources Canada, the Polar Continental Shelf Project and the Government of the Northwest Territories and the assistance of Mr. M Dahl.

REFERENCES

1. Holden, A. V., Monitoring organochlorine contamination of the marine environment by the analysis of residues in seals. In *Marine Pollution and Sea Life*, ed. M. Ruivo. Fishing News Books Ltd, UK, 1972, pp. 266–72.
2. Muir, D. C. G., Norstrom, R. J. & Simon, M., Organochlorine contaminants in arctic marine food chains: Accumulation of specific polychlorinated biphenyls and chlordane-related compounds. *Environ. Sci. Tech.*, **22** (1988) 1071–9.
3. Norstrom, R. J., Simon, M., Muir, D. C. G. & Schweinsburg, R. E., Organochlorine contaminants in arctic marine food chains: Identification, geographical distribution and temporal trends in Polar Bears. *Environ. Sci. Tech.*, **22** (1988) 1063–70.
4. Muir, D. C. G, Grift, N. P., Ford, C. A., Reiger, A. W., Hendzel, M. R. & Lockhart, W. L., Evidence for long range transport of toxaphene to remote arctic and subarctic waters from monitoring of fish tissues. In *Long Range Transport of Pesticides*, ed. D. Kurtz. Lewis Publishers, Chelsea, Michigan, 1990, pp. 329–46.
5. Norstrom, R. J., Simon, M. & Muir, D. C. G., Polychlorinated dibenzo-*p*-dioxin and dibenzofurans in marine mammals in the Canadian North. *Environ. Pollut.*, **66** (1990) 1–19.
6. Muir D. C. G., Wagemann, R., Lockhart, W. L., Grift, N. P., Billeck, B. & Metner, D., Heavy metal and organic contaminants in arctic marine fishes. *Environmental Studies* No. 42, Indian and Northern Affairs, Ottawa, Canada, 1987, 64 pp.
7. Bidleman, T. F., Patton, G. W., Walla, M. D., Hargrave, B.T., Vass, W. P., Erickson, P., Fowler, B., Scott, V. & Gregor, D. J., Toxaphene and other organochlorines in Arctic Ocean fauna: Evidence for atmospheric delivery. *Arctic*, **42** (1989) 307–13.
8. Oehme, M., Fürst, P., Krüger, Chr., Meemken, H. A. & Groebel, W., Presence of polychlorinated dibenzo-*p*-dioxins, dibenzofurans and pesticides in arctic seal from Spitzbergen. *Chemosphere*, **17** (1988) 1291–1300.
9. Muir, D. C. G., Wagemann, R., Hargrave, B. T., Thomas, D. J., Peakall, D. B. & Norstrom, R. J., Arctic marine ecosystem contamination. *Sci. Total Environ.*, (in press).
10. Wong, M. P., Chemical residues in fish and wildlife species harvested in northern Canada. *Environmental Studies Program* Report No. 46, Department of Indian and Northern Affairs, Ottawa, Canada, 1985.
11. Kinlock, D., Kuhnlein, H. & Muir, D., Inuit foods and diet: An assessment of benefits and risks. *Sci. Total Environ.*, (in press).

12. Dewailly, E., Nantel, A., Weber, J.-P. & Meyer, F., High levels of PCBs in breast milk of Inuit women from arctic Quebec. *Bull. Environ. Contam. Toxicol.*, **43** (1989) 641–6.
13. Barrie, L. A., Gregor, D., Hargrave, B., Lake, R., Muir, D., Shearer, R., Tracey, B. & Bidleman, T., Arctic contaminants: Sources, occurrence and pathways. *Sci. Total Environ.*, (in press).
14. Jensen, S., Johnels, G., Olsson, M, & Otterlind, G., DDT and PCB in marine animals from Swedish waters. *Nature*, **224** (1969) 247–50.
15. Reijnders, P. J. H., Organochlorine and heavy metal residues in harbour seals from the Wadden Sea and their possible effects on reproduction. *Neth. J. Sea Res.*, **14** (1980) 30–65.
16. Tanabe, S., PCB problems in the future: Foresight from current knowledge. *Environ. Pollut.*, **50** (1988) 5–28.
17. Holtz, A., Sharpe, M. A., Constable, M. & Wilson, M., Removal of contaminants from distant early warning sites in Canada's Arctic. Environment Canada, Environmental Protection Sevice, Western and Northern Region, Edmonton, Alberta (W&NR-86/87-CP(EP)-16), 1986, 111 pp.
18. Deo, P. G., Hasan, S. B. & Majumder, S. K., Isomerization of β-HCH in aqueous solution. *J. Environ. Sci. Health*, **B15** (1980) 147–64.
19. Voldner, E. C. & Ellenton, G., Production, usage and atmospheric emissions of priority toxic chemicals with emphasis on North America. Report ARD-88-4 prepared for the Canada/US International Joint Commission by Atmospheric Environment Service, Environment Canada, Downsview, Ontaro, Canada, 1987.
20. International Registry of Potentially Toxic Chemicals (IRPTC). Unpublished report based on a computerized search of Registry files, United Nations Environmental Programme, Geneva, Switzerland, 1988.
21. Canadian Council of Resource and Environment Ministers (CCREM), Canadian water quality guidelines. Environment Canada, Ottawa, Canada, 1987.
22. Rapaport, R. A. & Eisenreich, S. J., Historical atmospheric inputs of high molecular weight chlorinated hydrocarbons to eastern North America. *Environ. Sci. Tech.*, **22** (1988) 931–41.
23. Tanabe, S., Tatsukawa, R., Kawano, M, & Hidaka, H., Global distribution and atmospheric transport of chlorinated hydrocarbons: HCH (BHC) isomers and DDT compounds in the western Pacific, eastern Indian and Antarctic Oceans. *J. Oceanogr. Soc. Japan*, **38** (1982) 137–48.
24. Agarwal, H. C., Kaushik, C. P. & Pillai, M. K. K., Organochlorine insecticide residues in the rainwater in Delhi, India. *Water, Air Soil Pollut.*, **32** (1987) 293–300.
25. Von Rumker, R. E., Lawless, E. W. & Meiners, A. F., Production, distribution, use and environmental impact potential of selected pesticides. U.S. Environmental Protection Agency, EPA/540/1-74-001, 1974.
26. Eriksson, G., Jenson, S., Kylin, H. & Strachan, W., The pine needle as a monitor of atmospheric pollution. *Nature*, **341** (1989) 42–4.
27. Buchel, K. H. (ed.), *Chemistry of Pesticides* (translated by G. Holmwood). Wiley Interscience Publications, New York, 1983.
28. Rapaport, R. A. & Eisenreich, S. J., Atmospheric deposition of toxaphene to

eastern North America derived from peat accumulation. *Atmos. Environ.*, **20** (1986) 2367–79.
29. International Registry of Potentially Toxic Chemicals (IRPTC) Toxaphene. No. 32 in a series '*Scientific Reviews of Soviet Literature on Toxicity and Hazards of Chemicals*', ed. N. F. Izmerov. USSR Academy of Medical Sciences, Centre for International Projects, GKNI, Moscow, 1983, 15 pp.
30. Courtney, K. D., Hexachlorobenzene (HCB): A review. *Environ. Res.*, **20** (1979) 225–66.
31. Heit, M., Tan, Y. L., Klusek, C. S. & Burke, J. C., Anthropogenic trace elements and polycyclic aromatic hydrocarbons in sediment cores from 2 lakes in the Adirondack acid lake region. *Water Soil Air Pollut.*, **15** (1981) 441–64.
32. Barrie, L. A., Olson, M. P. & Oikawa, K. K., The flux of anthropogenic sulphur into the Arctic from mid-latitudes in 1979/80. *Atmos. Environ.*, **23** (1989) 2505–12.
33. Bidleman, T. F., Atmospheric processes, wet and dry deposition of organic compounds are controlled by their vapor–particle partitioning. *Environ. Sci. Tech.*, **22** (1988) 361–7.
34. Eisenreich, S. J., Looney, B. B. & Thornton, J. D., Airborne organic contaminants in the Great Lakes ecosystem. *Environ. Sci. Tech.*, **15** (1981) 30–58.
35. Murphy, T. J., Atmospheric inputs of chlorinated hydrocarbons to the Great Lakes. In *Toxic Contaminants in the Great Lakes*, ed. J. O. Nriagu and M. S. Simmons. Wiley Interscience, Toronto, Canada, 1984, pp. 53–80.
36. Slinn, W. G. N., Hasse, L., Hicks, B. B., Hogan, A. W., Lal, D., Liss, P. S., Munnich, K. O., Sehmel, G. A. & Vittori, O., Some aspects of the transfer of atmospheric trace constituents past the air–sea interface. *Atmos. Environ.*, **12** (1978) 2055–87.
37. Murphy, T. J. & Rzeszutko, C. P., Precipitation inputs of PCBs to Lake Michigan. *J. Great Lakes Res.*, **3** (1977) 305–12.
38. Doskey, P. & Andren, A., Modelling the flux of atmospheric polychlorinated biphenyls across the air/water interface. *Environ. Sci. Tech.*, **15** (1981) 705–11.
39. Duinker, J. C. & Bouchertall, F., On the distribution of atmospheric polychlorinated biphenyl congeners between vapor phase, aerosols, and rain. *Environ. Sci. Technol.*, **23** (1989) 57–62.
40. Barrie, L. A., Arctic air pollution: An overview of current knowledge. *Atmos. Environ.*, **20** (1986) 643–63.
41. Patton, G. W., Hinckley, D. A., Walla, M. D. & Bidleman, T. F, Airborne organochlorines in the Canadian high Arctic. *Tellus*, **41B** (1989) 243–55.
42. Bidleman, T. F., Organochlorines in the arctic atmosphere and ocean. In *Proceedings of the 8th International Conference of the Comité Arctique International on Global Significance of the Transport and Accumulation of Polychlorinated Hydrocarbons in the Arctic*, Oslo, Norway, September 18–22, 1989, ed. R. Shearer and A. Bartonova (in press).
43. Larsson, P. & Okla, L., An attempt to measure the flow of chlorinated hydrocarbons, such as PCBs, from water to air in the field. *Environ. Poll.*, **44** (1987) 219–25.
44. Atlas, E„ Foster, R., & Giam, C. S., Air–sea exchange of high molecular

weight organic pollutants: Laboratory studies. *Environ. Sci. Tech.*, **16** (1982) 283–6.
45. Larsson, P., Transport of ^{14}C-labelled PCB-compounds from sediment to water and from water to air in laboratory model systems. *Water Res*, **17** (1983) 1317–26.
46. Baker, J. E. & Eisenreich, S. J., Concentration and fluxes of polycyclic aromatic hydrocarbons and polychlorinated biphenyls across the air–water interface of Lake Superior. *Environ. Sci. Technol.*, **24** (1990) 342–52.
47. Gregor, D. J., Dominik, J. & Vernet, J.-P., Recent trends of selected chlorohydrocarbons in the Agassiz Ice Cap, Ellesmere Island, Canada. In *Proceedings of the 8th International Conference of the Comité Arctique International on Global Significance of the Transport and Accumulation of Polychlorinated Hydrocarbons in the Arctic*, Oslo, Norway, September 18–22, 1989, ed. R. Shearer and A. Bartonova (in press).
48. Hargrave, B. T., Vass, W. P., Erickson, P. E. & Fowler, B. R., Atmospheric transport of organochlorines to the Arctic Ocean. *Tellus*, **40B** (1988) 480–93.
49. Bidleman, T., Widequist, U., Jasson, B. & Soderlund, R., Organochlorine pesticides and polychlorinated biphenyls in the atmosphere of Southern Sweden. *Atmos. Environ.*, **21** (1987) 641–54.
50. Edwards, N. T., Polycyclic aromatic hydrocarbons (PAHs) in the terrestrial environment—a review. *J. Environ. Quality*, **12** (1983) 427–41.
51. Tuominen, J., Salomaa, S., Pyysalo, H., Skytta, E., Tikkanen, L., Nurmela, T., Sorsa, M., Pohjola, V., Sauri, M. & Himberg, K., Polynuclear aromatic compounds and genotoxicity in particulate and vapor phases of ambient air: Effect of traffic season and meteorological conditions. *Env. Sci. Tech.*, **22** (1988) 1228–34.
52. Oehme, M. & Mano, S., The long range transport of organic pollutants to the Arctic. *Fresenius Z. Anal. Chem.*, **319** (1984) 141–6.
53. Oehme, M. & Ottar, B., The long range transport of polychlorinated hydrocarbons to the Arctic. *Geophys. Res. Lett.*, **11** (1984) 1133–6.
54. Pacyna, J. M. & Oehme, M., Long-range transport of some organic compounds to the Norwegian Arctic. *Atmos. Environ.*, **22** (1988) 243–57.
55. Larsson, P. & Okla, L., Atmospheric transport of chlorinated hydrocarbons to Sweden in 1985 compared to 1973. *Atmos. Environ.*, **23** (1989) 1699–1711.
56. Patton, G. W. & Bidleman, T. F., Organochlorine and polycyclic aromatic compounds in the High Arctic. Paper presented at 82nd Annual Meeting & Exhibition, Air & Waste Management Association, Anaheim, California, June 25–30, 1989.
57. Hinkley, D. A. & Bidleman, T. F., Air–sea exchange of hexachlorocyclohexane in the Bering and Chukchi seas. Paper presented at 82nd Annual Meeting & Exhibition, Air & Waste Management Association, Anaheim, California, June 25–30, 1989.
58. Hoff, R. M. & Chan, K.-W., Atmospheric concentrations of chlordane at Mould Bay, N. W. T., Canada. *Chemosphere*, **15** (1986) 449–52.
59. Daisey, J. M., McCaffrey, R. J., & Gallagher, R. A., Polycyclic aromatic hydrocarbons and total extractable particulate organic matter in the arctic aerosol. *Atmos. Environ.*, **15** (1981) 1353–63.

60. Bjorseth, A., Lunde, G. & Lindskog, A., Long-range transport of polycyclic aromatic hydrocarbons. *Atmos. Environ.*, **13** (1979) 45–53.
61. Reichardt, P. B. & Reidy, S. K., Atmospheric polycyclic aromatic hydrocarbons: An aspect of air pollution in Fairbanks, Alaska. *Arctic*, **33** (1980) 316–25.
62. Stengle, T. R., Lichtenberg, J. A. & Houston, C. S., Sampling of glacial snow for pesticide analysis on the high plateau glacier of Mount Logan. *Arctic*, **26** (1973) 335–6.
63. Lunde, G., Gether, J., Gjos, N. & May-Berit, S. L., Organic micropollutants in precipitation in Norway. *Atmos. Environ.*, **11** (1977) 1007–14.
64. Paasivirta, J., & Paukku, R., Study of organochlorine pollutants in snow at North Pole. In *The Huure Expedition to the North Pole at 1984*, ed. Olli-Pekka Nordlund. Technical Research Centre of Finland, Research Notes #685, Espoo, Finland, February 1987, pp. 24–31.
65. Gregor, D. & Gummer, W., Evidence of atmospheric transport and deposition of organochlorine pesticides and PCBs in Canadian arctic snow. *Environ. Sci. Tech.*, **23** (1989) 561–5.
66. McNeely, R. & Gummer, W. D., A reconnaissance survey of the environmental chemistry in east-central Ellesmere Island, N.W.T. *Arctic*, **37** (1984) 210–23.
67. Arctic Laboratories Ltd, A study of the distribution of chlorinated hydrocarbon pesticides and PCBs in the Arctic Ocean using a Canadian Ice Island station. An unpublished report prepared for the Department of Supply and Services, Department of Indian and Northern Affairs and Department of Fisheries and Oceans by Arctic Laboratories Ltd, 2045 Mills Road, Sidney, B.C., Canada V8L 3S1, 1988, 115 pp.
68. Bobovnikova, T. I. & Dibtseva, A. V., Organochlorine in precipitation. *Meteorologiya i Gidrologiya*, **8** (1980) 46–51.
69. Erickson, P. & Fowler, B., The flux of suspended particulates, petroleum related hydrocarbons, trace metals and nutrients from the Mackenzie River during the winter season: A pilot study of the East Channel. *Environmental Studies* Report No. 48, Department of Indian and Northern Affairs, Ottawa, 1987.
70. Gregor, D. J., Water quality research. In *Northern Hydrology: Canadian Perspectives*, ed. S. Ommanney and T. Prowse. NHRI Science Report no. 1, National Hydrology Research Institute, Environment Canada, 11 Innovation Blvd., Saskatoon, Saskatchewan, Canada, 1990, pp. 163–86.
71. Lockhart, W. L., Gregor, D. J., Wagemann, R., Tracey, B., Sutherland, D. & Thomas, D. J., Presence and implications of chemical contaminants in the freshwaters of the Canadian Arctic. *Sci. Total Environ.*, (in press).
72. McCrea, R. C. & Fischer, J. D., Heavy metal and organochlorine contaminants in the five major Ontario rivers of the Hudson Bay Lowland. *Wat. Poll. Res. J. Can.*, **21** (1986) 225–34.
73. Langlois, C., Etude préliminaire de la qualité des eaux de surface de 15 cours d'eau majeurs du nouveau-Québec. *Wat. Poll. Res. J. Canada*, **22**, (1987) 530–44.
74. Stevens, R. J. J. & Neilson, M. A., Inter- and intralake distributions of trace

organic contaminants in surface waters of the Great Lakes. *J. Great Lakes Res.*, **15**, (1989) 377–93.
75. Lockerbie, D. M. & Clair, T. A., Organic contaminants in isolated lakes of southern Labrador, Canada. *Bull. Environ. Contam. Toxicol.*, **41** (1988) 625–32.
76. Windholz, M. (ed.), *The Merck Index: An Encyclopedia of Chemicals, Drugs and Biologicals.* Merck & Co., New Jersey, 10th edn, 1983, 1606 pp.
77. Schrimpff, E., The relationship between relief and deposition of some organic and inorganic contaminants in snow of northern Bavaria, F. R. G. In *Proceedings International Conference Ecological Impact Acid Precipitation,* ed. D. Drabløs & A. Tollan, Norwegian Interdisciplinary Research Programme, Oslo, Norway, 1980, pp. 130–1.
78. Giger, W., Leuenberger, C., Czuczwa, J. & Tremp, J., Organic micropollutants in the atmosphere: Determination, origins and behavior in rain, snow and fog. *EAWAG-NEWS* **22/23** (1987) 13–15.
79. Woo, Ming-ko, Heron, R., Marsh, P. & Steer, P., Comparison of weather station snowfall with winter snow accumulation in high arctic basins. *Atmosphere-Ocean*, **21** (1983) 312–25.
80. Barrie, L. A. & Hoff, R.M., Five years of air chemistry observations in the Canadian Arctic. *Atmos. Environ.*, **19** (1985) 1995–2010.
81. Barrie, L. A., Fisher, D. & Koerner, R. M., Twentieth century trends in arctic air pollution revealed by conductivity and acidity observations in snow and ice in the Canadian High Arctic. *Atmos. Environ.*, **19** (1985) 2055–63.
82. Woo, Ming-ko & Marsh, P., Determination of snow storage for small eastern high arctic basins. In *Proceedings of the Thirty-Fourth Annual Eastern Snow Conference*, Belleville, Ontario, Feb. 3–4, 1977, pp. 147–62.
83. Strachan, W. M. J., Organochlorine substances in the rainfall of Lake Superior: 1983. *Environ. Toxicol. Chem.*, **4** (1985) 677–83.
84. Strachan, W. M. J., Toxic contaminants in rainfall in Canada: 1984. *Environ. Toxicol. Chem.*, **7** (1988) 871–7.
85. Murphy, T. J. & Rzeszutko, C. P., Precipitation inputs of PCBs to Lake Michigan. *J. Great Lakes Res.*, **3** (1977) 305–12.
86. Swain, W. R., Chlorinated organic residues in fish, water and precipitation from the vicinity of Isle Royale, Lake Superior. *J. Great Lakes Res.*, **4** (1978) 398–407.
87. Peel, D. A., Organochlorine residues in Antarctic snow. *Nature*, **254** (1975) 324–5.
88. Peterle, T. J., DDT in Antarctic snow. *Nature*, **224** (1969) 620.
89. Rapaport, R. A., Urban, N. R., Capel, P. D., Baker, J. E., Looney, B. B., Eisenrich, S. J. & Gorham, E., 'New' DDT inputs to North America. *Chemosphere*, **14** (1985) 1167–73.
90. Hoff, R. M. & Barrie, L. A., Air chemistry observations in the Canadian Arctic. *Water Sci. Tech.*, **18** (1986) 97–107.
91. Rahn, K. A. & Shaw, G. E., Sources and transport of arctic pollution aerosol: A chronicle of six years of ONR research. *Naval Research Reviews*, **37** (1982) 3–26.
92. Robbins, J. A., Geochemical and geophysical applications of radioactive lead.

In *The Biogeochemistry of Lead in the Environment*, ed. J. O. Nriagu. Elsevier, Amsterdam, 1978, pp. 285–393.
93. Addison, R. F. & Zinck, M. E., PCBs have declined more than DDT-group residues in arctic ringed seals (Phoca hispida) between 1972 and 1981. *Environ. Sci. Technol.*, **20** (1986) 253–6.

Chapter 9

Local Sources of Pollution in the Arctic: From Prudhoe Bay to the Taz Peninsula

DANIEL A. JAFFE

Geophysical Institute and Department of Chemistry
University of Alaska Fairbanks
Fairbanks, Alaska 99775-0800, USA

9.1 INTRODUCTION: PRISTINE WILDERNESS OR PARADISE LOST?

> 'The polar atmosphere is fairly uncontaminated by impurities and, apart from fog, only precipitation or blowing snow will reduce visibility.'[1]

> 'On the Kola Peninsula, in an area of superphosphate plants, the entire winter's dust covers the grass and lichen in such a thick layer that it blocks photosynthesis and the plants die if a heavy rain does not wash the dirt from them.'[2]

These two passages give very different images of the Arctic. In fact both of these environments can be found in the Arctic. The vast majority of the arctic icepack remains uninhabited, undeveloped, and untouched by humans. However on the arctic landmasses, small and large population centers exist and certain regions have been extensively developed for fossil fuel and/or mineral resources both on and offshore. Some of the world's largest oil, natural gas and mining operations are found in the Arctic.

The influence of these sources on observed atmospheric pollution is exacerbated by a set of conditions which are unique to the Arctic. Principally these conditions are:

(1) The persistence of strong thermal inversions in the arctic boundary layer which traps pollutants and permits ground level concentrations to reach, at times, hazardous levels. These inversions are strongest and most persistent due to radiational cooling and the lack of solar insolation during winter. However, boundary layer inversions are also frequently found during summer.

(2) The relative scarcity of frontal systems penetrating into the Arctic results in a very dry climate. In lower latitudes, precipitation acts as a cleansing mechanism for many atmospheric pollutants.

An understanding of arctic pollution, both locally produced and that which is transported in, will be necessary to fully comprehend human influences on this important region. The long-range transport of pollutants to the Arctic, primarily during winter and spring, results in a distribution of pollutants which is relatively uniform over large horizontal distances. The concentrations of pollutants produced locally in the Arctic are not uniformly distributed, and give rise to pollution 'hot spots', whose concentrations are, at times, several orders of magnitude greater than that which results from long-range transport. In contrast to pollution transported into the Arctic, local sources are present in the Arctic year-round. Thus a picture of the Arctic is evolving which suggests a relatively uniform distribution of pollutants transported seasonally into the Arctic, punctuated by a number of large local arctic pollution 'hotspots'.

To date, most studies of arctic pollution have dealt principally with transport and chemistry, whereas little is known about the climatic and ecological impacts of this pollution. In a few well-documented cases, pollution levels in arctic cities can, at times, reach levels which are acutely hazardous to human health. However, little is known about the long-term impacts to humans or arctic ecosystems which may result from the continuous insult of lower concentrations of pollutants. There is no evidence that known (non-urban) concentrations of gas and aerosol species in the Arctic are detrimental to human or ecosystem health. However snow, in some heavily impacted areas, has been shown to contain elevated concentrations of many trace metals, sulfate and nitrate, sufficient to raise questions concerning ecosystem impacts of arctic pollution.

To some extent, development has occurred in all arctic nations. However, without question the USSR has the most populated and extensively developed arctic regions. Its arctic population is approximately 1–2 million people, with a total of six million people north of latitude 60°N, this is greater than the arctic population of all other regions combined.[3]

9.2 THE ARCTIC BOUNDARY LAYER

To simplify the picture, meteorology in the Arctic can be broken down into summer and winter patterns, with transitional periods occurring during spring and fall. During winter there are few differences between meteorology in polar ocean and land regions (including sub-arctic regions), both receive little solar insolation, are snow-covered, and generally possess strong thermal inversion gradients in the boundary layer. However, during summer the polar oceans remain cool, and the arctic landmasses warm up considerably and receive greater amounts of precipitation. A more detailed look at synoptic meteorology in the Arctic can be found in Chapter 2.

As a result of the large radiative cooling at the surface and the infrequent nature of convective frontal systems, temperature inversions are a common occurrence in the Arctic and sub-arctic. These inversions act as strong barriers to mixing and dilution of atmospheric pollutants, resulting in substantially elevated concentrations of the emitted species. During winter these inversions are found 80–100% of the time at all hours of the day.[4] Despite the presence of extended hours of daylight, inversions are still a common occurrence during the summer.

Two types of thermal inversions are generally found in the Arctic. In the first, radiative cooling of the surface is the primary cause. This is due to the fact that the surface of the Earth radiates in the infrared as nearly a black body, whereas the atmosphere radiates selectively in the infrared (mainly due to CO_2 and H_2O emission) causing substantial net heat loss at the surface in the infrared. If this heat is not compensated for by (short-wave) solar insolation (which is generally the case at night, or during the arctic winter) substantial cooling of the surface occurs and temperature inversions may result (in the absence of frontal activity, for instance). This is the primary type of inversion found in the Arctic during winter, and generally results in very steep thermal gradients which begin at the surface.[5,6]

Inversions can also form when a warm air mass moves in over cooler air. This is a common inversion type in coastal cities, such as Seattle or Los Angeles, when warm maritime air overrides a cooler continental air mass, trapping pollutants below. This type of inversion is generally observed to have an inversion base height (the elevation where temperatures begin to *increase* with height) of several hundred meters and less steep thermal gradients in the inversion layer. These type of inversions are also observed in the Arctic, usually during summer, when warmer maritime air may intrude into the polar regions and form a capping layer over the cooler arctic air.

Examples of both types of inversions are shown in Fig. 9.1 (after Ref. 5).

Generally, the temperature inversion during winter extends to almost 1 km, with much steeper thermal gradients found near the surface. The surface gradients found in winter are among the steepest observed anywhere in the world, reaching values as high as 30°C/100 meters for 30–50 meters, and 200°C/100 meters in the lowest two meters.[5,6] The

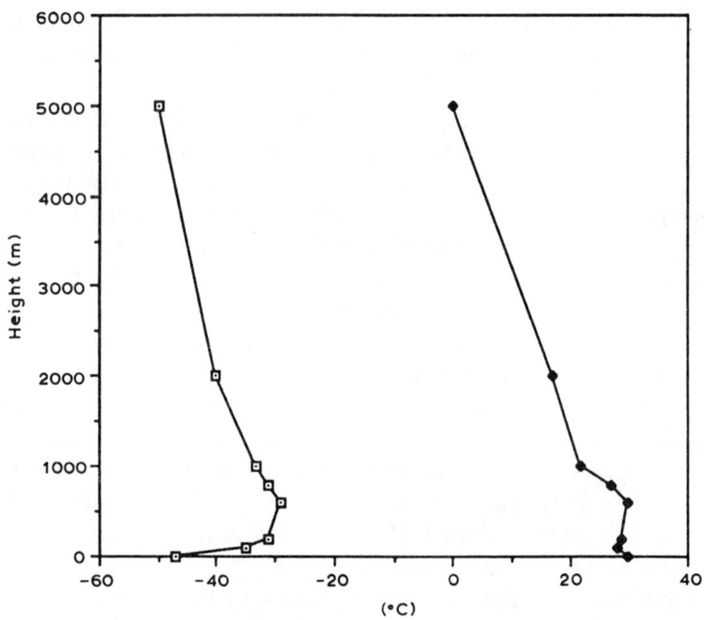

Figure 9.1 Comparison of a typical winter surface inversion in Fairbanks, Alaska, and a typical summer inversion in Los Angeles. Note that the Fairbanks inversion is much steeper and begins at the surface. —□—, Fairbanks; —◆—, Los Angeles.

steep gradients found near the surface are a strong barrier to boundary layer mixing. During summer, the gradients and inversion thickness tend to be much less than wintertime inversions.

Several factors are responsible for the extreme nature of wintertime arctic inversions, and therefore pollution episodes in arctic and sub-arctic locations. These are:

(1) With very low temperatures, and little or no solar insolation, net radiative heat loss at the surface is very large.

(2) The lack of solar heating also results in the frequent 'carryover' of inversion conditions from one day to the next. In lower latitude cities, afternoon solar heating is often the key to breaking down nocturnal inversions, and dilution of accumulated pollution.

(3) The fact that these inversions generally begin at the surface results in very small dilution volumes for pollution.

These factors primarily account for the fact that Fairbanks, Alaska, (*c.* 30 000 residents) can have CO concentrations comparable to much larger cities.[5]

9.3 INDUSTRIAL SOURCES

Principally, arctic development has occurred for one reason: resource extraction. As the industrial world exploits and grows short of resources which are closer to major markets it has increasingly looked north for major mineral and fuel supplies. Early arctic entrepreneurs came looking for gold, and worked with picks and shovels. Today, the entrepreneurs are more likely to be governments or multi-national corporations equipped with armies of diesel bulldozers, helicopters and gas turbines, and prepared to develop large portions of the arctic landmasses.

9.3.1 Coal, gas and oil

Large oil and natural gas fields in Alaska and the Soviet Arctic, account for a substantial fraction of these countries' current production and future reserves. New oil and gas projects are likely in these countries as well as Canada and the Norwegian Arctic. These industrial regions produce substantial quantities of air pollutants which result from combustion of fuels to power the complex, and possibly also from flaring of unwanted gas. Although arctic production of coal is not as significant as oil or gas, coal combustion is a very important source of air pollution in the Arctic, especially SO_2 and particulate trace metals. Much of this coal is imported into the arctic regions from other areas, and is used in thermal power plants and smelters. The emissions from coal combustion in the USSR are probably the largest source of air pollutants located in the Arctic.

In Alaska the major industrial development at Prudhoe Bay currently produces about two million barrels of oil per day, and accounts for approximately 25% of the total domestic production of crude oil in the USA, or 12% of its total consumption.[7] In the USSR, arctic and sub-arctic fields provide approximately half (about 250 billion cubic meters) of that country's supply of natural gas and 60% (about seven million barrels per day) of its oil.[8] The drilling, pumping, operations and maintenance of these facilities is generally powered by locally produced fuels, resulting in substantial emissions of SO_2, NO_x, hydrocarbons, and trace metals.

The oil field at Prudhoe Bay, on Alaska's arctic coast, was discovered in 1968. This region of the Arctic is covered by snow for 8–9 months of the year, and typically has an ice free harbor for only one month each year. During summer, the tundra is covered by shallow ponds and boggy, tundra vegetation. Thus, Prudhoe Bay's location in the Arctic made production difficult and delayed development until a means of transporting the oil to major markets could be found. In 1973 construction of the Alaskan pipeline was started and in 1977 oil production from Prudhoe Bay began.

Associated with this production, the Prudhoe Bay region has developed into a major industrial complex. The entire development spans an area of several thousand square kilometers, covered by about 500 kilometers of gravel roads. All major facilities and drilling sites must be built on gravel pads of about one-meter depth to prevent heating, melting and subsidence of the underlying permafrost layer. Major facilities include approximately 100 drilling sites, the Central Gas Facility (which separates gas and oil), the Cental Power Plant (which produces 160 megawatts of power), a small refinery for local fuels, several airports, a seawater treatment plant, and several large housing, dining and recreation complexes for more than 5000 Prudhoe Bay workers.[9] Figure 9.2 shows a map of the Prudhoe Bay region.

A substantial amount of natural gas is also obtained at Prudhoe Bay, but currently this is not marketed commercially because of lack of transportation to markets. The gas is separated from the crude, and utilized (locally) for energy production or reinjected into the drill holes to increase oil yields. Occasionally, equipment failures at Prudhoe Bay

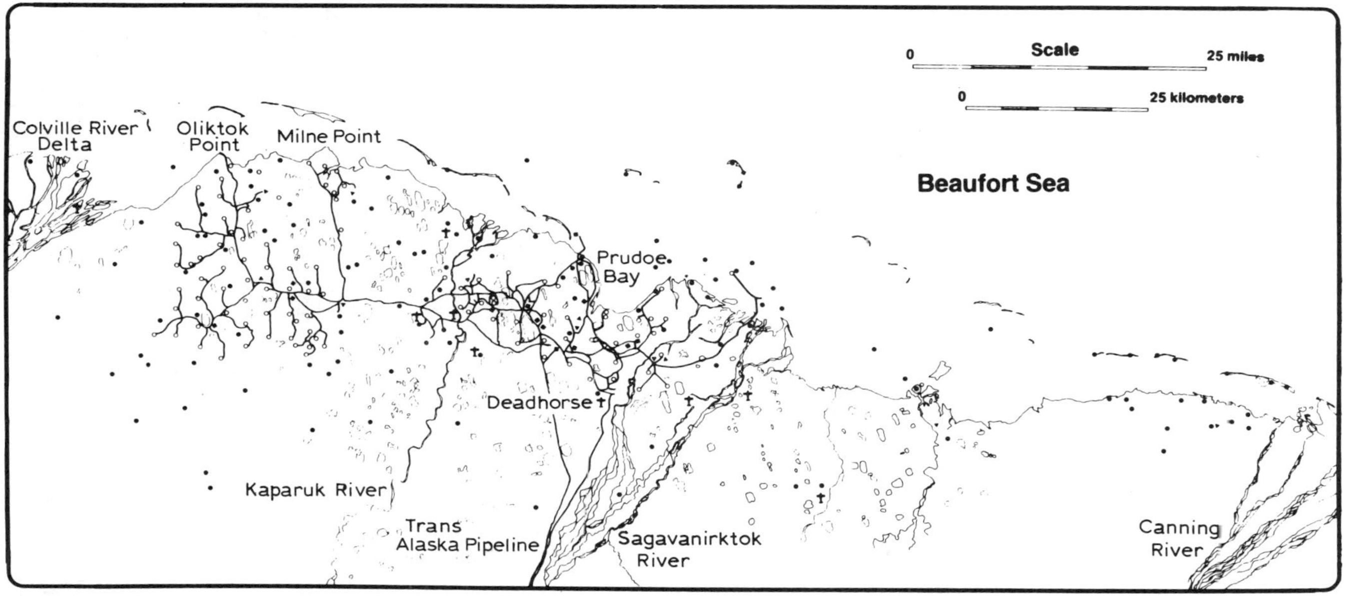

Map by Margie Gibson

- ◦ **Drilling pad**
- · **Exploratory site**
- ≻ **Road and/or pipeline system**
- · **Material site**
- · **Support facility**
- † **Airstrip**

Figure 9.2. Map of the Prudhoe Bay oil producing region. (From Ref. 9, reproduced by permission of the National Resources Defense Council, Trustees for Alaska, National Wildlife Federation.)

will lead to a situation where it is necessary to flare large amount of gas, leading to major black smoke plumes.[10] A separate natural gas pipeline has been proposed several times in the past decade; however to date construction of this project has not begun, and start-up is at least several years away.

Located at Prudhoe Bay is the world's largest collection of gas-fired turbines utilized for power generation. Associated with these turbines are substantial NO_x emissions. Permitted NO_x emissions are approximately 75 000 tons/year, although actual NO_x emissions are in the range 15 000–20 000 tons/year (Jack Coutts, Alaska Department of Environmental Conservation, 1988, pers. comm.). Emissions of hydrocarbons, sulfur dioxide, and carbon monoxide are much lower than NO_x. For comparison, the Prudhoe Bay emissions (actual) are approximately the same as the NO_x emissions from Washington, D.C., and represent about 0·1% of the total NO_x emissions of the USA.[9,11] Note, however, that the numbers given above include only the major operations and do not include smaller units, vehicles, or housing facilities.

The Prudhoe Bay facility is probably the best documented, large air pollution source in the Arctic. Besides the emissions estimates given above, ambient air monitoring has been required by the US Environmental Protection Agency (US EPA) and the Alaska Department of Environmental Conservation.[10] The ambient monitoring indicates that *annual average* NO_2 levels at Prudhoe Bay are substantially below the National Ambient Air Quality Standards (NAAQS) set by the US EPA (the standard for NO_2 is based on an annual average), though at times the concentrations can reach very high levels. The fact that the annual average is well below air quality standards, despite very large emissions, is likely due to the persistent winds found at Prudhoe Bay. Annual average wind speeds at Prudhoe Bay are of the order of six meters per second (13 mph), so that despite the presence of frequent inversions, dispersion and/or removal of pollutants from the immediate vicinity generally occurs.[10] Incidents of very high NO_2 levels are observed during infrequent periods of inversions and calm winds.

At the other end of the Trans Alaskan Pipeline System lies the port of Valdez, which is the tanker loading point for the majority of Alaskan petroleum. In addition to the recent major oil spill at Valdez, local air emissions associated with tanker operations have been cited as a significant environmental problem there.[12] During the loading process substantial quantities of hydrocarbon vapors, especially benzene, escape from tanker holds by displacement during the loading of crude oil. Since benzene is a known carcinogenic agent, this problem is of concern to local residents and has received attention from state environmental officials.

Without question, the most productive fossil fuel producing region in the Arctic is in Western Siberia, USSR. There, enormous gas and oil fields make a major contribution to the Soviet economy, by producing approximately 50% of the natural gas and 60% of the oil used throughout the country.[8] In addition a major pipeline begins at Vuktyl, and provides natural gas to Eastern and western Europe. In general, Soviet Arctic fuel sources are providing an increasingly large share of the total Soviet fuels, and will continue to increase in importance as fields closer to the major population centers decline in productivity.[8]

Although little concrete information is available, we can be reasonably certain that the Soviet oil and gas producing regions consume large amounts of the locally produced fuel (oil or gas) to power the operations, and therefore emit NO_x, SO_2, hydrocarbons, and particulates in varying, but probably substantial, amounts. In addition, natural gas flaring occurs at oil fields which are unable to process the natural gas.[13] These gas flares have been observed from space, and are perhaps the largest such flares on the planet.[14] It has been suggested that at some facilities, substantial quantities of natural gas (CH_4) may be released directly into the atmosphere without flaring (W. Sackinger, 1989, pers. comm.), contributing substantially to the atmospheric burden of this important greenhouse gas.

Coal is mined in several Soviet Arctic locations including the Pechora Basin and the region near Noril'sk. Coal combustion in the USSR is most certainly the largest source of SO_2 and many particulate trace metals to the arctic atmosphere.[15] Pechora Basin coal provides fuel for electricity and smelters (especially iron) in many northern Soviet locations, including the Kola Peninsula. According to Ref. 15 (NILU, 1984), about 550×10^3 metric tons of SO_2 is produced in the arctic regions of the USSR, principally due to coal combustion. This represents about 4% of the total SO_2 emissions from the USSR (see Table 9.1).

Recent data from the Prudhoe Bay region suggest that there may be impacts associated with these substantial air emissions as a result of fossil fuel production in the Arctic, including dust transport from gravel roads, potential for photochemical ozone impacts, and deposition of NO_3^-. Although similar information for air emissions in the Soviet Arctic is not available, comparable or worse impacts are to be expected. These issues will be discussed in Section 9.5 of this chapter.

Table 9.1 Soviet Arctic trace metal and SO_2 emission data. (Data from Ref. 15)

Region and source	*As*	*Cd*	*Cu*	*Cr*	*Mn*	*Ni*	*Pb*	*SO_2*	*Sb*	*Se*	*V*	*Zn*
The Kola Peninsula												
Copper–nickel production	154	15	173	2	2	535	412	26	14·0	16·0	—	61
Fossil fuel combustion	2	1	10	34	30	40	54	190	2·0	4·0	—	10
Steel and iron	3	1	29	47	37	44	13	30	2·3	1·5	—	19
Phosphate fertilizer	—	10	21	—	—	21	3	—	—	2·4	—	80
Wood processing	6	2	2	39	37	5	26	10	4·8	1·7	—	10
Gasoline combustion	—	—	—	—	—	—	237	—	—	—	—	
Total	165	29	235	122	106	645	745	256	23·1	25·6	122	180
The Pechora Basin												
Coal mining and combustion	8·5	2·1	34	60	55	62	27	70	7	17·5	49	40
Wood processing	3	1	15	21	19	11	13	30	4·8	4·7	17	16
Gasoline combustion							158					
Total	11·5	3·5	49	81	74	73	198	100	11·8	22·2	66	56
The Norilsk area												
Copper–nickel production	242	24	312	3	2·5	900	649	75	22	24·9	—	235
Fossil fuel combustion	1	0·5	7	8	7·0	11	27	100	1	3	113	8
Wood processing	3	1·0	83	20	18·5	24	13	5	2·4	4·7	17	19
Gasoline combustion							53					
Total	246	25·5	402	31	28	935	742	180	25·4	32·6	130	262
Total for all regions above (arctic and sub-arctic)	422·5	58	686	234	208	1653	1685	536	60·3	80·4	318	498
Percentage of total USSR emissions which are arctic or sub-arctic	15·2	6·7	13·0	7·0	6·8	35·5	5·5	3·8	17·2	13·2	4·7	2·4

Emissions of air pollutants in the arctic and subarctic USSR (in metric tons per year except SO_2 in 10^3 metric tons per year).

9.3.2 Mining and smelting

Mining, and the associated processing of the ore, emit substantial amounts of important air pollutants in certain regions of the Arctic. Mining activities occur in northwestern Alaska (near Kotzebue),[16] Canada (Northwest Territories), on Svalbard, in Eastern and Western Siberia, and most importantly on the Kola Peninsula.[3] A variety of minerals are mined and smelted in the Arctic including lead, zinc, apatite, copper, nickel, iron, gold, and coal. In general, the mining operations themselves will produce modest amounts of air pollution, principally exhaust from diesel equipment and dust. More important are the gaseous and particulate emissions from smelters and other processing operations. These sources will produce substantial amounts of SO_2, NO_x, and other gaseous and particulate species. Processing of apatite, a phosphate mineral, will likely produce the highly toxic gas hydrogen fluoride.

Table 9.1 shows estimated emissions for 11 trace metals and SO_2 from Soviet emission sources, including mining (data taken from Ref. 15). As these data show, emissions of these pollutants in the arctic regions of the Soviet Union are a substantial fraction of total Soviet emissions, and mining activities account for a substantial fraction of the total Soviet Arctic trace metal emissions.

9.4 NON-INDUSTRIAL SOURCES

From a health perspective, non-industrial sources of air pollution, such as automobiles and residential heating, may pose a greater, more immediate, health concern to some residents of the Arctic and sub-arctic than industrial sources. This is simply due to the fact that many of the large arctic developments are remote from substantial population centers. In the *urban arctic environment*, CO (carbon monoxide), Pb (lead) and polycyclic aromatic hydrocarbons (PAHs), from automobiles, municipal power plants and residential heating (especially wood and coal), present a clear and direct health hazard.

9.4.1 Automobiles

The importance of the automobile in the developed world cannot be overstated. In many urban arctic and sub-arctic locations, the role of the automobile takes on even greater significance as temperatures plummet. Unfortunately one important fact is painfully apparent: when operated in cold temperatures, virtually all automobiles and trucks produce much greater amounts of CO, especially during the intial warm-up period.[17] When combined with typical surface inversions which occur during winter, the CO concentrations in small cities can reach levels which are generally only seen in much larger cities, occasionally violating federal air quality standards. Automobiles also release substantial amounts of nitrogen oxides, lead, and other particulates. Although not strictly an 'arctic' issue, automobile emissions are an important concern to many high latitude residents and so will be discussed in this chapter.

Measured carbon monoxide concentrations in Fairbanks, Alaska, during winter are comparable to levels observed in much larger cities like Los Angeles. Violations of the 8 hour 9 ppm federal standard are common, with maximum eight-hour concentrations occasionally 2–3 times higher. Figure 9.3 shows the number of CO violations occurring each winter season (exceeding the 8 hour 9 ppm standard) in Fairbanks and the maximum eight-hour CO concentration recorded that year. The general declining trend can probably be attributed to the general improvement in new car CO emissions as older cars are slowly phased out. This decline in the number of violations has occurred despite an approxi-

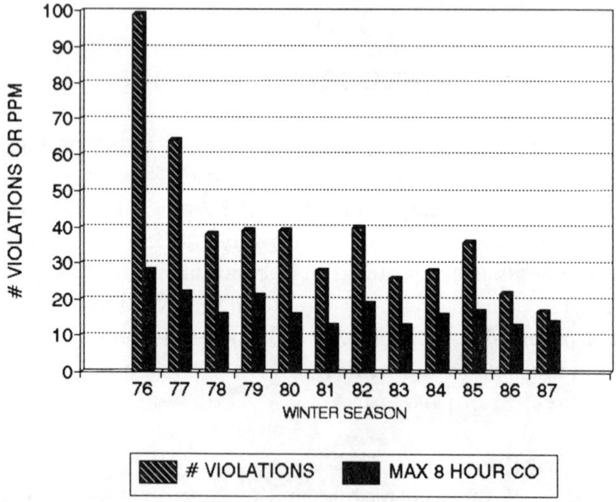

Figure 9.3 Number of CO violations each winter in the Fairbanks area and the maximum eight-hour average CO concentration recorded during that season. The year refers to the season beginning with December of the named year. (Data provided by the Fairbanks North-Star Borough.)

mately 50% increase in the number of vehicle miles travelled during this period.

In general, CO concentrations tend to be worst on days when the ambient temperature is in the range −6 to −30°C. In urban areas, temperatures lower than this result in the formation of 'ice fog' which acts to decrease the thermal gradient in the mixed layer, resulting in a larger mixing volume and therefore lower CO concentrations.

The CO problem has two facets which, because both are functions of temperature, are difficult to separate:[18]

(1) CO emissions are inversely related to temperature, and vary on a vehicle by vehicle basis; and

(2) the inversion strength (or in other words the thermal gradient) is also related to temperature, although no simple relationship is known.

This problem has been documented in several high latitude cities, incuding Fairbanks[19,20] and Anchorage,[21] Alaska, and Whitehorse, Yukon Territories.[22] Certainly it must be a serious problem in numerous other high latitude cities as well. Two approaches have been used to help alleviate the CO problem in Fairbanks: mandatory inspection and maintenance (IM) programs for automobile owners in impacted regions; and the use of electric engine pre-heaters. The IM program is aimed at identifying the relatively small number of vehicles which produce a disproportionate share of the emissions. Owners of faulty vehicles must then take corrective action to repair the problem. Although this program is very similar to other IM programs operating in other cities, it is not clear that testing vehicles after the initial 'cold-start' phase is an effective measure to control overall CO emissions.

In 1986 the Fairbanks North-Star Borough instituted an IM program which requires car owners to have their vehicles tested once per year. The improvements in CO emissions which are a result of the IM program, shown in Fig. 9.4, suggest that less carbon monoxide is being released into the local atmosphere per vehicle; however the data on violations per year (Fig. 9.3) do not show a significant decrease in the number of violations since the IM program began (1986). Presumably the increase in the number of vehicle miles and other factors, particularly meteorology, have masked the effect of the IM program.

Electric engine pre-heaters are designed to be used for at least several hours (or overnight) to warm up the engine prior to initial start up. This approach has been shown to reduce CO emissions from essentially all vehicles.[23]

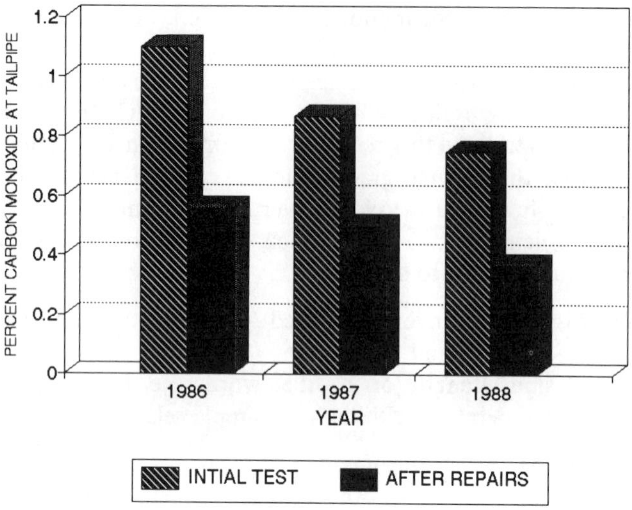

Figure 9.4 Reduction of automotive CO emissions as a result of the Fairbanks North Star Borough's Inspection and Maintenence program. This figure shows the average tailpipe CO concentrations (%) from all vehicles during the initial test and average concentrations during subsequent tests on those vehicles which failed the initial test. (Data provided by the Fairbanks North-Star Borough.)

The proportion of automotive CO which comes from a few worst offenders continuously, versus that which comes from all vehicles under cold-start conditions is unknown and a controversial question. One interpretation of the existing 'cold-start' test data is that although colder temperatures do produce additional CO, most CO is produced by a small fraction of heavily polluting vehicles.[24] If this is the case then the presence of a strong inversion (and consequently a small mixing volume) is the primary cause for elevated CO concentrations during winter. An alternative view is that it is the elevated emissions of all cars during cold-start conditions, combined with the meteorology, which causes the problem. These two views suggest different courses of action to remedy the problem. Proponents of the first view seek out better means of identifying those vehicles which are the worst emitters, whereas proponents of the second view seek legislation which will require manufactures to reduce the CO emissions emanating from new vehicles at low temperatures.[24] Amendments to the US Clean Air Act have been proposed which will require automobile manufacturers to test their vehicles for CO emissions under realistic cold-start conditions, although Stedman[24] argues that this is an unnecessary step and will not alleviate problems of urban CO violations.

Other species which have been measured in the urban Fairbanks include nitrogen oxides and lead. Lead was measured by Winchester *et al.*[25] and Holty[26] who found concentrations up to 6 $\mu g\, m^{-3}$, exceeding the federal standards. Winchester *et al.* also measured halogens and found that the Pb/Cl ratio was similar to that found in leaded gasoline, presumably the source.

Nitrogen oxides and ozone were measured by Coutts[27] in the Fairbanks area who found that while total NO_x was frequently very high, NO_2 concentrations were not very high and never exceeded the national standard (0·05 ppm). Ozone concentrations were generally very low and frequently went to 'zero'. This is due to the fact that nitric oxide (NO), produced by combustion, reacts rapidly with ozone to produce NO_2. In the presence of sunlight, the NO_2 will be photolyzed back to NO, regenerating ozone in the process. This cycle ultimately is responsible for producing significant urban ozone in sunny cities like Los Angeles. In Fairbanks during winter, the sun is only just above the horizon for 4–6 hours. The low solar angle yields essentially no UV light reaching the surface, and therefore essentially no photochemical reactions, resulting in the near complete destruction of ozone by nitric oxide within the surface inversion. High NO concentrations (as much as 0·3 ppm) and high NO/NO_2 ratios (approximately 10) are observed in the urban area. Benson[5] has argued that the high NO/NO_2 ratio is evidence for a 'reducing atmosphere'. On a relative scale it is probably true that the Fairbanks winter atmosphere is much less oxidizing than say Los Angeles by virtue of much lower ozone concentrations. However, in the chemical sense, it is incorrect to refer to any atmosphere with 21% oxygen as reducing.

Since federal standards for nitrogen oxides are based on NO_2 concentrations, Fairbanks is not in violation. Although NO is believed to be less hazardous than NO_2, the hazards associated with exposures to these high levels is not well known.

To summarize, automobiles produce substantially greater CO emissions during start up in cold weather. This fact combined with the steep surface inversions which are observed in arctic and sub-arctic cities often results in unhealthy levels of CO. In Fairbanks, CO concentrations have improved in recent years, possibly due to the imposition of an inspection and maintenance program, the overall improvement in new car CO emissions, or changes in surface inversion patterns. Other automobile emissions, especially Pb and nitrogen oxides, can also be very elevated in urban areas.

9.4.2 Residential heating

Arctic and sub-arctic residents utilize fossil and biomass fuels (coal, oil, gas, wood and peat) for residential heating, thus adding to the pollutant loading in many regions. The combustion of wood and coal is generally less efficient than oil and natural gas, therefore producing larger quantities of air pollutants. Wood and coal combustion also generate a greater quantity of hazardous pollutants, including sulfur dioxide (especially from coal) and a host of organics, including polycyclic aromatic hydrocarbons (PAH), a group of compounds which are known carcinogens.

Given the romantic image of the lonely arctic, it is hard to understand how a few wood stoves can produce unhealthy air quality. However, in numerous arctic and sub-arctic cities and towns, especially those in low-lying areas subject to frequent inversions, the presence of many coal and wood stoves in close proximity is a serious problem. On cold and clear nights, when these stoves are being used to capacity and inversions are most common, the stove emissions frequently result in substantial concentrations of hazardous organic compounds. Reichardt & Reidy measured the concentrations of polyaromatic hydrocarbons in Fairbanks, Alaska, during the winter of 1976–77. The concentrations they observed of these compounds were generally higher than the levels that are found in much larger cities. Although they attributed the polyaromatic hydrocarbons mainly to automotive emissions, wood combustion (which was not considered in their study) was certainly at least partially responsible.[28]

9.4.3 Biomass burning (forest and tundra fires)

Every summer, a substantial region of arctic and subarctic lands burn. These fires produce substantial emissions of NO_x, CO, hydrocarbons, soot and other particulates. Plumes from these fires can be transported long distances, and result in occurrences of very contaminated air away from the source region.

During the summer of 1988, approximately 8500 km^2 of land burned in Alaska (about the combined area of Rhode Island and Delaware). The plume from several of these fires frequently reached Fairbanks. During a period when the plume from a major fire in the Livengood area passed through Fairbanks, the Fairbanks North-star Borough's monitoring program recorded its highest ever concentration of PM-10 (aerosols below 10 μm diameter); 158 $\mu g\,m^{-3}$. Presumably other combustion pollutants were elevated as well, such as nitrogen oxides; however, these are not routinely measured in Fairbanks. Although forest fires during the summer of 1988 burned a large amount of land area, no records were set (the record is approximately $2 \times 10^4\,km^2$ burned in one year). On average, approximately 2000 km^2 is burned in Alaska each summer.

When arctic biomass burns, photochemically active gases such as carbon monoxide, hydrocarbons and NO_x are produced. Combined with the available sunlight during summer, this ensures that photochemical reactions will certainly ensue, producing ozone, nitric acid and other oxidants. Although no comprehensive studies of arctic pollutants formed during biomass fires have been reported, several studies have observed incidents of elevated concentrations of photochemically produced products which were attributed to forest fires. During collection of precipitation samples on Alaska's arctic coastal plain,[29] one summertime sample during 1986 was found to contain a nitrate concentration of 15 mg $litre^{-1}$ and pH 3·3. This high nitrate concentration and low pH was attributed to NO_x emissions from a large tundra fire burning at the same time. The NO_x emissions were thought to have photochemically produced HNO_3.[29] In a separate experiment, aircraft measurements over Alaska showed the existence of atmospheric layers and plumes containing elevated levels of O_3 and nitrogen oxides. These regions of photochemical activity were attributed to tundra or forest fires.[30]

9.5 UNIQUE FEATURES OF HIGH LATITUDE POLLUTION RESULTING FROM THE PHYSICAL AND METEOROLOGICAL CONDITIONS

As a result of the physical and meteorological conditions, arctic pollution behaves very differently from pollution in lower latitudes. Most importantly, the frequent presence of a strong, thermal inversion concentrates anthropogenic emissions at ground level. The dramatic solar cycle in the Arctic gives rise to long periods of photochemical activity followed by periods of essentially no photochemistry. Chemical reactions which create and process pollutants are fairly well understood at moderate temperatures but are rather poorly understood at low temperatures. Finally, the presence of a snow surface for 6–12 months each year (depending on location) can act to concentrate some pollutants and release these suddenly during a brief melting period in the late spring.

9.5.1 Removal and deposition of pollutants

Atmospheric gas and aerosols can be removed by wet or dry deposition to the Earth's surface. The mechanisms involved in this removal process are discussed in Chapters 3 and 5.

During the arctic winter, precipitation amounts are generally low and locally generated pollutants frequently do not mix above the boundary layer inversion, so that at cloud heights, where snow forms, locally generated pollutants are probably not incorporated into snow crystals. Therefore, deposition of locally generated atmospheric pollutants which does occur during winter is probably a result of dry deposition rather than wet deposition.

Precipitation scavenging by snow or ice crystals is quite different from scavenging which occurs duing cloud droplet formation. Whereas liquid cloud droplet formation is a very efficient scavenging mechanism for soluble pollutant aerosol species, such as H_2SO_4 or NH_4NO_3,[31] ice crystal formation is much less efficient as a scavenging mechanism and probably fails to incorporate much of the sulfate aerosol.[32] On the other hand, snow appears to be very efficient at removing gaseous HNO_3, either while in the atmosphere or at the snow surface.[33,34]

During summer, photochemical processes will generate gas and aerosol species which are more efficiently scavenged than the precursor, such as H_2SO_4 from SO_2 and HNO_3 from NO_x. The pollutants are no longer confined to the boundary layer as a result of substantial convective activity which occurs over much of the arctic landmasses, dispersing locally generated pollutants over a wider area. These systems also produce the majority of the annual precipitation. Therefore, during summer, wet deposition of pollutants in the Arctic must be more significant than during winter.

There are relatively few measurements of snow or rain chemistry in the Arctic. Measurements of precipitation chemistry were conducted by Dayan *et al.*[35] in Alaska. In addition, measurements of snow chemistry in Alaska were carried out during the spring of 1988. Thirty snowpack samples were collected from seven locations along the Dalton Highway from central Alaska across the Brooks range to the arctic coast[36] (see Fig. 9.5). Snowpack samples differ from individual event samples in that both cumulative, wet and dry, deposition will be incorporated into the observed concentrations, as well as any affects from sublimation of the snowpack.

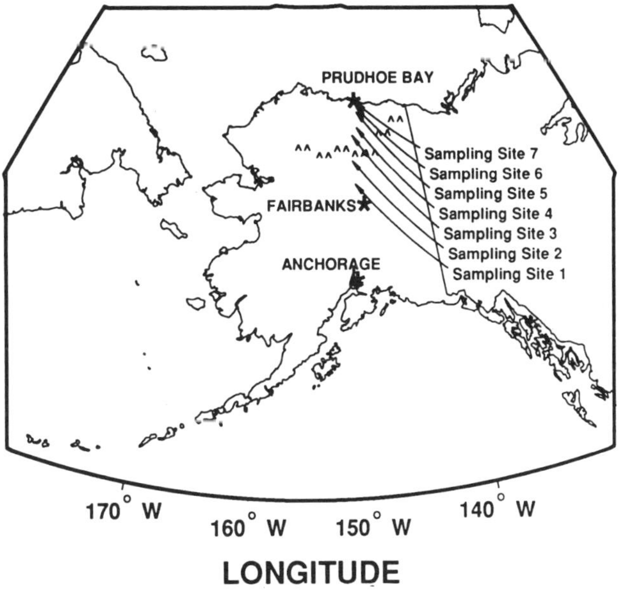

Figure 9.5 Map of Alaska showing snow sampling locations. (From Refs 40 and 41.)

The snowpack samples were analyzed for pH, conductivity, F^-, Cl^-, SO_4^{2-}, NO_3^-, Na, K, NH_4^+, Ca, Mg, Mn, Fe and Ba. Plots of pH, SO_4^{2-}, NO_3^- and Ca concentrations versus latitude are shown in Fig. 9.6. In general, samples collected south of the Brooks range had chemistries which were comparable to snow collected in other remote northern hemispheric regions. On the arctic coastal plain, concentrations of most measured species increased significantly, and pH decreased. Factor analysis on the data demonstrated that elevated SO_4^{2-}, Ca, pH and conductivity were correlated with each other to a high degree of confidence, as were Na and Cl. Although the increase in NO_3^- concentration was not as great as the other species, its increase was also statistically significant and it appeared on a single factor, suggesting that it has a source unique from the other species.

The increase in pH, conductivity, Cl^- and SO_4^{2-} is most likely a result of road dust associated with the Dalton Highway. Although all samples were collected 1–2 km from this gravel road the common presence of strong winds on the treeless arctic coastal plain results in a far greater distribution of dust than would be expected in the less windy interior of

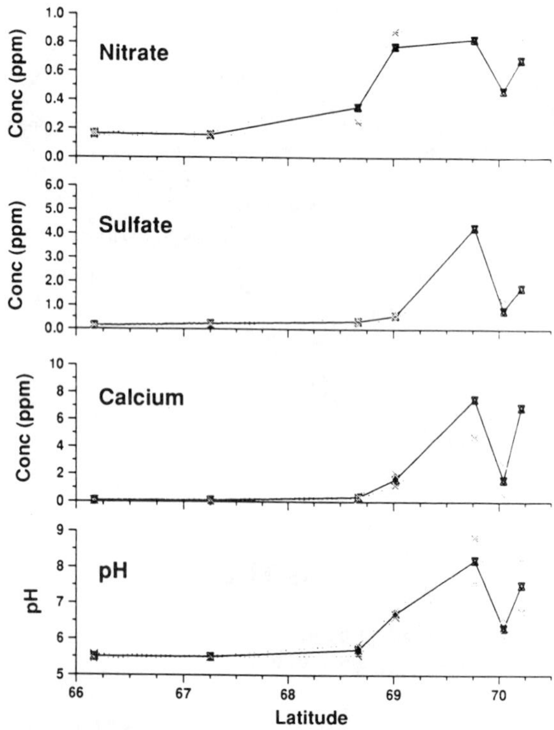

Figure 9.6 Average and standard deviation SO_4^{2-}, NO_3^- and Ca concentrations and pH for snowpack samples collected in northern Alaska during the spring of 1988. Sampling sites are shown in Fig. 9.5. Sampling site 1 is near 66°N latitude and sampling site 7 is near 70°N latitude, in the vicinity of Prudhoe Bay.

Alaska. Samples of snow collected within 5 m of the Dalton Highway (near sampling location 3) showed high pH values (8–9) and very high concentrations of all measured species except NO_3^-.

A possible source for the increase in NO_3^- concentrations seen on the coastal plain is the large gaseous emissions of NO_x from Prudhoe Bay. For this to occur, either the NO_x must be non-photolytically converted to HNO_3 which can then be efficiently removed to the snow surface, or possibly NO_2 could have a much higher deposition velocity at very cold temperatures than expected from measurements at moderate temperatures. Once deposited to the snow a slow oxidation process could oxidize the NO_2 to HNO_3. In either case, it is necessary to invoke a rather poorly understood mechanism to account for the enhanced snowpack NO_3^- concentrations seen on the coastal plain, assuming that Prudhoe Bay NO_x emissions are in fact responsible. Snowpack chemistry from other arctic locations, reflecting long range transported pollution rather than pollution from arctic sources, is discussed in Chapter 3.

In general, deposition processes act to cleanse the atmosphere by removing anthropogenic pollutants. In mid-latitudes, these removal processes generally yield atmospheric lifetimes of the order of a few days to a week for common sulfur, nitrogen and aerosol species. However, in the Arctic most of these removal processes are much less efficient, and this is one of the important factors which exacerbates arctic air pollution. However, as mentioned above there is evidence that under certain conditions some nitrogen species can be removed more quickly, via poorly understood deposition processes, than we would otherwise predict.

9.5.2 Photochemical and non-photochemical processes

The annual solar cycle is a dominant feature of the arctic environment. It affects all living things in the Arctic, plays a major role in meteorology, and also plays an important role in the cycling of chemical constituents in the atmosphere. During summer months the sun does not drop below the horizon, however it also does not get directly overhead either. This gives rise to the situation where solar insolation is continuous but of relatively low intensity (as compared to summer noon at mid-latitude locations). Of course in the winter, there is no solar insolation in the Arctic.

The most important atmospheric oxidant is the OH radical,[35] produced principally via:

$$O_3 + h\nu \Rightarrow O_2 + O^1D \quad h\nu < 320\,\text{nm}$$

$$O^1D + H_2O \Rightarrow 2OH$$

When the sun is shining, the OH radical is the primary oxidizer in the atmosphere, responsible for oxidizing organics and ultimately responsible for photochemical smog, oxidized organics such as formaldehyde, and urban ozone. Since the solar UV flux decreases to zero during winter in the Arctic, very little OH can be produced. During winter, other oxidants, such as O_3 and NO_3, probably are responsible for any oxidation that does occur in the Arctic.

There are very few documented cases of in-situ photochemical ozone production occurring in the Arctic. During one intensive airplane and ground-based measurement campaign in Alaska,[30] certain airmasses were observed to contain enhanced concentrations of O_3 and photochemical precursors to it such as nitrogen oxides and hydrocarbons which were attributed to emissions from arctic forest and tundra fires. It is surprising that there is only one report of photochemical ozone production considering the long hours of solar insolation during summer. However, three factors can account for this observation:

(1) Because of the substantial atmospheric absorptions, ground level UV is substantially diminished by high solar zenith angle.
(2) Temperatures will frequently get above 20°C over arctic landmasses, but rarely will they get much above 0°C over the arctic ocean during summer.
(3) Atmospheric inversions, which often occur during summer as well, have less steep gradients than in the winter and frequently have inversion bottoms which are 500 m or more above the surface.

One would expect that in polluted regions, photochemical 'smog' is possible in the Arctic, provided a plume of pollution remains intact. Elevated temperatures and solar zenith angles well above the horizon will act to enhance the reaction rates, and thus increase the likelihood that photochemical products will be produced before the plume can disperse.

Since 1973 the surface ozone concentration recorded at Barrow during summer has shown a steady 2% per year increase.[37] Increased photochemical ozone production, as a result of large increases in Prudhoe Bay NO_x emissions during this time period, has been suggested as a possible cause of this long-term increase based on the results of a photochemical box model.[38] The Prudhoe Bay facility, approximately 300 km to the ESE of Barrow is a large source of NO_x. Although an historical emissions inventory is not available, NO_x emissions are likely to have approximately followed oil production since 1977 when oil was first produced. Figure 9.7 shows the Prudhoe Bay oil production rate in million barrels of oil per day.

Figure 9.8 shows the results of a photochemical box model[38] for clear sky and cloudy (50% insolation) conditions in an assumed 500 km × 500 km box which includes both Barrow and Prudhoe Bay. Other parameters (temperature, background ozone, natural hydrocarbon emissions, etc.) were chosen to be appropriate for arctic summer conditions. The predicted steady state ozone concentration is shown as a

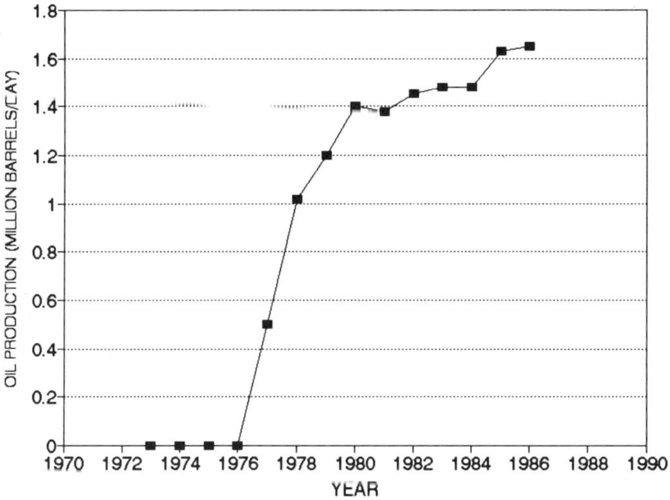

Figure 9.7 Prudhoe Bay oil production in millions of barrels per day versus year.

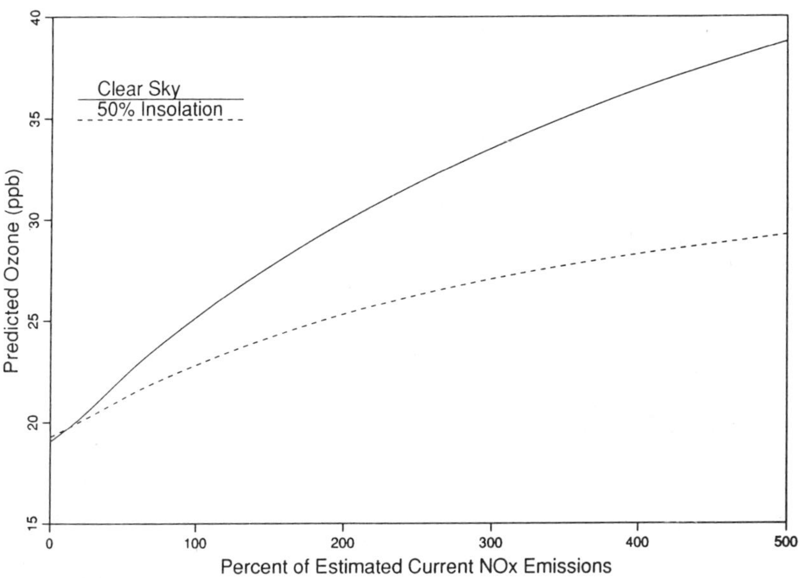

Figure 9.8 Steady state ozone concentrations calculated with an arctic photochemical box model for clear sky and cloudy (50% insolation) conditions as a function of estimated industrial NO_x emissions at Prudhoe Bay. Using this model, current NO_x emissions (100%) result in an ozone enhancement of 3–6 ppb.

function of the estimated current NO_x emissions from Prudhoe Bay (100% represents the current emissions). From this figure one observes that:

(1) Even at the low solar zenith angles found in the Arctic, sufficient light exists during summer to photochemically produce ozone from NO_x.
(2) The steady state ozone concentration is very dependent on the NO_x emissions and that this function is a non-linear function, with greatest production of ozone per unit of NO_x emissions at low concentrations.
(3) The ozone enhancement (over background concentrations) predicted by this model is between 3 and 6 ppb.

In Fig. 9.9 is shown the observed average summer (June, July and August) surface ozone concentrations at Barrow versus year (data from Ref. 37). These data show a 2% per year increase in the summer surface ozone concentrations, or an increase of approximately 5 ppb over this data record, a result not inconsistent with the previously mentioned box model. The use of a box model, however, will underestimate O_3 concen-

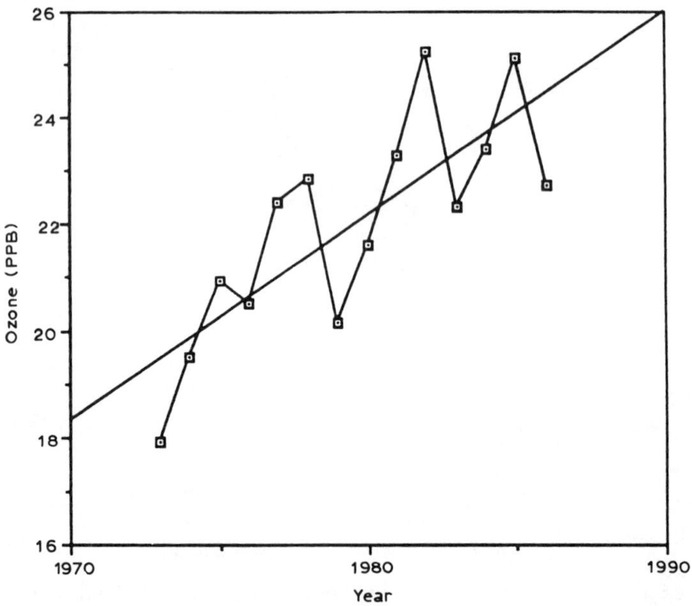

Figure 9.9 Average summer (June, July and August) surface O_3 concentrations measured at the Barrow NOAA-GMCC site with a linear regression fit to the data. (Data from Ref. 36.)

trations in the Prudhoe Bay plume. Future measurements of O_3 precursors (NO_x) concurrently with O_3 at Barrow should be able to clearly identify this process if it is indeed occurring. An alternative hypothesis is that the observed Barrow ozone increase is due to hemispheric scale pollution and ozone production and transport of the elevated ozone to the Arctic.

Measurements of nitrogen oxides conducted during the summer of 1988 did not show evidence for transport of the Prudhoe Bay plume to Barrow.[39] However the local meteorology during this period was rather atypical.

Measurements conducted during the spring of 1989 indicate that the plume of the Prudhoe Bay industrial area does, on occasion, reach Barrow.[40] Figure 9.10 shows the observed concentrations of NO and

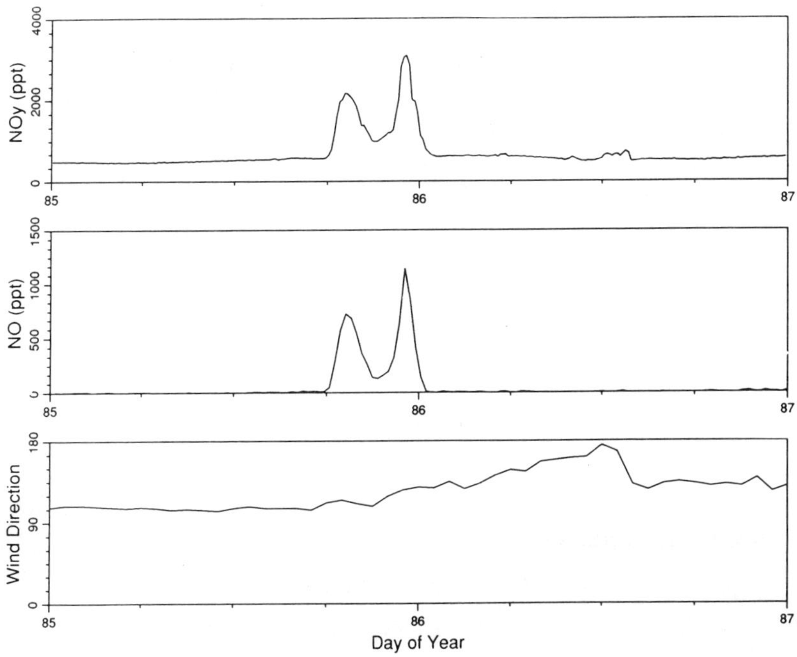

Figure 9.10 NO, NO_y concentrations measured at the Barrow NOAA-GMCC site along with the observed local wind direction during late March, 1989. Day of year is the Julian date (1 January is day 1, 31 December is day 365) and the time is Universal Time (UT). Note that UT is approximately nine hours ahead of Alaska Standard Time, so that local sunrise during March occurs at approximately 1800 UT. The Prudhoe Bay industrial area is located about 300 km east (118°) of Barrow.

NO_y (sum of reactive nitrogen oxides, $NO_x + HNO_3 + PAN + \cdots$) along with the observed Barrow wind direction for a two-day period. The occurrence of the large NO_y enhancement during day 85 coincides with a local wind direction from approximately 118°; the direction of Prudhoe Bay. Note that NO is present when the sun is up and therefore NO_x must represent a substantial fraction of the NO_y. This observation indicates that the NO_y is not 'aged' and has only recently been released into the atmosphere as NO_x. Concurrent measurements of SO_2 and sulfate aerosol during this period showed low total sulfur levels (Li, S. M., pers. comm.), consistent with the estimated Prudhoe Bay emissions. Since these measurements were conducted during the spring, when solar zenith angles are still very low in the Arctic, only slow photochemical ozone production is to be expected. Further confirmation of the role of Prudhoe Bay emissions on Barrow ozone concentrations must await further studies occurring during the summer months.

During the long dark arctic night, photochemistry must be totally absent. During this period, the usual oxidative mechanisms in the atmosphere (principally oxidation by the OH radical) are slowed considerably, although probably not eliminated entirely. The presence of 30–40 ppb ozone during winter[36] allows several possible oxidative pathways to continue, such as:

$$C_2H_4 + O_3 \Rightarrow C_2H_4O + O_2$$

$$SO_2 + O_3 \Rightarrow SO_3 + O_2$$

$$SO_3 + H_2O \Rightarrow H_2SO_4$$

$$NO + O_3 \Rightarrow NO_2 + O_2$$

$$NO_2 + O_3 \Rightarrow NO_3 + O_2$$

The nitrate radical (NO_3) is itself a strong oxidizer and can react with many other species, such as:

$$NO_3 + \text{organics} \Rightarrow HNO_3 + R^{\cdot}$$

$$NO_3 + NO_2 \Rightarrow N_2O_5$$

followed by:

$$N_2O_5 + H_2O \Rightarrow 2HNO_3$$

Many of these reactions are considered to be minor oxidative pathways for mid-latitude pollutants during daylight hours. During nighttime, some of these reactions may represent dominant oxidative pathways.[11]

The presence of four months of continuous darkness in the Arctic represents a unique opportunity to study many of these important processes, although to date little has been attempted in this regard.

9.5.3 Ice fog

Probably the most celebrated and unique local arctic air pollution phenomenon is that of ice fog. Ice fog begins to form in many urban locations when temperatures reach approximately $-30°C$, and becomes most intense at temperatures below $-40°C$. At these temperatures water vapor, from local combustion sources (automobiles, power plants, industrial, residential heating, etc.) and other sources, freeze out to form a fine mist of suspended ice particles. The ice particles are very fine (3–10 μm) and can obscure visibility to less than 50 m. Although little is known about the exact chemistry of ice fog, when allowed to collect on a clean collector surface the ice fog residue is basic and has an odour of gasoline. Not only is ice fog a problem in urban centers, but also in much smaller villages and industrial centers. Any location which has temperature inversions and sources of liquid water will be subject to this problem.

Surprisingly, non-combustion sources of water vapor are important contributors of water to the urban total. For example, in Fairbanks an analysis of water inputs into the urban atmosphere concluded that the warm water being discharged into the Chena River from a municipal thermal power plant allowed a substantial section of the river to remain unfrozen for the entire winter, and thereby contribute substantially to the ice fog problem. Benson has conducted an analysis of water inputs into the Fairbanks atmosphere using the stable isotope O^{18}/O^{16} ratio which distinguishes water which is old or recycled in the hydrological system versus new water created by the combustion of fossil fuels. In this analysis, approximately 50% of the water input was observed to be from the Chena River. (Benson, C. S, unpublished data, 1989).

Potentially these fine aerosols could be serious health concerns due to their size, which makes them ideal for inhalation into the lungs, and also their (unknown) chemical composition. Although the ice fog contains principally water, it is by no means pure. At these temperatures, various pollutants could be adsorbed onto the ice surface, including hydrocarbons, Pb compounds, CO, NO_x, and SO_2. When ice fog forms, the steep thermal gradient occurring at the surface is reduced, and a new radiative equilibrium is established which incorporates a greater mixing height. The ground level sources of CO are then diluted into a larger mixing volume and the elevated sources of SO_2 (smoke stacks) can now be included in the new mixed layer, resulting in an increase in the SO_2 concentration and a decrease in the CO concentration when ice fog forms.[5] However, some of the observed concentration decrease for reactive gases, such as SO_2, could also be a result of aerosol adsorption of gases.

9.6 IMPLICATIONS

9.6.1 Health implications

From the point of view of human health, the most important types of local air pollution in the Arctic are probably:

(1) CO emissions associated with automobiles;

(2) polyaromatic hydrocarbons and other compounds from automobiles and residential coal and wood burning; and

(3) lead compounds, though these are probably decreasing with the increased use of unleaded gasoline.

Probably less important, but still potentially hazardous, are the unknown constituents attached to ice fog aerosols.

CO concentrations in Fairbanks have improved since the imposition of mandatory inspection and maintenance programs for automobiles in the area. Additional improvements have occurred as a result of tighter new car CO standards and local programs to encourage utilization of electric 'plug-in' pre-heaters by area residents, though it is impossible to assess the relative importance of these. Also meteorology certainly impacts the area CO concentrations, however again, its contribution to recent improvements is difficult to assess.

9.6.2 Ecological implications

Impacts of arctic air pollutants on area ecosystems have only recently been considered, and so here too, little is known with certainty. Air pollution impacts can occur in two ways:

(1) The direct attack of oxidative gases (such as SO_2) on vegetation surfaces can damage plant tissue.

(2) Wet or dry deposited pollutants can accumulate on non-vegetation surfaces during winter, and be released quickly during the spring melt and/or concentrated as it is processed up the food chain.

In many cases, ground level concentrations of air pollutants are below those which are known to cause direct vegetation damage, so that only pollutants which are deposited to the surface represent potential problems.

Oxidative gases, especially SO_2, NO_2, and O_3, in high concentration are known to cause deleterious effects on numerous types of vegetation. These types of impacts are well known in non-arctic locations, but the author knows of no documented cases of this effect in arctic or sub-arctic locations. However, there is no inherent biological, physical or chemical reason why these impacts cannot occur in the Arctic. Since SO_2 is a direct industrial emission (as compared to NO_2 and O_3 which are photochemically generated and therefore produced more slowly at the high latitudes) it seems the most likely candidate for vegetation impacts in the Arctic. As mentioned previously, substantial SO_2 emissions are produced from power plants, smelters and pulp mills, and one can certainly envisage the situation where a large industrial plant with substantial SO_2 emissions operates in a confined local valley resulting in substantial SO_2 concentrations and vegetation impacts. For instance, several pulp mills operating in southeastern Alaska or smelters in the Soviet Union fit this description.

Because of the relatively short growing season, and generally low atmospheric concentrations, deposited pollutants may present a more serious hazard to arctic ecosystems. These ecosystems operate on a limited nutrient budget and therefore may be very susceptible to modest changes in atmospheric inputs. Barsdate & Alexander[41] reported that an arctic tundra ecosystem operates with a nitrogen budget which is approximately 1–10% of a typical nitrogen cycle for a mid-latitude ecosystem. Based on this observation they stated that:

> The general paucity of nitrogen and the high degree of its retention may pose unusual problems in the economic development of arctic tundra areas.

Their study was conducted prior to the substantial North Slope oil development at Prudhoe Bay. Since that time, our observations of snow NO_3^- indicate that it has increased approximately four-fold in the region extending out approximately 80 km from Prudhoe Bay (see Section 9.5.1 in this chapter) as compared to the snowpack NO_3^- concentrations reported by Barsdate & Alexander.[41]

The pollutants which are concentrated in the snowpack during the arctic winter are subsequently released during a relatively short spring thaw. The melting and elution of these snow contaminants further acts to concentrate many ions even further. Measurements at Toolik Lake[42] on the Alaskan arctic slope demonstrated that snowpack NO_3^- concentrations are concentrated by approximately a factor of five in the earliest meltwater as compared to the bulk mean concentration. Although the Arctic ecosystem impacts of this process are unknown, this affect has been found to be an important contributor to acid precipitation ecosystem impacts at lower latitudes.

9.6.3 Major uncertainties and research needs

Overall, the state of knowledge with respect to arctic air pollution is far below what we know about pollution in other regions. We know that pollution is often present at low concentrations but occasionally reaches much higher levels. We know that photochemical smog formation is much less pronounced in the Arctic, though not totally absent. We know that severe thermal inversions frequently act to concentrate pollutants into very limited mixing volumes, resulting in concentrations for certain pollutants, especially CO, which are unhealthy and exceed air quality standards.

However what we do *not* know constitutes a much longer list, for instance:

(1) What is the chemical processing of atmospheric constituents during the arctic night?

(2) What are the constituents of 'ice fog', and what are the physical and chemical mechanisms which are responsible for the presence of these pollutants in ice aerosols?

(3) How are pollutants removed to the snow surface during winter?

(4) Are there any effects on arctic ecosystems which occur as a result of long term changes in atmospheric deposition?

(5) What is the relative contribution of various factors (meteorology, cold start emissions versus continuous emissions) which contribute to the urban CO problem?

(6) Are urban polyaromatic hydrocarbon concentrations increasing or decreasing as a result of increased use of woodstoves, and what combustion factors are responsible for any changes?

(7) What are the health impacts of elevated NO concentrations?

(8) What are the natural contributions to air pollutants in the Arctic (e.g. fires, volcanoes, tundra)?

(9) How different is photochemistry in the Arctic during summer as compared to mid-latitudes?

9.6.4 Summary

Not to lose our perspective, pollution in the Arctic is generally found at concentrations which are well below those found in more populated and industrialized regions. But to maintain vital arctic ecosystems and prevent further degradation of urban arctic environments, it is vital that we understand the physical and chemical processes which cause arctic air pollution and the biological responses to this pollution. We know that long-range transport of pollution to the Arctic is responsible for a widely dispersed background concentration, particularly during winter and spring. Superimposed on this background are hotspots of industrial and urban pollution centers which impact the immediate vicinity and often wider areas around them. With further increases in arctic population and industrialization on the immediate horizon it is appropriate, and indeed essential, that we consider the impacts of anthropogenic emissions.

REFERENCES

1. Orvig, S., *World Survey of Climatology, Vol. 14: Climates of the Polar Regions.* Elsevier Pub. Co., New York, 1970.
2. Komarov, B., *The Destruction of Nature in the Soviet Union.* M. E. Sharpe, White Plains, NY, 1980
3. Central Intelligence Agency, *Polar Regions Atlas.* US Govt. Printing Office, Washington, DC, May 1978.
4. Bilello, M., Survey of Arctic and subarctic temperature inversions. Cold Regions Research and Engineering Laboratory, Tech. Report 161, Hanover, N.H., 1966.
5. Benson, C., Problems of air quality in localized areas. In *Arctic Air Pollution*, ed. B. Stonehouse. Cambridge University Press, NY, 1986.
6. Bowling, S., Climatology of high latitude air pollution as illustrated by Fairbanks and Anchorage, Alaska. *J. Clim. App. Met.*, **25** (1986) 22–34.
7. *Alaska Economic Trends*, **7** (10), 1987, 1–6.
8. Central Intelligence Agency, *USSR Energy Atlas.* US Govt. Printing Office, Washington, DC, January 1985.
9. National Resources Defense Council, *Oil in the Arctic: The Environmental Record of Oil Development on Alaska's North Slope.* NRDC, Washington, DC, 1988.
10. Standard Alaska Production Co., *Assessing the Impacts of Oil Development on Alaska's North Slope.* Standard Alaska Prod. Co., P.O. Box 196612, Anchorage, AK 99519-6612, 1988.
11. Finlayson-Pitts, B. J. & Pitts, J. N., Jr., *Atmospheric Chemistry—Fundamentals and Experimental Techniques.* Wiley Interscience Pub., NY, 1986.
12. Epler, P., Tankers inspected for vapors. *Anchorage Daily News*, 5 November 1989, p. B1.
13. Dienes, L. & Shabad, T., *The Soviet Energy System: Resource Use and Policies.* V. H. Winston and Sons, Washington, DC, 1979.
14. Croft, T. A., Nighttime images of the Earth from space. *Sci. Amer.*, July (1978) 94.

15. NILU, Emission sources in the Soviet Union. NILU TR 0-8147, Norwegian Inst. for Air Research, PO Box 130, N-2001 Lillestrøm, Norway, 1984.
16. Gale, L., Red Dog Mine. *Fairbanks Daily News Miner*, 13 August 1989.
17. Coutts, H. J., Low temperature automotive emissions. Alaska Dept. of Environmental Conservation, Report No. AK-AP-83-1, Volumes 1 and 2, 1983.
18. Chang, T. Y., Norbeck, J. M. & Weinstock, B., Ambient temperature effects on urban CO air quality. *Atmos. Environ.*, **14** (1980) 603–8.
19. Gilmore, T. M. & Hanna, T. R., Regional monitoring of ambient air carbon monoxide in Fairbanks, AK. *J. Air Poll. Control Assn.*, **24** (1974) 1077–9.
20. Coutts, H. J., A study of winter air pollutants at Fairbanks, AK. Corvallis Environmental Research Laboratory, US EPA, Corvallis, OR 97330, 1979.
21. Schweiss, J. W., Anchorage carbon monoxide study. US EPA, Region 10, Seattle, WA, 1983.
22. Lafleur, R. J., Wintuschek, E. P. & Emslie, J. H., Cold weather carbon monoxide survey at Whitehorse, Yukon Territory. Paper presented at Pacific Northwest Section of Air Pollution Control Assn., Anchorage, AK, September 1976.
23. Coutts, H. J., Automotive cold-start carbon monoxide emissions and preheater evaluation. Corvallis Environmental Research Laboratories, US EPA, Corvallis, OR 97330, 1981.
24. Stedman, D., Automobile carbon monoxide emissions. *Environ. Sci. Tech.*, **23** (1989) 147–9.
25. Winchester, J. W., Zoller, W. H., Duce, R. A. & Benson, C. S., Lead and halogens in pollution aerosols from Fairbanks, Alaska. *Atmos. Environ.*, **1** (1967) 105–19.
26. Holty, J. G., Air quality in a subarctic community, Fairbanks, Alaska. *Arctic*, **25** (1973) 292–302.
27. Coutts, H. J., A study of winter air pollutants at Fairbanks, Alaska. Corvallis Environmental Research Laboratory, Office of Research and Development, US EPA, Corvallis, OR 97330, 1979.
28. Reichardt, P. & Reidy, S., Atmospheric polycyclic aromatic hydrocarbons: an aspect of air pollution in Fairbanks, Alaska. *Arctic*, **33** (1980) 316–25.
29. Everett, K., Seasonal precipitation chemistry at Immavait Creek, Alaska. Presented at 40th Arctic Science Conference, Arctic Division AAAS, Fairbanks, September 1989.
30. Wofsy, S. C., Harriss, R. C., Talbot, R. W., Bradshaw, J., Sandholm, S., Browell, E., Gregory, G., Barrick, J., Shipman, M., Blake, S. D. & Blake, D. Enhanced trace gas and aerosol concentrations in the free troposphere over Alaska, July–August 1988. Presented at American Geophysical Union Conference, Baltimore, May 1989.
31. Junge, C., *Air Chemistry and Radioactivity*. Academic Press, New York, 1963.
32. Borys, R. D., Snow crystal riming and arctic snowpack chemical composition. In *Precipitation Scavenging, Dry Deposition and Resuspension*, ed. H. R. Pruppacher, R. G. Semonin and W. G. N. Slinn. Elsevier Pub. Co., NY, 1983, 1281–7.
33. Huebert, B. J., Fehsenfeld, F. C., Norton, R. B. & Albritton, D., The scavenging of nitric acid vapor by snow. In *Precipitation Scavenging, Dry*

Deposition and Resuspension, ed. H. R. Pruppacher, R. G. Semonin, and W. G. N. Slinn. Elsevier Pub. Co., NY, 1983, 293–300.

34. Chang, T. Y., Rain and snow scavenging of HNO_3 vapor in the atmosphere. *Atmos. Environ.*, **18** (1984) 191–8.
35. Dayan, U., Miller, J. M., Keene, W. C. & Galloway, J. N., An analysis of precipitation chemistry data from Alaska. *Atmos. Environ.*, **19** (1985) 651–7.
36. Zukowski, M. D., Jaffe, D. A., Utley, P. L. & Benson, C. S., Spatial trends in northern Alaskan snow chemistry. Presented at 39th Arctic Science Conference, American Association for the Advancement of Science, Fairbanks, Alaska, October 1988.
37. Oltmans, S. & Komhyr, W. P., Surface ozone distributions and variations from 1973–1984 measurements at the NOAA Geophysical Monitoring for Climatic Change baseline observations. *J. Geophys. Res.*, **91** (1986) 5229–36.
38. Jaffe, D. A. & Honrath, R. E. Ozone in the arctic troposphere. Presented at American Geophysical Union Conference, San Francisco, Dec. 1988.
39. Honrath, R. E. & Jaffe, D. A., Measurements of nitrogen oxides in the Arctic. *Geophys. Res. Lett.*, **17** (1990) 611–14.
40. Jaffe, D. A., Honrath, R. E., Herring, J. A., Li, S. M. & Kahl, J. D., Measurements of nitrogen oxides at Barrow, Alaska during spring: evidence for regional and northern hemispheric sources of pollution. *J. Geophys. Res.*, **96** (1991) 7395–405.
41. Barsdate, R. J. & Alexander, V., The nitrogen balance of arctic tundra: Pathways, rates and environmental implications. *J. Environ. Qual.*, **4** (1975) 111–17.
42. Zukowski, M. D., A study of northern Alaskan snow chemistry. MS thesis, University of Alaska Fairbanks, Fairbanks, Alaska, 1990.

Chapter 10

Potential Climate Change from Arctic Aerosol Pollution

JEAN-PIERRE BLANCHET

Canadian Climate Centre, Atmospheric Environment Service, 4905 Dufferin Street, Downsview, Ontario, Canada M3H 5T4

10.1 INTRODUCTION

One potential consequence of the increasing level of global air pollution is the alteration of climate. Both natural and anthropogenic modifications to the optical characteristics of the atmosphere can alter the radiation balance of the Earth's atmosphere and surface. The diabatic solar heating and terrestrial cooling terms from radiative processes are the ultimate energy source and sink, respectively, of the Earth's biosphere. They control its temperature.

An intimate relationship exists between radiative processes and most other atmospheric processes. This close interaction makes the problem of assessing climate changes a very difficult one. To understand climate change it is not sufficient to examine only the change of the radiation balance resulting from a modification of the atmospheric composition. The multiple relationships between physical processes within the climate system are the source of both positive and negative feedbacks. These feedbacks amplify or damp an initial change within the system. This process greatly affects the net resulting climate change. For air pollution the initial change is called an external forcing since it does not depend *a priori* on the climate state.

The degree of climatic change depends on the strength and nature of the external forcing, on its regional extent, and its temporal persistence. In the Arctic, two kinds of climate forcing associated with anthropogenic activities are generally considered. The 'greenhouse gases' dominated by carbon dioxide and a multitude of other trace gases like methane, nitrous oxide and CFCs are part of the first kind. Many studies have been dedicated to that topic.[1] The second kind is the aerosols or the particulate form of pollution. These two types of pollutants can be distinguished by their influences on solar and terrestrial radiation, their atmospheric residence times and their spatial distribution. In this chapter we discuss the second kind, aerosols, in relation to climate change as obtained by numerical simulations from climate models.

10.2 ARCTIC AEROSOL

Particulate matter, originating from mid-latitude source regions via long-range atmospheric transport, is periodically injected into the arctic air mass. As discussed in Chapter 5, the low efficiency of removal processes in the Arctic during winter and spring results in long residence times and a high accumulated aerosol concentration in the arctic air mass. These particles originate from natural and anthropogenic sources. With increasing human activities at high latitudes, the fraction of anthropogenic components has been steadily increasing since the middle of this century. To a large extent production of sulfate, soot and carbon dioxide are all closely related to fossil fuel combustion. In that respect, the climate change associated with carbon dioxide and arctic haze should be considered as one problem. However, from the analysis point of view, and due to their distinct characteristics, they are normally examined individually. Here, we focus on the effect of aerosol components. The modification of the greenhouse effect due to increasing concentrations of radiatively active gases (RAGs) has been widely discussed in the literature.[1]

Unlike the case of radiatively active gases, such as CO_2, the study of the effect of aerosols on climate requires detailed investigations of the aerosol optical properties. Next, the investigation is extended to radiative transfer within the atmospheric column. This stage is used to evaluate the effect of arctic aerosol on reflection and absorption of radiation within the atmosphere. Finally, using these detailed calculations as a reference, the aerosol is parameterized for inclusion into a climate model or a general circulation model. In this approach, the aerosol characteristics and effects are studied at three scales ranging from the sub-micron (optical properties), to kilometers (radiative transfer) and to 10 000 km (climate change).

The first stage considers only the intrinsic optical properties of the particles; that is the absorption and scattering coefficients. Alternatively the single scattering albedo, defined as the ratio of scattering to total extinction coefficients, is usually preferred. Ultimately, one may consider the more detailed scattering phase function, but in view of other modeling uncertainties this refinement is not yet justified.

In the second stage, the radiative effects of a vertically distributed aerosol on the local radiation balance is investigated. The absorption of radiation within the atmosphere is a diabatic heat source that increases the local temperature. Also, the presence of atmospheric aerosol layers alters the net surface radiation balance and directly affects the surface temperature. The change of the net radiation balance at the top of the atmosphere is the net external forcing imposed on the column surface–atmosphere system.

Although some insight into the climatic effect may be estimated from optical and radiative properties of aerosols, the climate is affected by several non-linear interactions that render this estimate questionable. The final stage of the problem is examined by considering the global Earth–atmosphere system.

At each of these levels, the research effort has been approached by direct observations and by numerical modeling. Generally observations are of the sum result of several interacting processes, while modeling provides a means of analyzing the effect of combined processes. These approaches are complementary. In the following section, the three scales of research are discussed in the context of potential climate changes.

10.3 OPTICAL PROPERTIES

A knowledge of the scattering and absorptive properties of aerosols is required to evaluate the effect of particulate pollution on the radiation balance and climate. The two approaches used to evaluate them are direct observations and theoretical calculations from aerosol models. Observations of samples collected on filters, together with nephelometer measurements, provide scattering and absorption coefficients at one or more wavelengths of visible light. Observed values for dry arctic aerosols are summarized in Table 10.1 for different periods of the year. The second method computes those coefficients using Mie theory and some measurement of the aerosol size distribution with representative values of the

Table 10.1. Summary of mean volume absorption and scattering coefficients from various authors and periods of the year in the Arctic. Measurements are mostly for mid-visible wavelengths. [From Ref 24, reproduced by permission from Pergamon, 1989.]

Authors Time	*Wavelength* (*nm*)	*Absorption* (m^{-1})	*Scattering* (m^{-1})
Porch *et al.*[44]			
March	550		$1{\cdot}1 \times 10^{-5}$
Patterson *et al.*[12]			
winter–spring	633	$1{\cdot}9 \times 10^{-6}$	
summer	633	$3{\cdot}0 \times 10^{-7}$	
March (1978)	550	$3{\cdot}9 \times 10^{-6}$	$1{\cdot}3 \times 10^{-5}$
Rosen *et al.*[45]			
winter–spring	633	$2{\cdot}8 \times 10^{-6}$	
Heintzenberg[2,46]			
April–May 1979	450		$2{\cdot}2 \times 10^{-5}$
April–May 1979	550		$1{\cdot}5 \times 10^{-5}$
April–May 1979	700		$9{\cdot}6 \times 10^{-6}$
winter–spring	550	$8{\cdot}2 \times 10^{-7}$	$1{\cdot}5 \times 10^{-5}$
summer	550	$4{\cdot}4 \times 10^{-8}$	$1{\cdot}7 \times 10^{-6}$
Barrie *et al.*[47]			
winter–spring	550		$2{\cdot}5 \times 10^{-5}$
summer	550		3×10^{-6}
Bodhaine *et al.*[48]			
winter–spring	visible		$1{\cdot}9 \times 10^{-5}$
summer	visible		$3{\cdot}2 \times 10^{-6}$
Wendling *et al.*[49]			
March 1983	550	$2{\cdot}7 \times 10^{-6}$	$3{\cdot}5 \times 10^{-5}$
Clarke & Noone[14]			
March–April 1982	550	$2{\cdot}3 \times 10^{-6}$	$1{\cdot}6 \times 10^{-5}$
Porch & MacCracken[50]			
(model) spring	500	$4{\cdot}5 \times 10^{-6}$	

refractive index. Providing all data are consistent, the results from both methods should agree.

The key ingredients entering the calculations of optical properties include: particle size distribution, material composition, and type of mixture. The size distribution has been determined using particle counters and multi-wavelength integrating nephelometers,[2] and impactors for large particles[3] or diffusion batteries for small particles.[4] Composition is determined by chemical analysis.[5] From measurements at different sites it is evident that arctic aerosols have fairly homogeneous physical and chemical characteristics throughout the Arctic. The size spectrum is dominated by a narrow distribution in the accumulation mode. This mode is strongly influenced by anthropogenic sulfate and soot, and controls the optical properties of arctic haze. The large particle mode typically has a volume of particles similar to the accumulation mode[6] but composed mostly of natural materials (sea-salt, soil, pollen, etc.).

The distribution and composition of the particles are characteristic of aged aerosols with long residence times. This fact is important in two ways for potential climate impact. Firstly, it implies a long period of interaction with the radiation fields. Secondly, it also means a broad horizontal extent with somewhat uniform optical properties. These two aspects favor a large scale and persistent climate forcing. For these reasons the effect of arctic haze may be significant although the absolute concentration of pollutants is lower in the Arctic than in heavily industrialized regions of the mid-latitudes.

Another important aspect for radiative effects of the arctic haze is the state of the material mixture within the aerosol. This factor particularly affects the absorption coefficient and therefore the diabatic heating rate within the atmosphere. From electron microscopy of chemically treated plates, Bigg[7] reported that, unlike particles observed at baseline atmospheric monitoring stations outside the Arctic, the particles from Alaska contained more than half insoluble inclusion with almost all particles being coated with sulfuric acid. An internal and heterogeneous (type I) mixture of light-absorbing black carbon (soot) inclusion with a soluble hygroscopic shell leads to enhanced absorption coefficients by as much as 50% for sub-micron particles.[8,9] A second kind of mixture (type II) has soot particles embedded in the surface of large droplets or crystals. In type II, the absorption coefficients may be enhanced by as much as one or two orders of magnitude.[10,11] Due to their strength, the radiative effects of type II particles are likely to be transient (e.g. short lived), and their implications for climate have yet to be investigated. Type I is more relevant to arctic haze where particles in sub-saturated air can increase their volume by a factor of two or three by water absorption without being activated to form cloud droplets.

Knowledge of the optical properties required for computation of radiative transfer must cover the complete spectrum of solar and terrestrial radiation (0·3–100 μm). Since the arctic aerosol contains hygroscopic particles the optical properties also need to be known as a function of relative humidity. The most suitable way to find these optical properties is to construct an aerosol model. This model uses an observed size distribution, observed composition and information about the state of the mixture.

Patterson *et al.*[12] and Blanchet & List[9] independently developed arctic aerosol models based on Mie theory for spherical particles and spherical shell particles (i.e. type I mixture as described above). These models agree well with each other at a wavelength of 0·55 μm and also agree with observations. Figure 10.1 compares model results for various soot concentrations and relative humidities with observations. Similar results[9] are obtained for the infrared spectrum (2–40 μm). This figure shows that the absorption coefficient for aerosol with low soot concentrations (e.g. less than 0·1 $\mu m\, m^{-3}$) agrees well with cleaner air found near Spitsbergen. The two-year mean from Barrow, Alaska, agrees with the model values derived from observed composition. It may be concluded that observations of arctic aerosols are consistent with theory. In turn the aerosol model yields the basic information for evaluation of the radiative and climatic effects of arctic aerosols.

From the climate point of view, aerosols can act either to warm or cool the atmosphere. The sign of the temperature change depends primarily on the surface albedo and the absorption to back-scattering ratio. This is illustrated by a simple model[13] composed of one aerosol layer with an absorption to back-scattering ratio (a/b) above a surface of albedo A. It may be shown, by accounting for multiple scattering between the aerosol layer and the surface, that a critical value of absorption to back-scattering $(a/b)_c$ exists for which this system is unaffected by the aerosol. This is expressed by

$$(a/b)_c = (1 - A)^2/2A \tag{10.1}$$

and is illustrated by the solid line in Fig. 10.2. Values of (a/b) above that line result in warming of the system while values below result in cooling. This expression describes the fact that a scattering aerosol layer over a dark surface increases the reflectivity of the system (cooling effect), while an absorbing aerosol like soot over a bright snow surface reduces the reflectivity of the system (warming effect). In the Arctic the ground albedo is typically high, which favors warming. In arctic haze, soot is directly related to absorption whilst sulfate is responsible for the enhancement of scattering and visibility reduction (haziness). The absorption to back-

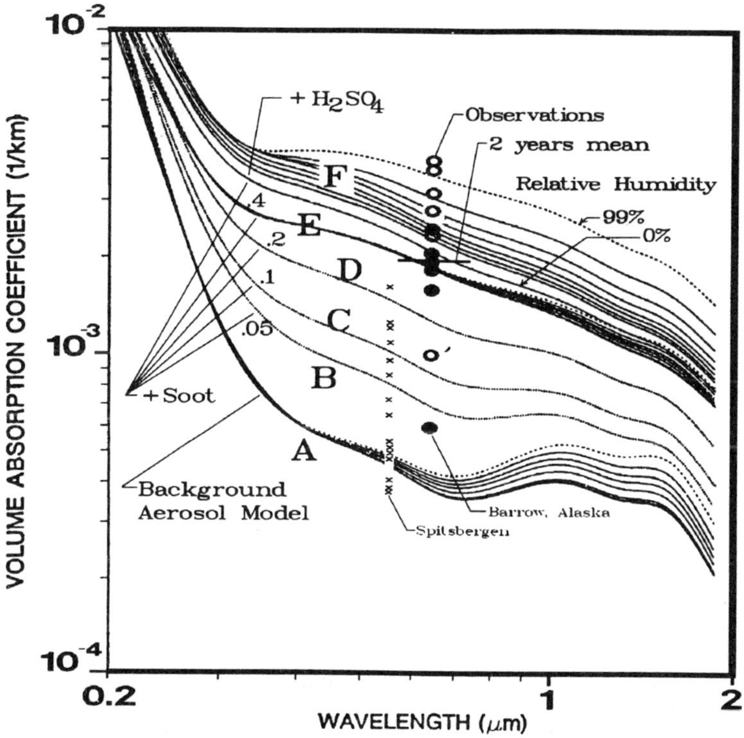

Figure 10.1. Volume absorption coefficients for the solar spectrum computed from an arctic haze model[9] as compared to measurements by Heintzenberg[46] (×), Rosen *et al.*[45] (open dots), and Patterson *et al.*[12] (black dots). The two-years winter–spring mean at Barrow is indicated by a bar. Curve A represents the 'background' aerosol model (no soot or sulfuric acid). Curves B–E contain increasing amounts of soot: 0·05, 0·1, 0·2, and 0·4 $\mu g\, m^{-3}$, respectively. Curves F are similar to E but also include 1·6 $\mu g\, m^{-3}$ of sulfuric acid in a shell, coating other externally mixed materials. Cases A, E and F were evaluated for various relative humidities (0%, 30%, 50%, 70%, 80%, 90%, 96%, 98% and 99%). Curve F, compared to E, shows the effects of an internal mixture of soot and sulfuric acid. [From Ref. 24, reproduced by permission from Pergamon, 1989.]

scattering ratio increases with soot amount and decreases with the amount of condensed water on hygroscopic aerosol (sulfate, sea-salt, etc.). Another consideration is the systematic low solar zenith angle in the Arctic that favors scattering as evaluated by a column radiation model (dashed lines in Fig. 10.2) for 60° and 84·3° respectively.

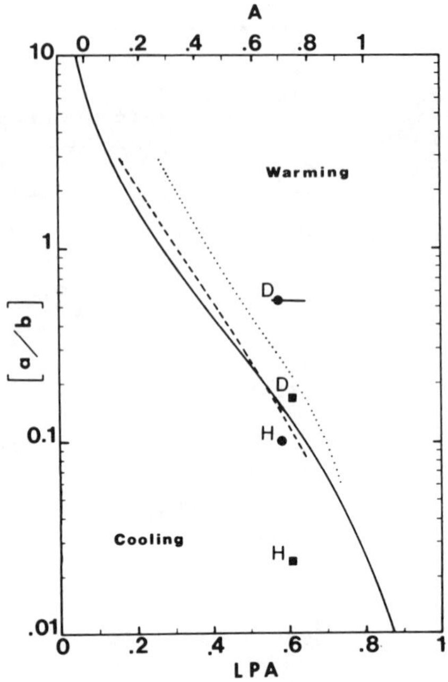

Figure 10.2. Absorption to backscattering ratio (*a*/*b*) as a function of planetary albedo (*LPA*) and ground albedo (*A*). The solid line represents the critical value at which the aerosol has no net effect on solar radiation absorption. Dashed and dotted lines were evaluated for solar zenith angles of 60° and 84·3°, respectively, using a radiation model to show the effect of the solar zenith angle. For points above those curves the aerosol has a warming effect, while below it has a cooling effect. Square points represent 'natural' aerosol in (H) hazy (99% relative humidity) and (D) dry (55%) conditions. Circles are the corresponding 'natural + anthropogenic' aerosol in hazy and dry conditions. [From Ref. 24, reproduced by permission from Pergamon, 1989, and from Tellus, 1987.]

Two quantities related to absorption and scattering coefficients, and of direct relevance to climate change, are the *single scattering albedo* (SSA), $\tilde{\omega}$, and the *aerosol optical depth*, τ. The first is defined as the ratio of scattering coefficient to extinction coefficient (scattering + absorption)

$$\tilde{\omega} = \sigma_{scat}/(\sigma_{scat} + \sigma_{abs}) \qquad (10.2)$$

It is somewhat analogous to the ratio (a/b) and determines the sign of the climate effect. The second quantity of interest is the optical depth of aerosol defined as

$$\tau = \int_0^\infty \sigma_{ext} \cdot dz \qquad (10.3)$$

where σ_{ext} is the volume extinction coefficient. It is proportional to the amount of aerosol present in the column. This quantity determines the strength of the perturbation to the radiation field and ultimately the climatic effect.

The evaluation of the SSA requires simultaneous measurement of the absorption and scattering coefficients. Clarke *et al.*,[53] showed that in the lowest 5 km of the arctic atmosphere a high correlation exists between those coefficients leading to a narrow range of SSA (at a wavelength of 0·55 μm) of 0·77–0·93 for dry aerosols. This result agrees with the earlier estimate of 0·8 from Patterson *et al.*[12] and is similar to values observed in urban and rural environments. The mean value of SSA of 0·86 also agrees well with the model evaluations[9] for dry particles. However, the models show that the SSA must approach unity as the particle water content increases with relative humidity (Fig. 10.3). In dense haze, the SSA becomes similar to that of cloud droplets with diameters of about 1 μm.[15] This haze is a transition between aerosol and cloud in terms of optical properties.

The integrated aerosol optical depth may be evaluated from measurements of the attenuation of solar radiation using a filter sun photometer.[16] During haze events, the total integrated optical depth due to aerosol is typically in the range 0·1–0·3,[17] a value comparable to many mid-latitude urban environments.[18] Arctic aerosols forming hazy layers are partly a natural phenomenon, but it is greatly enhanced by the addition of anthropogenic H_2SO_4 which can absorb large amounts of water, even in low relative humidities. Most of the optical depth is produced by water vapor condensed or deposited on existing aerosol particles. This fact was deduced by Leighton[19] from comparison between observations of direct and global irradiance at the surface,[17] and model results. The optical depth for absorption was then estimated to be about 10% of the total extinction optical depth, a value that agrees with observed SSA. In other words, light scattering by water condensed on particles accounts for most of the total light extinction by arctic haze.

Since condensed water adds little to the absorption, the ratio (a/b) decreases with increasing relative humidity. Drier aerosol layers therefore have more warming effect than very moist aerosol layers (Fig. 10.2) in

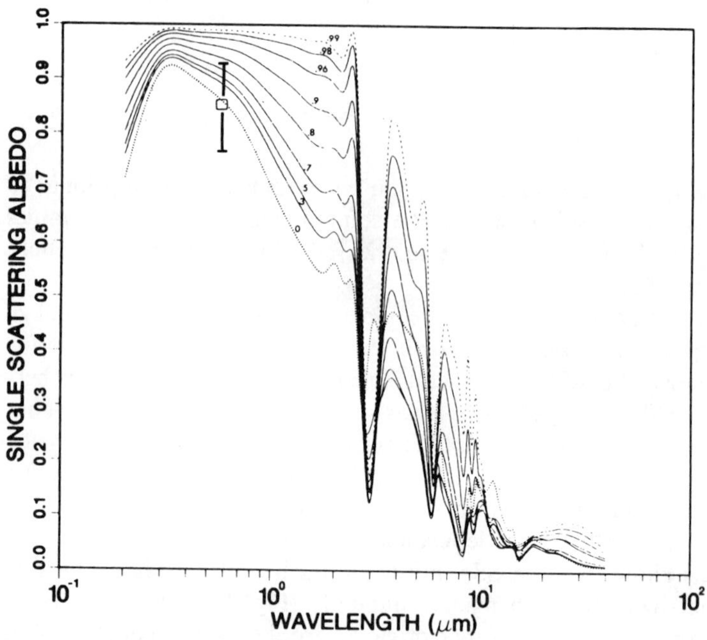

Figure 10.3. Single scattering albedo computed as a function of wavelength[9] for relative humidities (RH) ranging from 0 to 99% of saturation with respect to water. No freezing is assumed. The composition of the aerosol is similar to that observed at Barrow in March. The solar region (0·2–2 μm) is influenced by soot (0% RH) and liquid water (99%). The case of high relative humidity (99% RH) approaches the case of cloud droplets[15] of about 1 μm diameter. Mean observed[53] values at a wavelength of 0·55 μm of dried (from nephelometer) arctic aerosol are represented by the square point with a range indicated by the vertical bar (all dry aerosol), indicating that the model correctly predicted the optical properties of the dried arctic aerosol. [From Ref. 9, reproduced by permission from *Atmosphere-Ocean*, 1983.]

spite of the water shell enhancement of the absorption coefficient of sooty particles by as much as 50%.[8,9]

10.4 RADIATIVE PROPERTIES

To estimate the magnitude of the effect of arctic aerosol on the radiation field, one may use radiative transfer models applied to a column atmosphere. By dividing the atmosphere into several layers, the radiative fluxes can be computed for vertically inhomogeneous atmospheres. Calculations are performed for cases with aerosols and compared to cases either without aerosols or with a background natural aerosol as a reference. The difference between the two cases is the radiative forcing term to the climate system. The main output of these time-independent calculations is the net change of the radiative energy fluxes at the top of the system and at the ground. Results for solar radiation and from various investigators are shown in Table 10.2

Unlike CO_2, which affects long-wave terrestrial radiation, soot affects only solar radation. The additional absorption to that of pristine conditions is due to soot particles and takes place mostly in the visible light range.[20] At the top of the atmosphere the increase in the net radiation absorbed by the Earth–atmosphere system is largest during spring and for higher latitudes. The daily averaged estimate ranges from 1 to 25 W m^{-2} depending on aerosol concentration, month, and latitude. During winter, due to the limited insolation, only a small increase in solar heating is obtained near the Arctic southern limit. Unlike CO_2, which generates a global and continual forcing, soot generates a regional and seasonal forcing of an amplitude several times larger than that of CO_2. For instance, the doubling of CO_2 would produce a forcing of about 2 W m^{-2} at the top of the global atmosphere. The actual estimates suggest that this level of CO_2 will only be reached in more than a century hence. On the other hand, the forcing due to soot in the Arctic already exceeds the double CO_2 forcing by a factor of two or more during the spring season. Furthermore, the soot radiative forcing (RF) increases linearly with soot concentration while that of CO_2 increases only logarithmically with concentration. This has significant implications for a scenario of increasing anthropogenic pollutants in the Arctic.

Although sooty aerosols are found to be increasing in many parts of the world, they have their greatest impact on solar radiation in cloud-free regions. In the Arctic, soot becomes a significant contributor due to low cloudiness during spring. Other factors favoring the relative increase in soot RF in the Arctic are: the high surface albedo; the long period of sunlight during late spring; and the long path length through aerosol layers with low solar elevations. Another factor is the relative importance of soot with respect to water vapor in arctic clear air. Water vapor is the dominant absorber of radiation in the atmosphere but its concentration is limited by the low saturation point of the cold arctic air mass. In that context, soot becomes an important competitor for absorption of sunlight within the atmosphere. In other words, considering the lower absorption

Table 10.2 Summary of model evaluations of the change in the absorbed solar irradiance by the atmosphere ΔF_{atm} ($W m^{-2}$) and by the surface ΔF_{sfc} ($W m^{-2}$) from various authors; (SS) Shaw & Stamnes,[51] (PM) Porch & MacCracken,[50] (L) Leighton,[19] (C) Cess,[25] (W) Wendling *et al.*,[49] and (BL) Blanchet & List.[29] Groups show: (1) model intercomparisons, (2) diurnal average changes, (3) seasonal dependence, and (4) latitude dependence. Other columns represent the solar zenith angles (ϕ), the surface albedo (A_{sfc}) and the aerosol optical depth for absorption τ_{abs}, respectively. [From Ref 24, reproduced by permission from Pergamon, 1989.]

Authors	ΔF_{atm} (Wm^{-2})	ΔF_{sfc} (Wm^{-2})	ϕ (°)	A_{sfc}	τ_{abs}	*Remarks*
(1) *High solar elevation (model intercomparison)*						
PM	20·8	−4·1	72	0·80	0·0214	
C	21·4	−4·7	72	0·80	0·0214	
BL	20·2	−6·1	72	0·80	0·0214	
W	3·2	−0·5	75	0·74	0·0031	Low optical depth τ_{abs}[a]
(2) *Diurnal averaged (for Barrow, Alaska)*						
SS	20	−4·7	avg.	0·80	0·040	equinox (large τ_{abs})
L	10·9	−3·3	avg.	0·80	0·023	equinox, $k=0{\cdot}01$[b]
L	5·9	−2·8	avg.	0·80	0·013	equinox, $k=0{\cdot}005$
PM	5		avg.	0·80	0·021	late March
BL	7·4	−2·2	avg.	0·78	0·012	15 April, fixed albedo
BL	7·8	+0·8	avg.	var.	0·012	15 April, computed albedo[29]
(3) *Seasonal dependence at 65°N*						
C	2·1	−1·0	avg.	0·53	0·010	November
C	1·4	−0·4	avg.	0·63	0·037	December
C	3·1	−1·1	avg.	0·61	0·034	January
C	11·3	−4·9	avg.	0·58	0·044	February
C	12·4	−5·3	avg.	0·56	0·026	March
C	17·1	−7·7	avg.	0·50	0·025	April
C	8·2	−5·4	avg.	0.30	0·010	May
(4) *Latitude dependence in April*						
C	17·1	−7·7	avg.	0·50	0·025	Latitude = 65°N
C	19·1	−5·1	avg.	0·71	0·025	75°N
C	24·5	−5·7	avg.	0·77	0·025	85°N

[a]Scaled by the optical depth ratio ($0{\cdot}021/\tau_{abs} \approx 7\times$), ΔF_{atm} and ΔF_{sfc} are comparable to other authors (22·4 and −3·5 $W m^{-2}$ respectively).
[b]Evaluated[19] from bulk refractive index (k) for haze.

of sunlight in pristine arctic air due to low water content and high surface albedo, the inclusion of a minute amount of soot becomes relatively important for the radiation budget.

A second result of radiative transfer calculations consists of vertical profiles of the radiative heating rates. The heating rate is proportional to the radiative flux convergence within an atmospheric layer. This has been evaluated from observation of the net solar irradiance at various aircraft altitudes.[21–23] These results are readily comparable to model results. The daily average values reach 0·1–0·2°C day^{-1} in the lowest 5 km of the atmosphere[24] (Fig. 10.4(a), 1 ×).

The variation of the vertical distribution of aerosol above high albedo surfaces has little effect on the variation of net radiation balance at the top of the atmosphere and at the surface.[25] When considering the effect of radiative heating on atmospheric circulation and surface processes (sea-ice, snow, temperature, etc.) the vertical location of the aerosol layers is the determining factor. The increase in local radiative heating is directly proportional to the concentration of soot. For elevated aerosol layers, the diabatic heating is most effective in changing the generation of available potential energy which directly affects the formation of synoptic disturbances and the meridional transfer of heat, moisture and momentum from mid-latitude regions to polar regions. For an aerosol layer located close to the surface, the heating is influenced by several positive feedback processes leading to heating amplification.[24] These considerations will be further discussed in the next section.

Most soot particles reach the surface by wet and dry deposition processes[26] and have a lifetime of 2–4 weeks in the Arctic.[27] Soot is radiatively active as long as it remains in the atmosphere or in the top portion of the snow layer.[28] Simultaneous computations of radiation in aerosol and snow permit an investigation of the spectral signature of anthropogenic aerosols. Blanchet & List[29] showed that the radiative heating occurs in the 0·4–1·1 μm range. Soot is an effective absorber at short wavelengths where atmospheric gases are weakly absorbing and snow reflectively is highest. Above 1·3 μm snow albedo is very low and the aerosol becomes a cooling agent (Fig. 10.2) in the near infrared (Fig. 10.5).

Observations[30] and models[31] indicate that soot on the snow surface affects the radiation balance in a comparable manner to airborne particles. Clarke & Noone[14] showed that soot is widespread in the arctic snow and increases the absorption of solar radiation at the surface by 1–8 W m^{-2}. In the spring, as snow melts, soot often accumulates in the melting layer near the surface. This effect potentially compounds the reduction of snow albedo. The model shows that the cumulative effect can increase the surface absorption[31] by 5–40 W m^{-2}. The strong heating predicted by models during final melting occurs under ideal conditions. It is expected that soot has its greatest impact along the snow line because of the large increase of the local absorption of solar radiation when snow retreats.

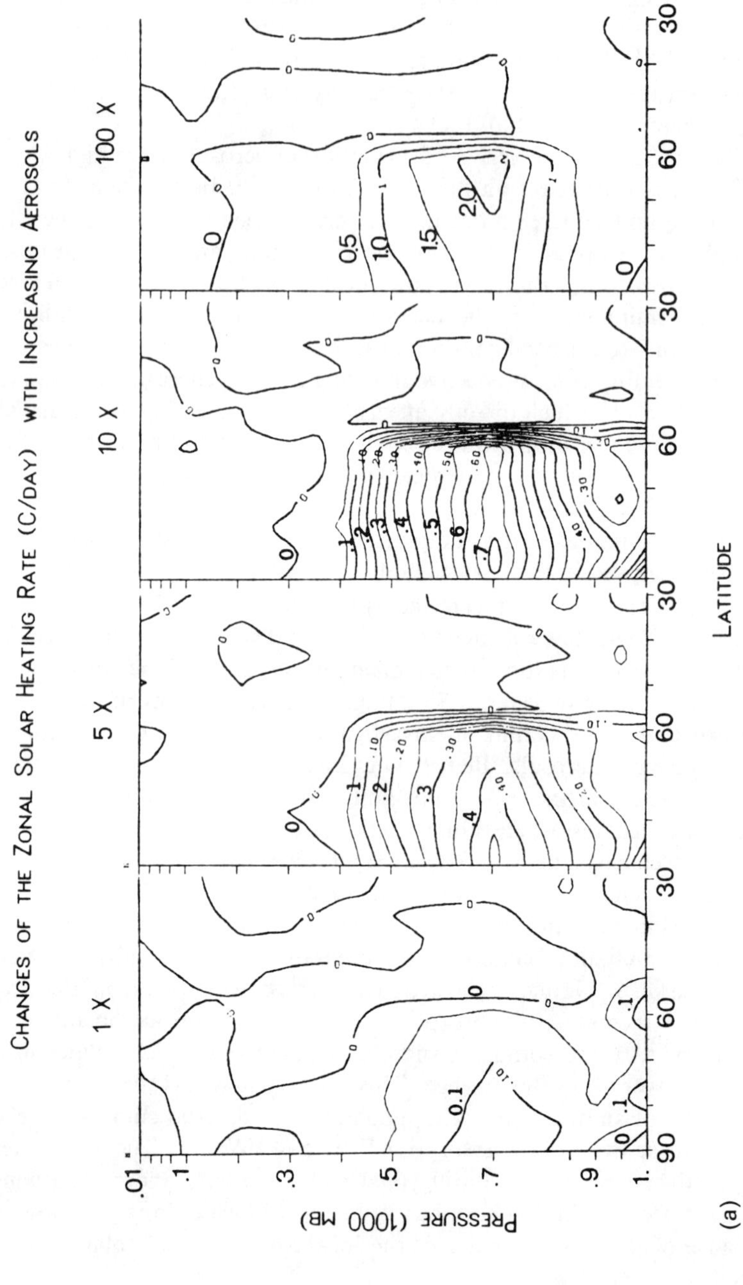
CHANGES OF THE ZONAL SOLAR HEATING RATE (C/DAY) WITH INCREASING AEROSOLS
1 X
5 X
10 X
100 X
PRESSURE (1000 MB)
LATITUDE
.01
.1
.3
.5
.7
.9
1.
90
60
30

(a)

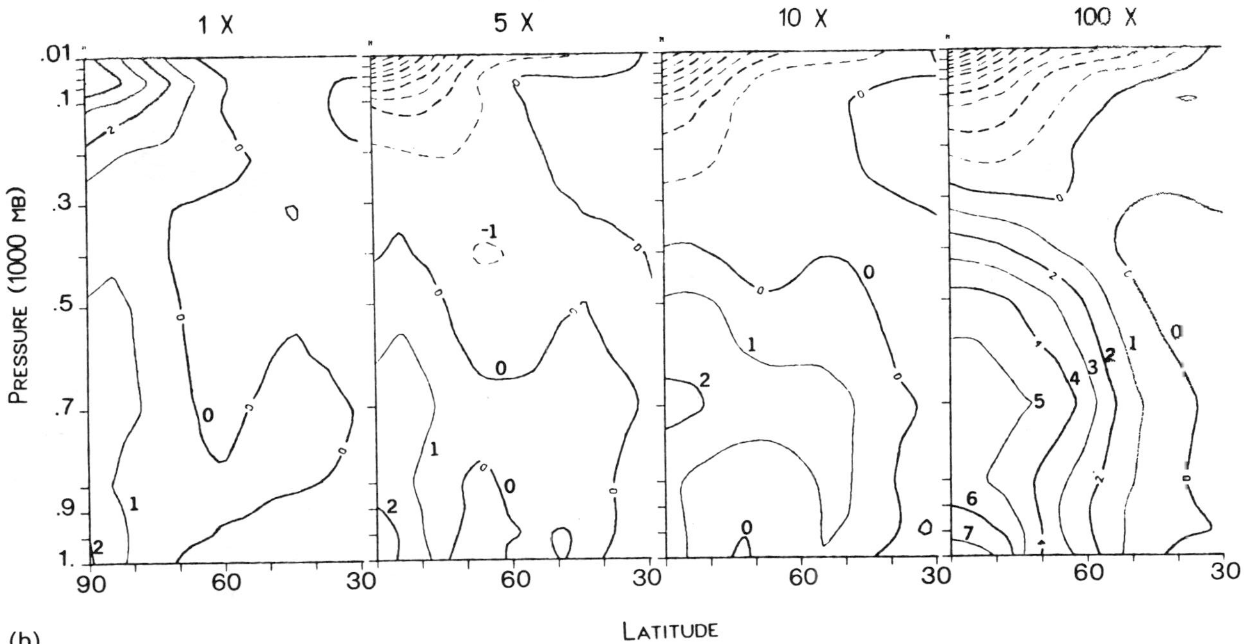

(b)

Figure 10.4. Zonal and seasonal changes in (a) solar radiative heating rates (°C day^{-1}), (b) air temperature (°C), and (c) terrestrial longwave radiative heating rates (0·1°C day^{-1}) for a GCM simulation.[39] The concentration of dry aerosol is set constant (6·54 μg m^{-3} for 1 × case) in the region poleward of 60°N and below 5 km altitude (500 mb). Aerosol particles have variable optical properties depending on the GCM's prognostic relative humidity. Four cases are shown with increasing aerosol concentration of 1 ×, 5 ×, 10 × and 100 × above the current pollution level. Solid contours represent increases while dashed lines are for decreases with respect to the control simulation without aerosols. [From Refs. 24 and 39, reproduced by permission from Pergamon, 1989, and Deepak Publishing, 1988.]

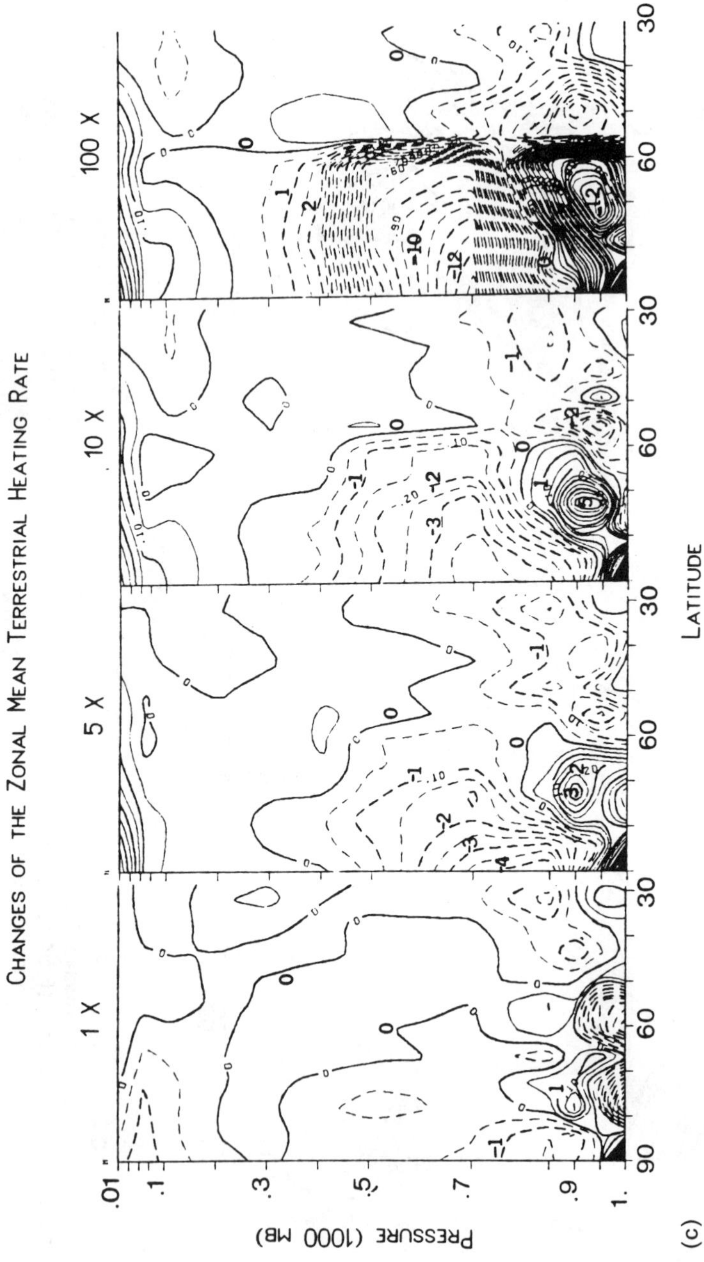
Changes of the Zonal Mean Terrestrial Heating Rate
1 X
5 X
10 X
100 X
Pressure (1000 MB)
.01
.1
.3
.5
.7
.9
1.
90
60
30
Latitude

(c)

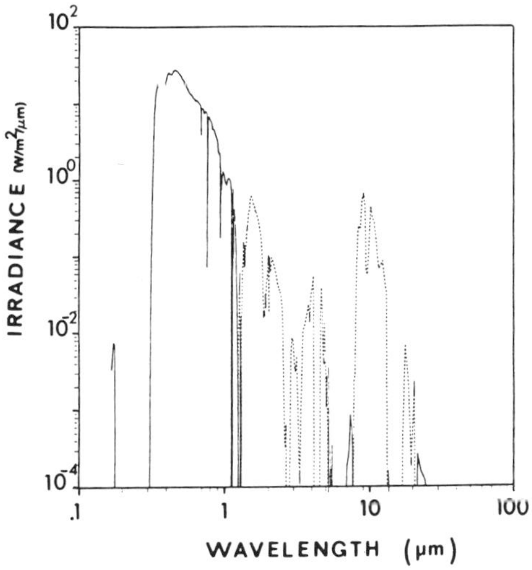

Figure 10.5. Spectral absorption at the top of the atmosphere due to arctic haze as computed by a radiative transfer model. The solid line is a gain by the surface–atmosphere system and the dotted line is a loss. Most of the warming occurs below 1·1 μm where the snow albedo is high. A peak around 10μm is produced by deliquescent aerosols in moist layers which exhibit a sharp transition in extinction coefficient. [From Ref. 24, reproduced by permission from Pergamon, 1989.]

Few results are available on the effect of arctic aerosols in the terrestrial infrared (IR) spectrum. Dry aerosol layers have a low optical depth at IR wavelengths. This implies little disturbance of the IR radiation field. In moist air, however, arctic haze forms small droplets or ice crystals that have a volume up to two orders of magnitude larger than dry aerosol particles. The consequence is a substantial increase of aerosol optical depth that is particularly effective in the 8–12 μm window region.[9]

The vertical temperature profile in the Arctic is characterized by a pronounced surface inversion. The top of the haze layer can be 10–20K warmer than the skin temperature of the radiating surface. Since haze layers act as grey bodies to thermal radiation, they can increase the outgoing long-wave radiation (OLR)[29] by 1–2 W m^{-2}. In essence this is an 'inverse greenhouse' effect. Increasing the haze opacity results in increasing the IR cooling rate in the planetary boundary layer. This is analogous to the effect of increasing carbon dioxide in the stratosphere where a temperature inversion also exists.

The optimum IR cooling is produced by wet aerosol (or cloud) layers that exhibit a sharp transition in opacity at its top (like a sharp cloud top). This is due to the fact that the increase of OLR at the cloud top from IR emission by particulates below the cloud top is not counteracted by a significant increase of the downward IR component from particulates above the cloud top. Sharp boundaries may be produced by ice crystals or deliquescent particles like ammonium sulfate. This is due to the fact that, at the saturation point or at the deliquescence point, particles exhibit a distinct phase change and consequently a change in size. In contrast, hygroscopic sulfuric acid has no deliquescence point and absorbs atmospheric water vapor to form diffuse haze layers at all relative humidities. In the absence of a sharp vertical gradient of the sulfuric acid constituent, a moist aerosol layer forms a hazy top with loose transition of its opacity. In that case both the upward (OLR) and downward IR nearly counteract each other. Calculations show that diffused haze layers result in gradual transitions of aerosol emissivity, favoring the internal exchange of radiative energy within the atmosphere. This is a redistribution of energy within the atmosphere as opposed to a genuine loss to space obtained for layers with sharp tops. This points to the fact that the morphology of aerosol layers is an important consideration for effects of arctic haze on climate.

10.5 ATMOSPHERIC RESPONSE AND CLIMATE CHANGE

The radiative transfer calculations reveal that the alteration of the radiation balance is sufficient to expect pollution to have some effects on arctic climate. Arctic climate is characterized by large variability; under these conditions the detection of climate change is difficult. Climate models predict that the positive feedbacks characteristic of the Arctic should result in that region experiencing the largest temperature changes on the globe[1] in response to changing atmospheric conditions. In the last decade, most regions of the world have experienced warming by a few tenths of a degree Celsius. Many arctic regions do not show the expected degree of warming, whilst some arctic stations have measured little change or even cooling on an annual basis. It is premature to attribute these changes to man's activities.

Kukla & Gavin[32] showed from observations that the period 1974–1978 was colder during fall, winter and early spring, while late spring and summers were slightly warmer than the mean for 1891–1978. Kukla &

Gavin[33] mention the good qualitative agreement between surface temperature along the belt of melting snow and the predicted increase of the temperature due to CO_2 increase.[34] As discussed above, this warming along the snow line is also consistent with reduction of the snow albedo from soot accumulation on the surface. Also, as suggested by Kukla & Gavin, the cooling is consistent with aerosol effects during the cold seasons. As mentioned above, the terrestrial IR cooling from sulfate wet aerosols can enhance OLR lost to space and cool the atmosphere during that season.

Jones & Kelly[35] examined the temperature data of the Northern Hemisphere and found a significant warming trend during the last 100 years. The observed trend in the Arctic is well correlated with the trend of the Northern Hemisphere temperature. The period from 1880 to 1940 warmed about 2°C in the Arctic and peaked near 1940. Subsequently, between 1940 and 1970, the trend was reversed with a cooling of about 1°C. These variations are representative of the natural climate variability. Since 1970 the trend is upward again. This change is less than the polar warming predicted by most climate models under the current increasing CO_2 trend.

The largest uncertainty in actual climate models, and particularly relevant for the arctic climate, is the treatment of the ocean. Most experiments to date have been done by taking account of the mixed layer only. In these models the ocean circulation is replaced by residual surface heat fluxes representing the ocean heat transport in the present climate state. This residual surface heat budget is evaluated by constraining the climate model, in its current climate state, to have the observed sea surface temperature and sea-ice amount. The same residual is then used in both the control (present climate) and experiment ($2 \times CO_2$) runs. In a warmer arctic climate, the decrease of the temperature gradient between the pole and mid-latitudes would likely result in a reduction of heat transported by the ocean. Therefore, the prescribed residual heat flux is likely to be over-estimated in the experiment resulting in excessive positive feedbacks via sea-ice and temperature.

Recent simulations by Washington & Meehl[36] with explicit ocean circulation showed a much reduced warming in the Arctic. Some regions of the North Atlantic were cooled substantially (up to 6°C during winter). These simulations are complex and must be further investigated.

Blanchet *et al.*[37] examined an unusually severe pollution episode over central Europe around 15 January, 1982. The aerosol observed during that period had many similarities with the arctic aerosol. This is in part due to central Europe being a major source for arctic air pollution. On 15 January, 1982, the city of Hamburg recorded concentrations of 24 $\mu g\, m^{-3}$ for soot and 400–600 $\mu g\, m^{-3}$ for sulfate. The ratio soot/sulfate is comparable to that of the arctic aerosol but at a concentration two orders of magnitude above that of the Arctic. The SSA (at 0·5 μm) of 0·92 is also similar to that of the arctic aerosol. This high concentration of pollutants was produced under an intense surface inversion and was confined to the lowest 200 m of the atmosphere. In the Arctic the aerosol is distributed over the lowest 5 km of the atmosphere. Other conditions for the case study that were comparable to arctic conditions during mid-March were: a snow covered surface; similar temperature profile; and solar elevation. For this extreme event, the model computed a diabatic heating due to soot (2·2°C) that agreed well with the observed diurnal warming. The observed diurnal variation of ambient relative humidity is explained mostly by diabatic heating from the aerosol. These results provide some evidence for the relevance of the model evaluations on the effect of pollution in the atmosphere.

MacCracken *et al.*[38] used a two-dimensional climate model to investigate the influence of soot on the arctic climate. They showed that the solar heating from soot enhances the downward atmospheric IR received at the surface. This effect overcompensates for the reduction of solar radiation received by the surface. (Table 10.2). This is due to increased air temperature and atmospheric water vapor. They estimated that the warming due to anthropogenic aerosol would reach 1–2°C in the lower 3 km of the atmosphere during April. The springtime warming due to soot acts in the same direction as the expected effect of increasing CO_2 and other radiatively active trace gases.

A series of general circulation model (GCM) experiments have been initiated by Blanchet.[39] Due to the large temperature variability in polar regions, a full GCM simulation for the case of a small perturbation requires a long time series to build up a statistically significant signal. Alternatively, by repeating a shorter seasonal simulation (March-April-May) at various aerosol concentrations (1, 5, 10 and 100 times actual values) to increase the signal to noise ratio, some consequences for the atmosphere can be deduced. In this experiment, the ocean surface temperature and the sea-ice are prescribed to climatological values. The consequences of these choices will be discussed later.

In this experiment, the aerosol optical properties are functions of the local relative humidity. This allows for the formation of haze layers in moist conditions. The vertical and horizontal aerosol concentrations are prescribed to 6·54 $\mu g\,m^{-3}$ (mass of dry aerosol) north of 60°N and below an altitude of 5 km. No aerosols are allowed elsewhere. This vertical profile of aerosol is similar to the mean observed concentration profile plus one standard deviation as reported by Barrie.[40] Spectral optical properties are those of Blanchet & List.[9]

Figure 10.4(a) shows the seasonal and zonal mean aerosol radiative forcing in the solar spectrum at 1 times (1 ×) to 100 times (100 ×) the present day aerosol concentration. The extreme case (100 ×) is comparable in concentration and heating rate to that found in extreme pollution events of mid-latitude cities.[37] As an example, this case could be produced by large-scale fires in mid-latitude regions. Figure 10.4(b) shows the corresponding change of the seasonal-zonal mean air temperature for each respective pollution concentration. Surprisingly, the temperature change of 1–2°C is insensitive to the aerosol concentration in the range 1–10 times the present level. In that range of concentration the diabatic heating is balanced by a reduction of the meridional heat transport. In other words, above a certain minimum pollution threshold, the regional arctic warming stabilizes around 1 or 2°C during spring. The absorbed heat in excess of that required to warm the air is efficiently used by the atmospheric system to alter the meridional circulation. This is shown in Fig. 10.4(b) by a tongue of warming centered at 700 mb and extending equatorwards from the Arctic Circle to the northern limit of the Hadley cell (≈35°N). This structure results from the weakening of the polar and Ferrel cells because of the reduction in the radiation deficit in the Arctic. The increasing aerosol concentration in the Arctic usually reduces the surface pressure and increases the mean temperature, specific humidity, cloud cover, cloud opacity and precipitation.

Figure 10.4c shows the corresponding change in terrestrial radiative cooling rate due to aerosols. The change of IR cooling rate may be divided into two components: (1) the change associated with the aerosol layer emissivity (direct effect), and (2) the change associated with the temperature change (indirect effect of solar heating). In the top portion of the aerosol layer (2–5 km) the direct effect dominates with weak cooling (−0·1°C day^{-1}) broadly increasing with concentration (up to −1·2°C day^{-1} at 100 ×), while in the lower atmosphere (below 1 km) the indirect effect dominates. The IR radiation shows a net tendency to propagate the effect of solar heating downwards.

Figure 10.6 shows a regional map of the aerosol radiative forcing in the solar spectral domain. This RF is produced by 0·4 $\mu g\,m^{-3}$ of soot (specific absorption of 5·0 $m^2\,g^{-1}$ at 0·55 μm) from the 1 × concentration

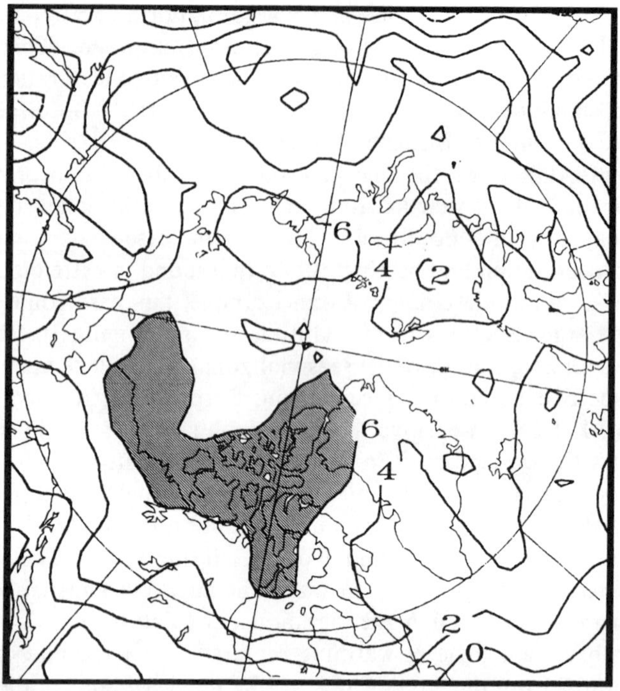

Figure 10.6. Changes in the spring (March-April-May) mean absorption of solar radiation due solely to arctic haze in a GCM simulation.[39] The contours ($W\,m^{-2}$) are the difference between the experiment and the control simulation (case 1 × in Fig. 10.4). The mean haze optical depth of 0·1 is comparable to observations. The shaded area indicates a region of maximum warming. Regional variations are affected by elevated clouds which attenuate solar radiation absorbed by aerosol located in the lower troposphere. [From Ref. 24, reproduced by permission from Pergamon, 1989.]

case. In spite of the uniform aerosol distribution within the Arctic Circle, the RF is strongly modulated. A maximum heating of 6–8 $W\,m^{-2}$ occurs in northern Alaska and over the Canadian Arctic islands. In contrast, minima of 2–4 $W\,m^{-2}$ are obtained in cloudier baroclinic regions of northern Europe and on the elevated Greenland plateau. It is also found that the change in solar radiation absorbed at the surface exhibits variations with an amplitude ($\pm 25\,W\,m^{-2}$) exceeding the direct attenuation due to aerosols by a factor of 10. The indirect effects of aerosols on clouds and moisture appear to be more important than the direct attenuation at the surface as shown in Table 10.2

A second series of experiments is presently being carried out with the Canadian Climate Centre general circulation model (CCC/GCM-2). The aim is to investigate the seasonal response of the arctic atmosphere and surface to anthropogenic aerosols. This new GCM version includes a mixed layer ocean, thermodynamic sea-ice and interactive cloud cover.[41] The aerosol model,[9] the long-wave radiation[42] and solar radiation[43] codes are the same as used in the first experiments. A thermodynamic sea-ice model implies that ice forms and expands in response to the local surface energy deficit. No horizontal transport of sea-ice is allowed. Figure 10.7 shows (a) the 10-year mean amount of sea-ice ($100\,kg\,m^{-2}$) during May of the control climate run and (b) the corresponding standard deviation ($kg\,m^{-2}$).

In this experiment, the arctic aerosol is assumed to have a 5× present-day concentration from November to May of each year, with the same horizontal and vertical distribution (c.f. Fig. 10.4(a)). Only a small background amount of aerosol ($0{\cdot}65\,\mu g\,m^{-3}$ of total dry material including $0{\cdot}04\,\mu g\,m^{-3}$ of soot) is assumed from June to October. The aerosol is allowed to form haze droplets in moist air. The simulation is 19 months long starting from 1 January of the 15th year of the model reference climate run. Seasonal and diurnal cycles are accounted for. Due to the limited length of this simulation, the 5× case is preferred in order to have a strong radiative forcing and climate response. As seen from Fig. 10.4(a), the RF may be linearly scaled to 1× and thus the response for current aerosol concentration may be estimated from the 5× case. The 5× pollution amount produces an RF that increases with insolation from $4\,W\,m^{-2}$ at 60°N in January to $40\,W\,m^{-2}$ near the pole in May, and down to near zero from June to October. The seasonal mean RF is about 5 times that shown in Fig. 10.6. Only background aerosols, with a 50 times lower concentration than in spring, are assumed from June to October inclusively.

One objective of this experiment is to assess the seasonal response of surface temperature and sea-ice to arctic haze. It is of particular interest to examine the phenomenology of changes occurring during the simulation. At the beginning (January of year 15) the temperature differences (experiment minus control runs) are small. During the first spring the temperature rises by 1–2°C (as in Fig. 10.4(b)), but without the apparent heat anomaly (Fig. 10.4(b)) extending into the mid-latitudes. In the first experiment, the ocean temperature and the sea-ice were prescribed and the excess of heat mostly altered the meridional heat flow. In this second experiment, most of the excess of heat interacts with the surface, melting

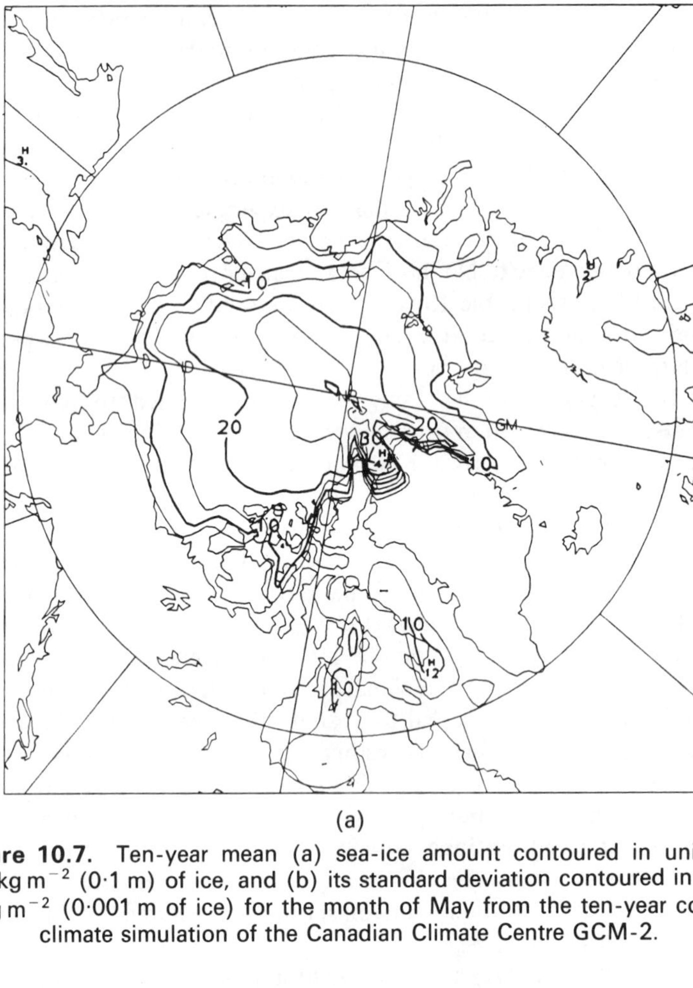

(a)

Figure 10.7. Ten-year mean (a) sea-ice amount contoured in units of 100 kg m^{-2} (0·1 m) of ice, and (b) its standard deviation contoured in units of kg m^{-2} (0·001 m of ice) for the month of May from the ten-year control climate simulation of the Canadian Climate Centre GCM-2.

snow and ice, as well as affecting the heat storage into the ocean, leaving little for modification of the meridional heat flow (during the first spring season).

In this experiment, most of the extra sea-ice melt (200–700 kg m^{-2}) occurs during July, at a time when the perturbed experiment has practically no air pollution present (0·04 μg m^{-3} of soot left as a background). The extra melting is a consequence of the warmer spring and reduced snow cover. Simultaneously the sea surface albedo is reduced and the solar heat storage in the Arctic Ocean increased (20–40 W m^{-2}

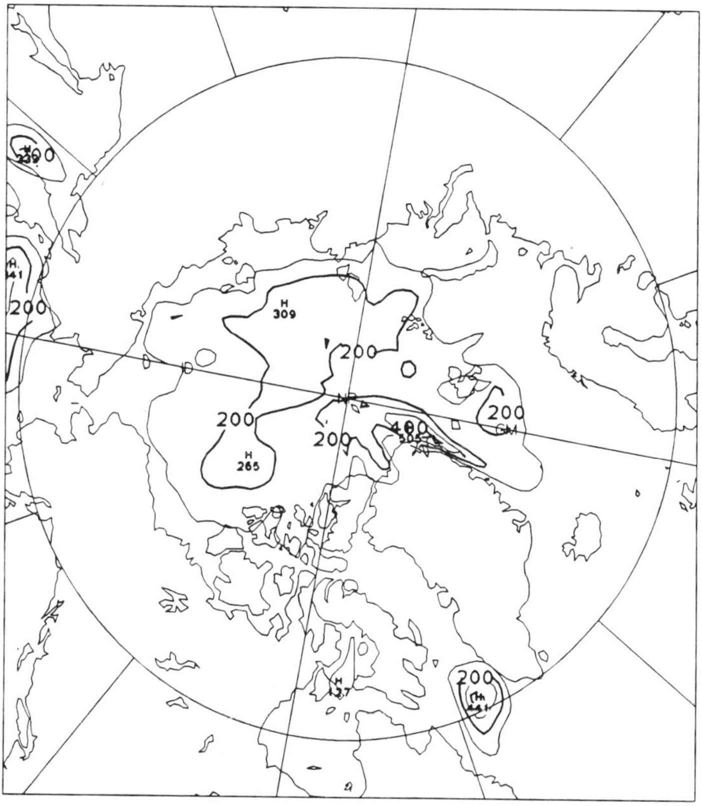

(b)

of net surface heating). During the following fall and early winter, the heat stored in the ocean is released to the atmosphere. The air temperature in the lower atmosphere is 2–4°C warmer than the control run. Due to the surface temperature inversion, the anomaly is at a maximum in January at ground level with values around 8–12°C (Fig. 10.8(a)). In the second spring the heat storage has been diminished and the surface temperature becomes gradually closer to the control run. Most of the sea-ice remains considerably (100–600 kg m^{-2}) below the control run. With the second spring the pollution RF again acts to increase the temperature in the lower troposphere by a few degrees. The second summer shows further depletions of the multi-annual sea-ice (an extra 200–400 kg m^{-2} with extreme total depletions reaching 800 kg m^{-2}). This shows that more than one year is needed to reach a new quasi-equilibrium of ice mass.

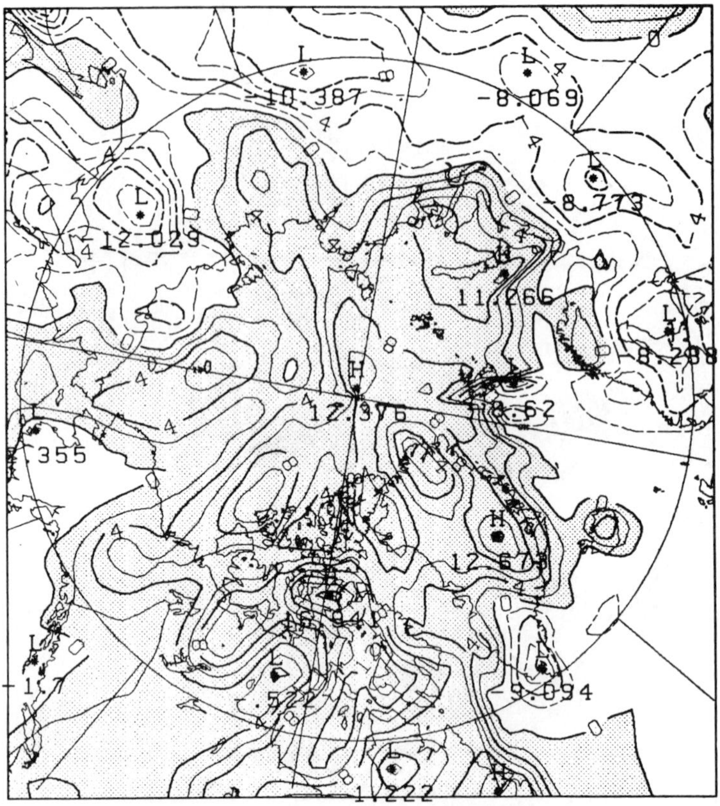

(a)

Figure 10.8. Change of the ground temperature (°C) between the arctic haze (5×) and the control simulation. These cases represent only one month of (a) January and (b) April (year 16 of the 20-year control climate run) in a 19-months transient simulation. The solid lines and shaded areas represent increases while dashed lines are decreases. Note the cool anomaly near sea-ice edges and at lower latitudes similar to the independent cases shown in Fig. 10.4(b) (1× to 10×). Apart from these cool anomalies, the January warming pattern of case (a) is very similar to the 10-year mean January temperature change pattern due to a doubling of CO_2 and generated by the same GCM (not shown). The 8–16°C warming over sea-ice regions is directly related to the reduction of sea-ice amounts in the model [e.g. Fig. 10.9] through a sea-ice–temperature feedback.

From this experiment appears an important result. The initial radiative forcing may exceed a certain threshold such that the amount of sea-ice depletions during summer is sufficient to allow for enough heat storage in the ocean to keep the sea-ice anomaly persisting over the winter.

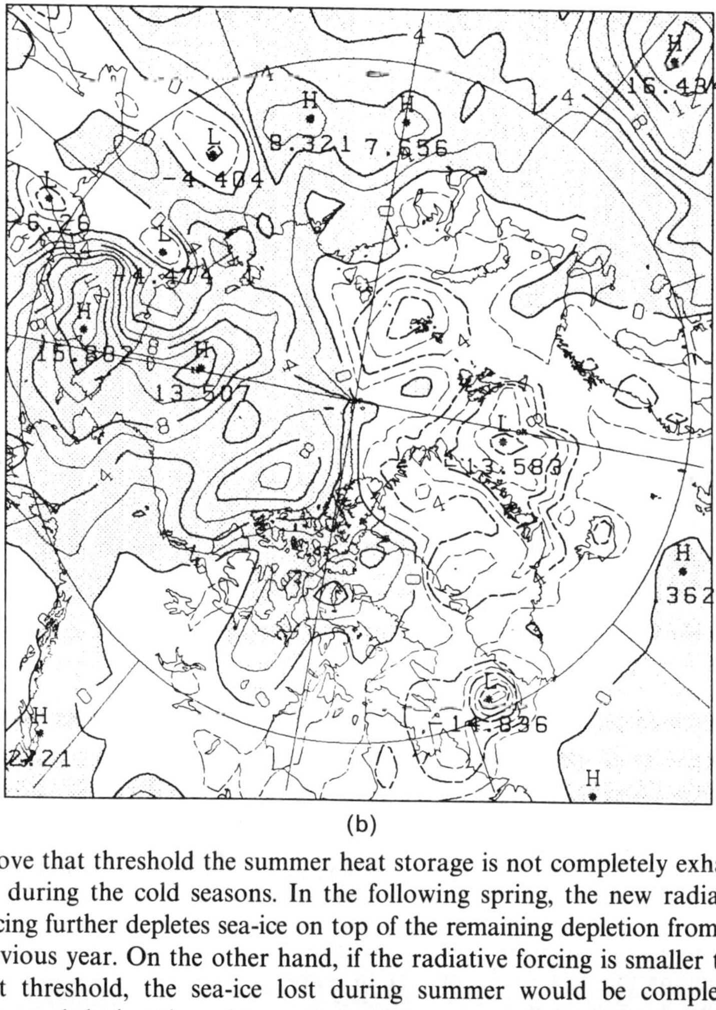

(b)

Above that threshold the summer heat storage is not completely exhausted during the cold seasons. In the following spring, the new radiative forcing further depletes sea-ice on top of the remaining depletion from the previous year. On the other hand, if the radiative forcing is smaller than that threshold, the sea-ice lost during summer would be completely recovered during the cold season and the second year would start in a state similar to that of the pevious year. In other words, the effect of a small amount of soot is practically forgotten during the cold season and the sea-ice amount adjusts to the radiative forcing within one year. The 5 × case examined here exceeds this threshold and the amount of sea-ice requires several years to adjust to a new equilibrium. This 'runaway' process would eventually be limited by leads and deformation of the sea-ice, not considered in the current model. Similar thresholds would exist for other types of radiative forcing, like increasing CO_2 and other radiatively active trace gases. This illustrates the importance of transient

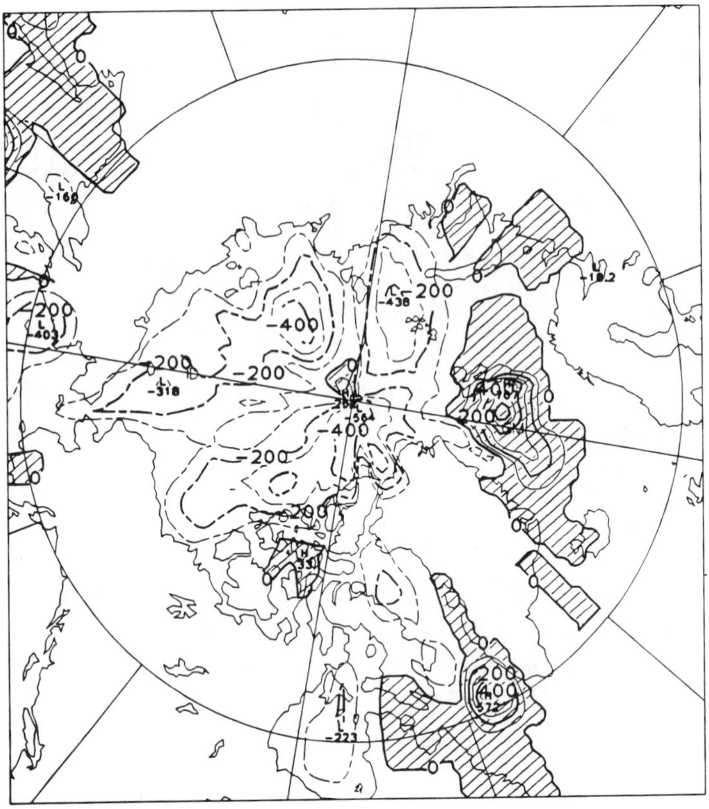

(a)

Figure 10.9. Change of the sea-ice amount contoured in units of kg m^{-2} (0·001 m of ice) between (a) arctic haze (5 ×) and control simulations, and (b) 2 × CO_2 and 1 × CO_2 simulations. The arctic haze case (a) represents only one month of May (year 16 of the 20-year control climate run) from a 19-months' transient simulation. The CO_2 case (b) represents a 10-year mean for the month of June. The solid lines and shaded areas represent increases while dashed lines are decreases. Note that the maximum depletion in (b) reached 1600 kg m^{-2} (1·6 m) (not contoured above 1000 kg m^{-2}) in the central arctic region.

climate experiments over the constant forcing type (like the resulting climate of a double CO_2 scenario).

An interesting result of the second spring season is shown in Fig. 10.9(a) by comparison to the change from doubling CO_2 from present level (10 years' mean) using the same model (Fig. 10.9(b)). In the case of

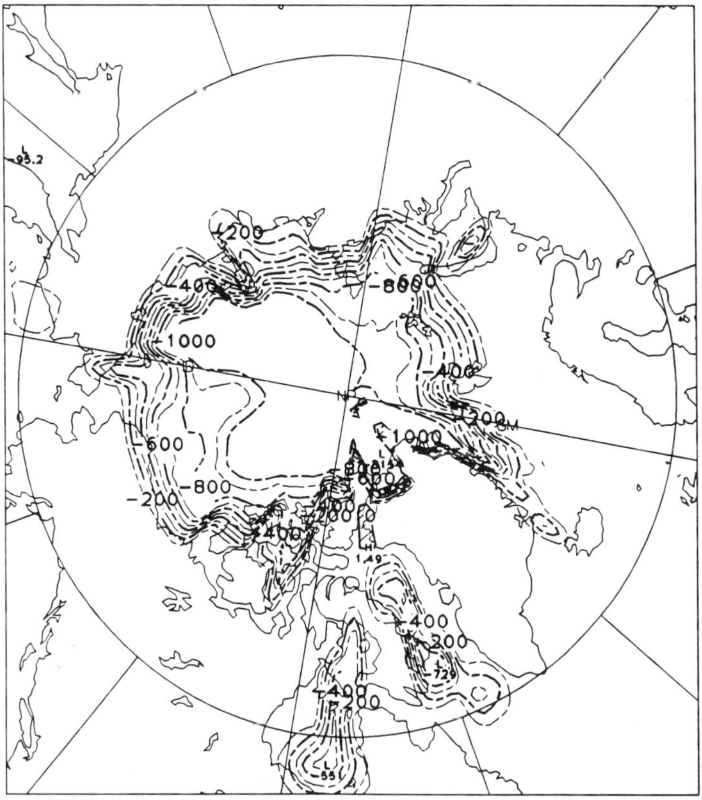

(b)

arctic haze the sea-ice decreases at high latitudes, but it increases along the southern ice edges: in the northern Atlantic, the Labrador Sea and the Sea of Okhotsk. This ice expansion at the southern edge of the Arctic is related to the negative temperature anomaly in the previous months (Fig. 10.8(b)). In turn this cold anomaly is produced by a change in the circulation. As a result of the positive temperature anomaly in the central Arctic during winter (Fig. 10.8(a)) the surface pressure is modified and the meridional heat flow to the Arctic is reduced. As in the first experiment (Fig. 10.4), the reduction in the pole to mid-latitude temperature gradient leads to a reduction in the meridional heat flow from mid-latitudes to the pole. This behavior is also shown in Fig. 10.4(b) for three of the simulations (1 ×, 5 ×, and 10 ×). At the surface and southern edge of the Arctic, a negative temperature anomaly is obtained (the 100 × shows only a trough in the low levels). The intermediate latitudes (50–70°N) normally benefit from warm air in transit to colder latitudes. As the pole to mid-latitude gradient diminishes the middle regions are cooled. As the ice expands southwardly, the albedo increases and the temperature further decreases (Fig. 10.8(b)) leading to a progression of the ice edge.

A similar result has been obtained by Washington & Meehl[36] for a transient CO_2 simulation with an explicit ocean circulation model. The CO_2 case is different from the aerosol experiment, yet some conditions are similar. The arctic aerosol produces a regional gradient of the temperature anomaly. The change of atmospheric circulation is analogous to that described by Washington & Meehl. A similar pattern was obtained by Jones & Kelly[35] from observations of the warm period at the beginning of this century (1917–1940). While the pole was 1–3°C warmer, a belt of cool anomaly or only slight warming was seen around the Arctic perimeter. These examples suggest that the investigation of climate changes in the Arctic requires particular attention to the interactions of the Arctic with mid-latitude regions via the atmospheric circulation.

10.6 CONCLUSION

The evaluation of climate change related to arctic air pollution involves the study of aerosol optics and micro-physics, the radiative transfer in column atmospheres and the study of regional and large-scale changes. For this effort, observations and models are complementary approaches.

The soot component of the aerosol is responsible for the increase in diabatic warming of the atmosphere and of the surface snow layer. At the present concentrations, the diurnally averaged heating rate is increased by 0·1–0·3°C day^{-1} above that without the presence of soot. GCM simulations at various aerosol concentrations indicate that the 'first' spring initial warming (i.e. that in the first year following the addition of aerosol in the model simulation) is nearly insensitive to aerosol concentration in the range $1\times$ to $10\times$ present day levels. When the sea surface temperature and sea-ice amount are prescribed, the excess of heat is spent in altering the heat transport via meridional circulation. When the ocean and the sea-ice are interactive with the atmosphere, the warmer spring affects the mass of sea-ice during late spring and summer.

The consequence is an increase of heat storage in the ocean that is released during the fall and winter seasons. Due to the surface inversion in the cold season, the temperature anomaly is largest at the surface. When the atmosphere is subjected to periodic annual pollution events and the associated radiative forcing, two situations may occur depending on the amount of pollutants. In the small perturbation range, the sea-ice and temperature anomalies are restored within one annual cycle. The system has no 'multi-annual memory' of the pollution events. For large perturbations, it takes more than one year for the sea-ice to reach a new equilibrium. The temperature–sea-ice feedback is amplified in the course of several years to produce a large wintertime warming. The criterion determining which type occurs is the threshold amount of soot that would cause sufficient oceanic heat storage in summer to prevent the sea-ice from recovering within the following cold season. Both cases investigated here, the 5 × aerosol amount and the double CO_2 scenario are in the large perturbation range and result in strong feedbacks. This threshold is yet to be determined, but at current levels the increase in both soot and CO_2 are likely to be in the small perturbation range without multi-annual amplification.

An interesting result of the reduction of the arctic heat deficit is the interaction with mid-latitudes via atmospheric circulation. The reduction of the meridional heat flux causes cool anomalies along the southern Arctic border during spring. Consequently, the ice margin may extend southwards during that period.

From the point of view of arctic climate, the increase of aerosol containing soot and radiatively active trace gases are of comparable importance. Both are warming factors and their effect would be additive (in the qualitative sense). Current indications suggest that anthropogenic sulfate aerosol would act as a cooling agent in the Arctic. The magnitude of long-wave radiative cooling is estimated to be less than the magnitude of the warming due to soot. However, sulfate may also play a significant role in the atmospheric water cycle during winter[24] and cloud albedo during summer.[52] All these processes act as cooling factors and deserve more investigation.

Many questions are still unresolved due to the extensive ramifications of the problem. Because of the unique conditions, the investigation of arctic pollutants leads to a better understanding of several related climate issues: urban pollution, volcanic eruptions, forest fires, desert aerosols and even the extreme case of extensive fires that are hypothesized to have caused the extinction of many living species in the past.

ACKNOWLEDGMENTS

The author is grateful to Barrie Maxwell for reviewing the manuscript.

REFERENCES

1. Schlesinger, M. E. & Mitchell, J. F. B., Climate model simulations of the equilibrium climatic response to increasing carbon dioxide. *Rev. Geophys.*, **25** (1987) 760–98.
2. Heintzenberg, J., Particle size distribution and optical properties of Arctic haze. *Tellus*, **32** (1980) 251–60.
3. Bailey, I. H., Airborne observations of Arctic aerosols. II Giant particles. *Geophys. Res. Lett.* **11**(5) (1984) 397–400.
4. Shaw, G. E., Microparticle size spectrum of Arctic haze. *Geophys. Res. Lett.*, **11**(5) (1984) 409–12.
5. Heintzenberg, J. & Covert, D. S., Chemically resolved submicronmetric size distribution and external mixing of the Arctic haze aerosols. *Tellus*, **39B** (1987) 374–82.
6. Radke, L. F., Lyons, J. H., Hegg, D. A., Hobbs, P. V. & Bailey, I. H., Airborne observation of Arctic aerosols. I: Characteristics of Arctic haze. *Geophys. Res. Lett.*, **11**(5) (1984) 393–6.
7. Bigg, E. K., Comparison of aerosol at four baseline atmospheric monitoring stations. *J. Appl. Met.*, **19** (1980) 521–33.
8. Mita, A., Light absorption properties of inhomogeneous spherical aerosol particles. *J. Meteor. Soc. Japan*, **60** (1982) 765–76.
9. Blanchet, J. -P. & List, R., Estimation of optical properties of Arctic haze using a numerical model. *Atmos-Ocean*, **21** (1983) 444–64.
10. Chylek, P., Ramaswamy, V. & Cheng, R. J., Effect of graphitic carbon on the albedo of clouds. *J. Atmos. Sci.*, **41** (1984) 3076–84.
11. Chylek, P., Srivastava, V., Cahenzli, L., Pinnick, R. G., Dod, R. L., Novakov, T., Cook, T. L. & Hinds, B. D., Aerosol and graphitic carbon content of snow. *J. Geophys. Res.*, **92** (1987) 9801–9.
12. Patterson, E. M., Marshall B. T. & Rahn, K. A., Radiative properties of the Arctic aerosol. *Atmos. Environ.*, **16** (1982) 2967–77.
13. Russell, P. B. & Grams, G. W., Application of soil dust optical properties in the analytical models of climate change. *J. Appl. Meteor.*, **14** (1975) 1037–43.
14. Clarke, A. D. & Noone, K. J., Soot in the Arctic snowpack: A cause for perturbations in radiative transfer. *Atmos. Environ.*, **19** (1985) 2045–53.
15. Irvine, W. M. & Pollack, J. B., Infrared optical properties of water and ice spheres. *Icarus*, **8** (1968) 324–60.
16. De Luisi, J. J. & Herman, B. M., Estimation of solar radiation absorption by volcanic stratospheric aerosols from Agung using surface-based observations. *J. Geophys. Res.*, **82** (1977) 3477–80.
17. Mendonca, B. G., DeLuisi J. J., & Schroeder, J. A., Arctic haze and perturbation in the solar radiation fluxes at Barrow, Alaska. *Proceedings from*

the 4th Conference on Atmospheric Radiation. American Meteorological Society, Toronto, Ontario, Canada, June 1981, pp. 95–6.

18. Rahn, K. A., Chemical and optical properties of the Barrow aerosol. *Geophysical Monitoring for Climatic Change*, No. 7. Summary report for 1978. NTIS PB80-165962, pp. 97–9.
19. Leighton, H., Influence of the Arctic haze on the solar radiation budget. *Atmos. Environ.*, **17** (1983) 2065–8.
20. Rosen, H. & Hansen, A. D. A., Role of combustion-generated carbon particles in the absorption of solar radiation in the Arctic haze. *Geophys. Res. Lett.*, **11** (1984) 461–4.
21. Valero, F. P. J., Ackerman, T. P. & Gore, W. J. Y., Radiative effects of the Arctic haze. *Geophys. Res. Lett.*, **10** (1983) 1184–7.
22. Valero, F. P. J., Ackerman, T. P & Gore, W. J. Y., The absorption of solar radiation by the Arctic atmosphere during the haze season and its effects on radiation balance. *Geophys. Res. Lett.*, **11** (1984) 465–8.
23. Valero, F. P. J., Ackerman, T. P, Gore, W. J. Y. & Weil, M. L., Radiation studies in the Arctic. *Aerosols and Climate*, ed. P. V. Hobbs and M. P. McCormick. A. Deepak Publishing, Hampton, Virginia, 1988 pp. 271–5.
24. Blanchet, J.-P., Toward estimation of climatic effects due to Arctic aerosols. *Atmos. Environ.*, **23** (1989) 2609–25.
25. Cess, C. R., Arctic aerosol: Model estimates of the interactive influences upon the surface–troposphere radiation budget. *Atmos. Environ.*, **17** (1983) 2555–64.
26. Noone, K. J. & Clarke, A. D., Soot scavenging measurements in Arctic snowfall. *Atmos. Environ.*, **22** (1988) 2773–8.
27. Ogren, J. A. & Charlson, R. J., Elemental carbon in the atmosphere: Cycle and lifetime. *Tellus*, **35B** (1983) 241–54.
28. Warren, S. G. & Wiscombe, W. J., A model for the spectral albedo of snow—II. Snow containing atmospheric aerosols. *J. Atmos. Sci.* **37** (1980) 2734–45.
29. Blanchet, J. -P. & List, R., On radiative effects of anthropogenic aerosol components in Arctic haze and snow. *Tellus*, **39B** (1987) 293–317.
30. Clarke, A. D., Charlson, R. J. & Radke, L. F., Airborne observations of Arctic aerosol, IV: Optical properties of Arctic haze. *Geophys. Res. Lett.*, **11** (1984) 405–8.
31. Blanchet, J. -P., Optical and radiative properties of the winter–spring Arctic aerosol with an assessment of the climatic implications. PhD thesis. School of Graduate Studies, University of Toronto, Toronto, Canada, 1984, 142pp.
32. Kukla, G. & Gavin, J., Cool autumn in the 1970's. *Mon. Wea. Rev.*, **109** (1981) 903–8.
33. Kukla, G. & Gavin, J., Summer ice and carbon dioxide. *Science*, **214** (1981) 497–503.
34. Ramanathan, V., Lian, M. S. & Cess, R. D., Increased atmospheric CO_2: Zonal and seasonal estimates of the effects on the radiation energy balance and surface temperature. *J. Geophys. Res.*, **84** (1979) 4949–58.
35. Jones, P. D. & Kelly, P. M., The spatial and temporal characteristics of Northern Hemisphere surface air temperature variations. *J. Climatol.*, **3** (1983) 243–52.
36. Washington, W. M. & Meehl, G. A., Climate sensitivity due to increasing

CO_2: Experiment with a coupled atmosphere and ocean general circulation model. *Climate Dynamics*, **4** (1989) 1–38.
37. Blanchet, J. -P., Heintzenberg, J. & Winkler, P., Radiative heating during an intense pollution episode in Hamburg, FRG. *Beitr. Phys. Atmosph.*, **59** (1986) 359–74.
38. MacCraken, M. C., Cess, R. D. & Potter, G. L., Climatic effects of anthropogenic Arctic aerosols: An illustration of climatic feedback mechanisms with one- and two-dimensional climate models. *J. Geophys. Res.*, **91** (1986) 14445–50.
39. Blanchet, J. -P., The effects of Arctic aerosols on a general circulation model simulation: A sensitivity study. *Aerosols and Climate*, ed. P. V. Hobbs and M. P. McCormick. A. Deepak Publishing, Hampton, Virginia, 1988, pp. 373–83.
40. Barrie, L. A., Arctic air pollution: An overview of current knowledge. *Atmos. Environ.*, **20** (1986) 643–63
41. McFarlane, N. A., Boer, G. J., Blanchet, J. -P. & Lazare, M., Simulation of equilibrium climate sensitivity to changes in atmospheric CO_2 with the Canadian Climate Centre second generation general circulation model. Part I: Description of the model and simulated control climate, *J. Climate* (1992).
42. Morcrette, J. -J., Smith, L. & Fouquart, Y., Pressure and temperature dependence of absorption in longwave radiation parameterization. *Beitr. zur Phys*, **59** (1986) 455–69.
43. Fouquart, Y. & Bonnel, B., Computation of solar heating of the Earth's atmosphere: A new parameterization. *Beitr. zur Phys.*, **53** (1980) 35–62.
44. Porch, W. M., Charlson, R. J. & Radke, L. F., Atmospheric aerosol. Does a background level exist? *Science*, **70** (1970) 315–17.
45. Rosen, H., Novakov T. & Bodhaine, B. A., Soot in the Arctic. *Atmos. Environ.*, **15** (1981) 1371–5.
46. Heintzenberg, J., Size-segregated measurements of particulate elemental carbon and aerosol light absorption at remote Arctic locations. *Atmos. Environ.*, **16** (1982) 2461–9.
47. Barrie. L. A., Hoff, R. M. & Daggupaty, S. M., The influence of mid-latitude sources on haze in the Canadian Arctic. *Atmos. Environ.*, **15** (1981) 1407–19.
48. Bodhaine, B. A., Harris, J. M. & Herbert, G. A., Aerosol light scattering and condensation nuclei measurements at Barrow, Alaska. *Atmos. Environ.*, **15** (1981) 1375–89.
49. Wendling, P., Wendling, R., Renger, W., Covert, D. S., Heintzenberg, J. & Moerl, P., Calculated radiative effects of Arctic haze during a pollution episode in spring 1983 based on ground-based and airborne measurements. *Atmos. Environ.*, **19** (1985) 2181–93.
50. Porch, W. M. & MacCracken, M. C., Parametric study of the effects of Arctic soot on the solar radiation. *Atmos. Environ.*, **16** (1982) 1365–71.
51. Shaw, G. E. & Stamnes, K., Arctic haze: Perturbation of the Polar radiation budget. *Ann. N. Y. Acad. Sci.*, **338** (1980) 533–9.
52. Coakley, J. A., Jr., Bernstein, R. L. & Durkee, P. A., Effect of ship-stack effluents on cloud reflectivity. *Science*, **237** (1987) 1020–22.
53. Clarke, A. D., Charlson, R. J. & Radke, L. F. Airborne observations of Arctic aerosol—IV. Optical properties of Arctic haze. *Res. Lett.*, **11** (1984) 405–8.

Index